Olive Mill Waste

Olive Mill Waste

Recent Advances for Sustainable Management

Editor

Charis M. Galanakis

Galanakis Laboratories, Chania, Greece

Academic Press is an imprint of Elsevier
125 London Wall, London EC2Y 5AS, United Kingdom
525 B Street, Suite 1800, San Diego, CA 92101-4495, United States
50 Hampshire Street, 5th Floor, Cambridge, MA 02139, United States
The Boulevard, Langford Lane, Kidlington, Oxford OX5 1GB, United Kingdom

Notices
Knowledge and best practice in this field are constantly changing. As new research and experience broaden our understanding, changes in research methods, professional practices, or medical treatment may become necessary.

Practitioners and researchers must always rely on their own experience and knowledge in evaluating and using any information, methods, compounds, or experiments described herein. In using such information or methods they should be mindful of their own safety and the safety of others, including parties for whom they have a professional responsibility.

To the fullest extent of the law, neither the Publisher nor the authors, contributors, or editors, assume any liability for any injury and/or damage to persons or property as a matter of products liability, negligence or otherwise, or from any use or operation of any methods, products, instructions, or ideas contained in the material herein.

Library of Congress Cataloging-in-Publication Data
A catalog record for this book is available from the Library of Congress

British Library Cataloguing-in-Publication Data
A catalogue record for this book is available from the British Library

ISBN: 978-0-12-805314-0

For information on all Academic Press publications
visit our website at https://www.elsevier.com/

Publisher: Nikki Levy
Acquisition Editor: Nancy Maragioglio
Editorial Project Manager: Billie Jean Fernandez
Production Project Manager: Caroline Johnson
Designer: Greg Harris

Typeset by Thomson Digital

Contents

List of Contributors

Evita Agrafioti
Department of Research & Innovation, Galanakis Laboratories, Chania, Greece

Christos S. Akratos
Department of Environmental and Natural Resources Management, University of Patras, Agrinio, Greece

Mercedes Ballesteros
Biofuels Unit, Energy Department-CIEMAT, Madrid, Spain

Pilar Bernal
Spanish National Research Council (CSIC). Campus Universitario de Espinardo, Murcia, Spain

Eulogio Castro
Department of Chemical, Environmental and Materials Engineering, Agrifood Campus of International Excellence, ceiA3, University of Jaén, Jaén, Spain

Rafael Clemente
Spanish National Research Council (CSIC). Campus Universitario de Espinardo, Murcia, Spain

Maria K. Doula
Benaki Phytopathological Institute, Department of Phytopathology, Kifissia, Greece

Abdelilah El-Abbassi
Food Sciences Laboratory, Department of Biology, Faculty of Sciences—Semlalia, Cadi Ayyad University, Marrakech, Morocco

Sonia Esposto
Department of Agricultural, Food and Environmental Sciences, University of Perugia, Perugia, Italy

Paris A. Fokaides
School of Engineering, Frederick University, Nicosia, Cyprus

Charis M. Galanakis
Department of Research & Innovation, Galanakis Laboratories, Chania, Greece

Giovanni Gigliotti
Department of Civil and Environmental Engineering, University of Perugia, Perugia, Italy

Abdellatif Hafidi
Food Sciences Laboratory, Department of Biology, Faculty of Sciences—Semlalia, Cadi Ayyad University, Marrakech, Morocco

Vassilis J. Inglezakis
School of Engineering, Department of Chemical Engineering, Environmental Science & Technology Group (ESTg), Nazarbayev University, Astana, Republic of Kazakhstan

Hajar Kiai
Food Sciences Laboratory, Department of Biology, Faculty of Sciences—Semlalia, Cadi Ayyad University, Marrakech, Morocco

Konstantinos Komnitsas
School of Mineral Resources Eng, Technical University of Crete, Chania, Crete, Greece

Kali Kotsiou
Department of Chemistry, Section of Industrial and Food Chemistry, University of Ioannina, Ioannina, Greece

George K. Lamprou
Laboratory of Organic Chemical Technology, School of Chemical Engineering, National Technical University of Athens, Zografou, Greece

Paloma Manzanares
Biofuels Unit, Energy Department-CIEMAT, Madrid, Spain

Jose Luis Moreno-Ortego
CEBAS-CSIC, Department of Soil and Water Conservation and Organic Resources Management, Espinardo, Murcia, Spain

Luigi Nasini
Department of Agricultural, Food and Environmental Sciences, University of Perugia, Perugia, Italy

María José Negro
Biofuels Unit, Energy Department-CIEMAT, Madrid, Spain

Enrico Novelli
Department of Comparative Biomedicine and Food Science, University of Padova, Legnaro, Italy

Maria Antónia da Mota Nunes
LAQV/Requimte, Faculty of Pharmacy, University of Porto, Rua Jorge Viterbo Ferreira, Porto, Portugal

Maria Beatriz Prior Pinto Oliveira
LAQV/Requimte, Faculty of Pharmacy, University of Porto, Rua Jorge Viterbo Ferreira, Porto, Portugal

Tania Pardo
Spanish National Research Council (CSIC). Campus Universitario de Espinardo, Murcia; Spanish National Research Council (CSIC). Av. De Vigo, Santiago de Compostela, Spain

Primo Proietti
Department of Agricultural, Food and Environmental Sciences, University of Perugia, Perugia, Italy

Luca Regni
Department of Agricultural, Food and Environmental Sciences, University of Perugia, Perugia, Italy

Francisca Rodrigues
LAQV/Requimte, Faculty of Pharmacy, University of Porto, Rua Jorge Viterbo Ferreira, Porto, Portugal

Encarnación Ruiz
Department of Chemical, Environmental and Materials Engineering, Agrifood Campus of International Excellence, ceiA3, University of Jaén, Jaén, Spain

Apostolos Sarris
Foundation for Research and Technology, Hellas, Institute for Mediterranean Studies, Laboratory of Geophysical-Satellite Remote Sensing and Archaeo-environment, Nik. Foka & Melisinou, Rethymnon, Crete, Greece

Sami Sayadi
Laboratory of Environmental Bioprocesses, Centre of Biotechnology of Sfax, University of Sfax, Sfax, Tunisia

Maurizio Servili
Department of Agricultural, Food and Environmental Sciences, University of Perugia, Perugia, Italy

Safa Souilem
Laboratory of Environmental Bioprocesses, Centre of Biotechnology of Sfax, University of Sfax, Sfax, Tunisia

Agnese Taticchi
Department of Agricultural, Food and Environmental Sciences, University of Perugia, Perugia, Italy

Athanasia G. Tekerlekopoulou
Department of Environmental and Natural Resources Management, University of Patras, Agrinio, Greece

Federico Tinivella
Center for Agricultural Experimentation and Assistance, Albenga (SV), Italy

Ioanna A. Vasiliadou
Department of Engineering and Architecture, University of Trieste, Trieste, Italy

Dimitrios V. Vayenas
Department of Chemical Engineering, University of Patras, Rio; Institute of Chemical Engineering Sciences (ICE-HT), Platani, Patras, Greece

Gianluca Veneziani
Department of Agricultural, Food and Environmental Sciences, University of Perugia, Perugia, Italy

Anestis Vlysidis
Laboratory of Organic Chemical Technology, School of Chemical Engineering, National Technical University of Athens, Zografou, Greece

Apostolos G. Vlyssides
Laboratory of Organic Chemical Technology, School of Chemical Engineering, National Technical University of Athens, Zografou, Greece

Preface

The cultivation of olives and the production of olive oil have deep roots in the history of Mediterranean region. This tradition represents a very important asset for many countries, not only in terms of culture and health, but also in terms of wealth. However, olive oil production is accompanied with the generation of huge amounts of by-products and waste that leave a congested environmental footprint. These materials are undesirable for the olive oil industry in terms of sustainability and environmental impact, but perhaps more important in view of high disposal costs. Therefore, they have been considered as a matter of minimization, prevention, and treatment for as many decades as olive oil industrial production exists. Indeed, the proposed treatment methods and the respective literature and references are endless. Despite this fact, olive oil industry remains unsustainable, with few opposite examples that confirm this rule of thumb.

Why is this happening? Is it a matter of inadequate treatment technologies or is it about cost? Olive oil is a sector challenged by many directions. Consumers demand extra virgin olive oils of ultra-high quality, product's final price varies a lot from time to time, and local authorities demand from production units to reduce their environmental impact. Under these conditions, even cheap solutions that promise the total treatment of olive mill waste may collapse financially olive oil industries. Consequently, most treatment solutions have been rejected in practice due to industries' denial that claim to close down production and society's tolerance that delays the enforcement of environmental legislations implementation. Can olive oil industries overpass environmental legislations forever? Does this consideration fall in the frame of the modern bioeconomy? Can they adapt any alternative strategies? The urgent need for sustainability within olive oil industry has turned the interest of researchers and professionals to investigate the management of olive mill waste with another perspective. This resource contains valuable components, such as water, organic compounds, and a wide range of nutrients that could be recycled. The prospect of recycling ingredients from olive mill waste is a story that started few decades ago. For instance, solvent extraction had been applied to recover oil from olive kernel, which is one of the by-products derived from olive oil production. Olive kernel is considered an established commodity similar to olive fruit, whereas scientists focus on the recovery of polyphenols, the reutilization of irrigation water, as well as the production of compost to be used as soil amendment. Subsequently, there is a need for a new guide covering the latest developments in this particular direction.

Following this trend, the current book covers the most recent advances of olive mill waste management in the name of sustainability. It aspires to fill in the gap of transferring knowledge between academia and industry by describing in details the viable industrial applications and scenarios. It highlights success stories and solutions that are already applied in some olive oil industries, whereas it explores the advantages, disadvantages, and real potentiality of relevant processes and products in the market. The ultimate goal is to inspire scientific community and producers that aspire to develop real commercialized applications.

The book consists of 12 chapters. Chapter 1 discusses olive oil production, its environmental effects, and the current sustainability challenges of the sector. Chapter 2 introduces the current advisable practices for the sustainable development of olive oil industry, focusing on olive mill waste, and two soil remediation methods, applied within the framework of European project. Chapter 3 deals with the industrial valorization of residues from olive oil industry within the integrated concept of biorefinery. Rest chapters focus on more specific applications. For instance, Chapter 4 presents the possibilities and alternative strategies to recover energy from olive oil processing residues. In Chapter 5, the benefits and risks of using olive mill waste as a soil amendment are discussed and recommendations on their proper application are provided, too. Chapter 6 describes industrial case studies for the detoxification of olive mill wastewater using Fenton oxidation process followed by biological processes for energy and compost production. Chapter 7 presents an integrated and commercialized approach for the treatment of olive mill wastewater and solid residue using only biological treatments (i.e., trickling filters, constructed wetlands, and composting). Chapter 8 deals with the cocomposting of olive mill waste as well as the design and operation of two case (pilot-scale and full-scale) studies. Chapter 9 reviews the use of the different olive mill by-products in phytoremediation strategies of contaminated soils. Chapter 10 denotes the different available technologies for the recovery of bioactive compounds from olive oil processing by-products and suggests an integral methodology that ensures the sustainability of the process. The applications of membrane processes for this purpose are thoroughly discussed, whereas detailed information for the current patented and commercialized methodologies are provided. Finally, Chapters 11 and 12 describe available and potential applications of compounds recovered from olive mill waste in food products and cosmetics, respectively.

Conclusively, the ultimate goal of the book is to provide a handbook for all the professionals and producers activated in the field, trying to optimize olive oil industry performance and reduce its environmental impact. It concerns chemical engineers and technologists working in the olive oil and food industry as well as researchers, specialists, and new product developers working in the edge of food and environmental sectors. It could be used as a textbook and/or ancillary reading in graduates and postgraduate level, and multidiscipline courses dealing with agricultural science, food and environmental technology, sustainability, and chemical engineering.

I would like to take this opportunity to thank all the authors and contributors of the book for their collaboration and high quality work in bringing together different topics and sustainable approaches in a comprehensive text. I consider myself fortunate to have had the opportunity to collaborate fruitfully with so many knowledgeable colleagues from Cyprus, Greece, Italy, Kazakhstan, Morocco, Portugal, Spain, and Tunisia. Their acceptance of book's concept and their dedication to editorial guidelines is highly appreciated. I would also like to thank the acquisition editor Nancy Maragioglio for our collaboration in this project and all the team of Elsevier, particularly Billie Jean Fernandez for her assistance during editing and Caroline Johnson during production. I would also like to acknowledge the support of Food Waste Recovery Group of ISEKI Food Association, which is the most relevant group worldwide in the particular field.

Last but not least, a message for the readers. In a collaborative project of this size, it is impossible for it not to contain errors. If you find errors or have any objections to its content, please do not hesitate to contact me.

Charis M. Galanakis
Research & Innovation Department
Galanakis Laboratories
Chania, Greece
e-mail: cgalanakis@chemlab.gr

Food Waste Recovery Group
ISEKI Food Association
Vienna, Austria
e-mail: foodwasterecoverygroup@gmail.com

CHAPTER

1

OLIVE OIL PRODUCTION SECTOR: ENVIRONMENTAL EFFECTS AND SUSTAINABILITY CHALLENGES

Safa Souilem*, Abdelilah El-Abbassi, Hajar Kiai**, Abdellatif Hafidi**, Sami Sayadi*, Charis M. Galanakis†**

**Laboratory of Environmental Bioprocesses, Centre of Biotechnology of Sfax, University of Sfax, Sfax, Tunisia; **Food Sciences Laboratory, Department of Biology, Faculty of Sciences—Semlalia, Cadi Ayyad University, Marrakech, Morocco; †Department of Research & Innovation, Galanakis Laboratories, Chania, Greece*

1.1 INTRODUCTION

Olive oil is produced from olive trees, each olive tree yielding between 15 and 40 kg of olives per year. Worldwide olive oil production for the year 2002 was about 2.5 million tons produced from approximately 750 million productive olive trees, the majority of which are in the Mediterranean region. In particular, there are about 25,000 olive mills worldwide. Mediterranean countries alone produce 97% of the total olive oil production, while European Union countries produce 80–84%. The biggest olive oil-producing countries are Spain, Italy, Greece, and Turkey (~0.9, 0.6, 0.4, and 0.2 million tons in 2002, respectively), followed by Tunisia, Portugal, Morocco, and Algeria. Outside the Mediterranean basin, olives are cultivated in the Middle East, the USA, Argentina, and Australia (Paraskeva and Diamadopoulos, 2006).

Olive oil production tends to increase over the last decades as a valuable source of antioxidants and essential fatty acids in the human diet and constitutes one of the most important dietary trends worldwide. The olive oil extraction systems could be classified in two main categories: traditional pressing process (Kapellakis et al., 2008), used for many centuries with minor modifications, and centrifugal processes, including two centrifugation systems, called three- and two-phase systems. The extraction of olive oil generates huge quantities of wastes that may have a great impact on land and water environments because of their high phytotoxicity. The most pollutant and phytotoxic wastes are known as olive mill waste (OMW). During the olive oil production, almost all the phenolic content of the olive fruit (~98%) remains in the olive mill by-products (Rodis et al., 2002). Besides being a serious environmental problem, OMW represents today a precious resource of useful compounds for recovery and valorization purposes (Hamza and Sayadi, 2015; Skaltsounis et al., 2015; Chiaiese et al., 2011).

Olive Mill Waste. http://dx.doi.org/10.1016/B978-0-12-805314-0.00001-7

1.2 OLIVE FRUIT PROCESSING AND OLIVE OIL PRODUCTION TECHNOLOGIES

Olive oil extraction involves different processes, such as leaf removal, olive washing, grinding, beating, and separation of the oil. The amount and physicochemical properties of the produced wastes and effluents depend on the method used for the extraction. Olive oil is extracted directly from the fresh fruit of olive tree (*Olea europaea* L.) using only mechanical methods, in order to maintain its natural organoleptic characteristics according to the European Commission Regulation No. 1513/2001 (EC, 2001). Olive fruits must be processed as quickly as possible after harvesting to minimize oxidation and preserve low acidity. All the operations of the olive mill can influence on olive oil quality (Gimeno et al., 2002; Kiritsakis et al., 1998). The mechanical processes used to extract olive oil from olive fruit include olives crushing, malaxation of resulting paste, and separation of the oily phase by pressure or centrifugation. In the latest one, three different systems are commonly used: the traditional discontinuous press process, the three-phase, and the two-phase decanter centrifuge methods (Fig. 1.1).

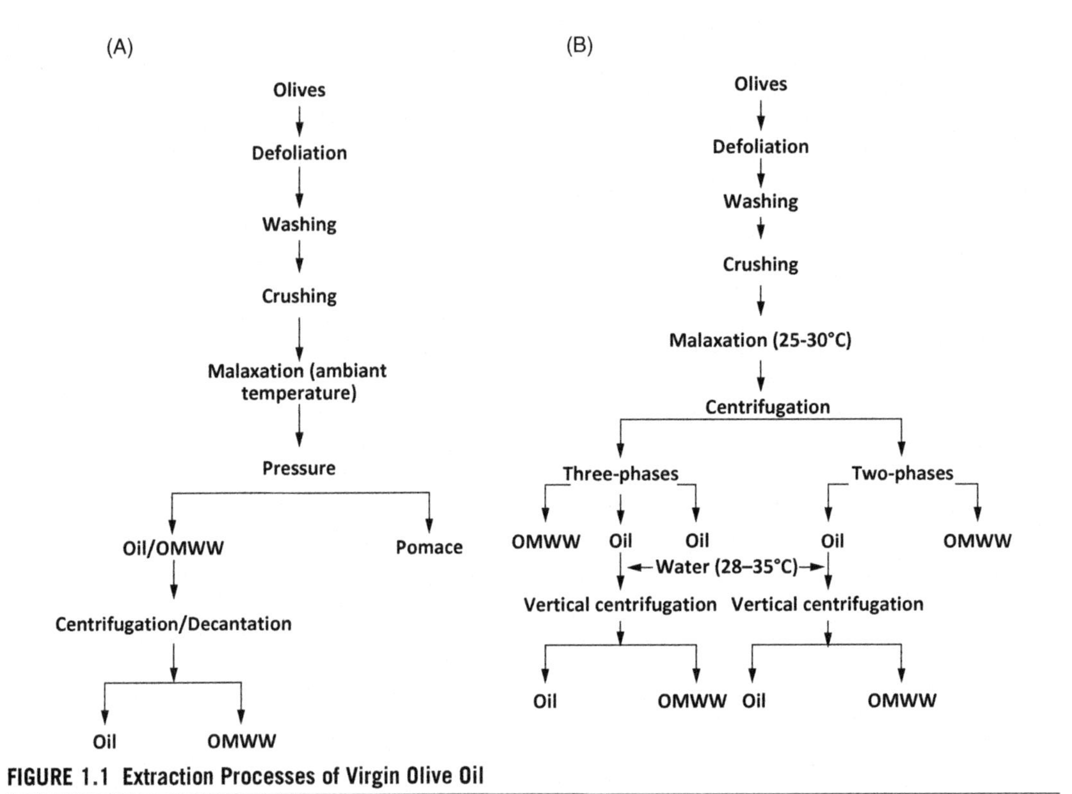

FIGURE 1.1 Extraction Processes of Virgin Olive Oil

(A) Traditional process. (B) Modern process.

1.2.1 TRADITIONAL DISCONTINUOUS PRESS SYSTEM

Traditional extraction press is still used in some mills. After grinding olives in stone mills, the paste is spread on fiber diaphragms (which are stacked on top of each other) and then placed into the press (Fig. 1.2A). Hydraulic pressure is applied on the disks, thus compacting the solid phase of the olive paste and percolating the liquid phases (oil and vegetation water). To facilitate separation of the liquid phases, water is run down the sides of the disks to increase percolation speed. Traditionally the disks were made of hemp or coconut fiber, but in modern times they are made of synthetic fibers, allowing easier cleaning and maintenance (Kapellakis et al., 2008).

This process generates a solid fraction called olive husk (or kernel), an emulsion containing the olive oil and a water phase. The olive oil is finally separated from the remaining wastewater by decantation or vertical centrifugation. The traditional process is reputed to produce a high quality olive oil. However, many precautions should be taken in consideration, for example, a proper cleaning of the disks and a rapid treatment of the paste to avoid fermentation. The latest generates unwanted flavors that lead to the production of defected olive oil. Nowadays, due to the need to process large amounts of olives and to obtain higher yields of olive oil, the evolution of the oil extraction process has led to the replacement of traditional pressure mills with modern continuous centrifugal extraction.

1.2.2 CONTINUOUS CENTRIFUGAL EXTRACTIONS

After washing, crushing, and malaxation steps, the mechanical oil extraction is mainly carried out using a continuous process based on centrifugation with a decanter. The decanter centrifuge is designed

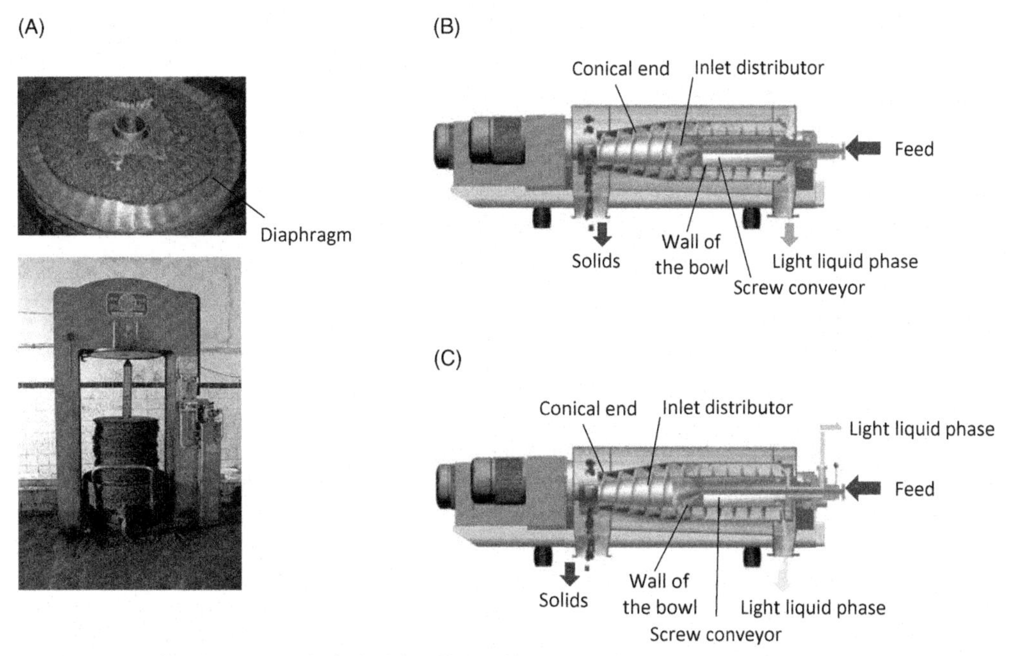

FIGURE 1.2 Extraction Processes of Virgin Olive Oil (VOO)

(A) Press system. (B) Decanter (two-phases). (C) Decanter (three-phases). Decanters photos courtesy of Flottweg Separation Technology, Germany flottweg.com.

with screw conveyer and rotating bowl, allowing to process large amounts of olives in short time (Catalano et al., 2003). Actually two types of centrifugal decanters are currently used: two-phase centrifugation and three-phase centrifugation (Fig. 1.2B and C).

The continuous three-phase centrifugation process was introduced during the 1970s, in order to increase processing capacity and extraction yield and reduce labor (Demicheli and Bontoux, 1997). During the three-phase process, an addition of hot water is required to wash the oil. The process yields three phases: oil phase, solid residue: olive cake (olive pulp and stones) and the olive mill wastewater (OMWW). Solid residue is separated from the other two phases in the decanter. The liquid phases are subsequently submitted to a vertical centrifugation in order to separate the olive oil from OMWW.

A disadvantage of this process is the large amounts of produced wastewaters that are due to the high consumption of water, 1.25–1.75 times more water than press extraction (Vlyssides et al., 2004). The failure to develop a suitable end-of-pipe wastewater treatment technology challenged technology manufacturers to develop the two-phase process. The latest uses only washing water and delivers oil (liquid phase) and a very wet substrate (semisolid phase called wet olive pomace or two-phase OMW) using a more effective centrifugation technology (Vlyssides et al., 2004). This process has reduced environmental impact due to the lower water demands and waste amounts produced, but it requires extra energy for drying prior olive kernel oil extraction. As the olive mills are becoming bigger, the chances of automation are also increasing, reflecting an improvement in the olive oil yield and quality. Therefore, a detailed knowledge of the whole extraction process is crucial in order to design the appropriate control strategies (Bordons and Nunez-Reyes, 2008).

1.2.3 FEEDING, LEAF REMOVAL, AND WASHING

The first step in the olive oil extraction process is olive fruit cleaning and removal of stems, leaves, twigs, and other debris left with the olives. Washing is also aiming to remove pesticides and dirt. Light contaminants are removed by a heavy air flow and heavy objects sinks in the water bath. Olive washing in closed loop systems is a critical control point at the olive mill due to microbiological cross-contamination and fruit physical damage. Furthermore, when the olives were short-term stored before oil extraction, sensory attributes of virgin olive oil (VOO) diminished due to changes in phenolic and lipoxygenase derived volatile compounds (Vichi et al., 2015).

1.2.4 CRUSHING

Crushing step aims to break down the cellular membranes of olive fruits and thus release small drops of oil from the vacuoles (Rodis et al., 2002). This operation produces a mixture of two distinct liquid phases (raw oil and water) and an extremely heterogeneous solid phase (pit, skin, and pulp fragments). Crushing can be considered a critical point affecting the quality of the produced olive oil, especially in terms of phenolic content and volatile compounds (Servili et al., 2015). The constitutive parts of olive fruit are pulp, stone, and seed. The pulp is rich in phenolic compounds while the seed contains large amount of endogenous oxidoreductases and low amounts of phenols (Servili et al., 2004, 2007). Two enzymes, polyphenol oxidase and peroxidase, are highly concentrated in the olive kernel. Thus, crushing operation allows a direct contact of the peroxidase and polyphenoloxidase (POD) enzymes with phenolic compounds and induces their oxidation, which results in lower phenolic concentration in the oil. At this phase, the main hydrophilic phenols of VOO (e.g., secoiridoid aglycons) are generated

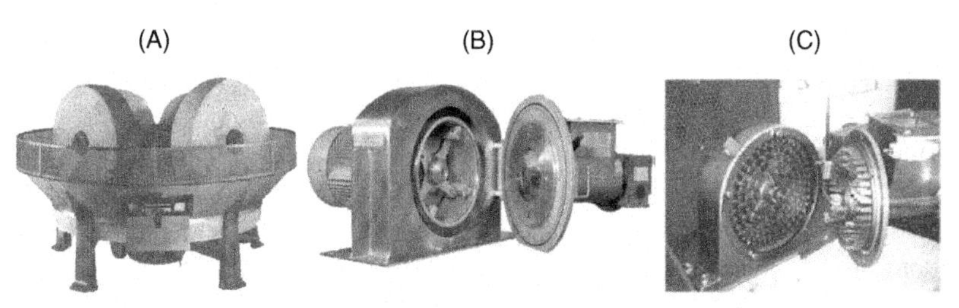

FIGURE 1.3 Types of Crushers (A) Stone Miller. (B) Disc Crusher. (C) Hammer Miller.

Photos courtesy of Alfa Laval, Italy alfalaval.com.

at the hydrolysis of oleuropein, demethyloleuropein, and ligstroside, and as catalyzed by endogenous β-glucosidases (Servili, 2014). The control of the enzymatic activity during the crushing step is a good strategy to maintain a high phenolic concentration in the resulting olive oil. For this purpose, mild seed crushing methods that reduce seed tissue degradation are used to limit the release of POD in the paste and decrease the rate of oxidation of hydrophilic phenolic compounds (Servili et al., 2007, 2015). Mechanical extraction from destoned olives has also been proposed to avoid enzymatic reaction catalyzed by POD, resulting in higher oxidative stability and nutritional value of the obtained olive oil (Del Caro et al., 2006). Moreover, oils extracted from destoned olives have a better sensory profile than the oils obtained from the traditional milling of entire fruits, although olive destoning lowers the oil yield (Ben Mansour et al., 2015). In another study aiming the evaluation of functional phytochemicals in destoned VOO, the oils showed higher contents of phenolic compounds, tocopherols, and aromas, whereas a potentially higher stability and shelf-life was reported (Ranalli et al., 2009).

On the other hand, olive seed oil is highly marketable in cosmetics and pharmaceutical industries while the destoned olive pomace containing a high phenolic concentration and high-value monounsaturated fatty acids, can be used in animal feeding (Servili et al., 2015). Crushing is generally carried out using a traditional stone mill or using hammer or disk crushers (Fig. 1.3). Hammer crushers produced oils with greater amounts of phenolic compounds as compared to stone crushers (Leone et al., 2015). Concerning the olive oil quality, Leone et al. (2016) reported that a flash thermal treatment of olive paste after crushing improves the phenolic and volatile profile of the oil significantly as compared to the traditional process. Since destoning process has led to better olive oil quality, destoner machines have been developed. The destoner consists of a cylindrical perforated stationary grill and a rotary shaft. The olives are pushed by centrifugal force toward the perforated grill (Amirante et al., 2006). The current trend in the olive oil market is the production of high quality monovarietal oils or mixing with the most widespread olives. The cocrushing of Picual variety with a Galician (Spain) local variety (80:20) was shown to improve the sensory and health properties of Picual extra VOO (Reboredo-Rodriguez et al., 2015).

1.2.5 MALAXATION

Malaxation consists of a slow and continuous kneading of the olive paste in order to facilitate the cohesion of small oil droplets obtained during crushing step, leading to separation of the oil from the water phase. This step is essential to achieve high yields of extraction. Several researchers have investigated the effect of malaxation parameters on the quality of produced oil. Malaxation time and temperatures found to affect greatly the olive oil quality especially the aroma and phenolic profiles (Jiménez et al., 2014;

Reboredo-Rodríguez et al., 2014). Aliakbarian et al. (2008) demonstrated that increasing the malaxation time from 90 to 150 min decreased the nutritional quality of olive oil, mainly due to the increased oxidation of phenolics. In order to compromise quality and yield of olive oil, a low malaxation temperature and a process time between 30 and 45 min are recommended (Clodoveo et al., 2014). The old type of malaxation machines are designed with a stainless steel grill closure that was reported to cause a loss of the oil's phenolic and volatile compounds, due to the wholes of the upper grill (Amirante et al., 2006). In fact, the higher presence of O_2 during malaxation activate the POD and PPO enzymes responsible for the oxidation of phenolic compounds and loss of flavors (Tamborrino et al., 2014). A hermetic sealing improves the heat transfer leading to a reduction of the malaxation time and lower loss of volatiles and perfect control of the atmosphere contacting the paste (Leone et al., 2014). Besides, an innovative system has been developed to monitor the oxygen concentration during malaxation (Catania et al., 2013, 2016). At this case, the composition of the obtained olive oil in volatile compounds was enhanced by blowing pure oxygen in the headspace of the machine to modify the atmosphere. Recently, Catania et al. (2016) have developed a system to control the atmosphere in the headspace of the malaxation machine and improve the fatty acid composition of olive oil. The same group has developed a software that allows the acquisition and recording of the oxygen concentration in the headspace of the malaxation machine. Using N_2 during malaxation extended induction time, raised phenolic and tocopherol contents, leading to a strong antioxidant potential of oils (Yorulmaz et al., 2011).

The malaxation efficiency can also be improved by the addition of coadjuvants to promote the breakage of oil/water emulsions and consequently increase the recovery yield of the oil (Espinola et al., 2015; Guermazi et al., 2015; Sadkaoui et al., 2015). The list of studied coadjuvant consists of enzymes, talc, vegetable fiber, calcium carbonate, and salt. Researchers reported an increase of oil yield by the addition of natural microcrystalline talc during olive processing. In addition, an increase of oil stability due to the increase of phenolic content and tocopherols was denoted (Espinola et al., 2015; Caponio et al., 2015, 2016). The micronized natural talc is authorized by the European Commission and can be used during malaxation step (Caponio et al., 2016). The utilization of solid carbon dioxide (CO_2) for the extraction of extra-VOO has been proposed, too (Zinnai et al., 2015). For instance, the addition of cryogen to the olives during premilling phase greatly increased the extraction yield (ranging from 1% to 21%) improving olive oil quality and increasing its shelf-life (Zinnai et al., 2015).

Another approach is the addition of pectolytic, hemicellulolytic, and cellulolytic enzymes. The addition of these enzymes enables the breakdown of cell wall structure of olives and reduced the complexation of hydrophilic phenols with polysaccharides (Aliakbarian et al., 2008). More free phenolic compounds are consequently released into the olive oil depending on the used enzyme formulation (De Faveri et al., 2008). The modification of the pH of malaxation had also an influence on the produced olive oil. Aliakbarian et al. (2009) assessed the feasibility of increasing the free phenols in the olive oil paste by simultaneous addition of citric acid during the malaxation step and control of the kneading time. An increase of phenolic compounds in the olive oil was correlated to a better hydrolyze of pectic polysaccharides, cellulosic, and hemicellulosic fractions in olive pulp which allowed better release of phenolics from the paste to the oil.

1.2.6 HORIZONTAL CENTRIFUGATION

The centrifugation process is based on the density differences of olive paste components (olive oil, water, and insoluble solids). Decanters are constituted of a cylindrical conical bowl drum where a screw

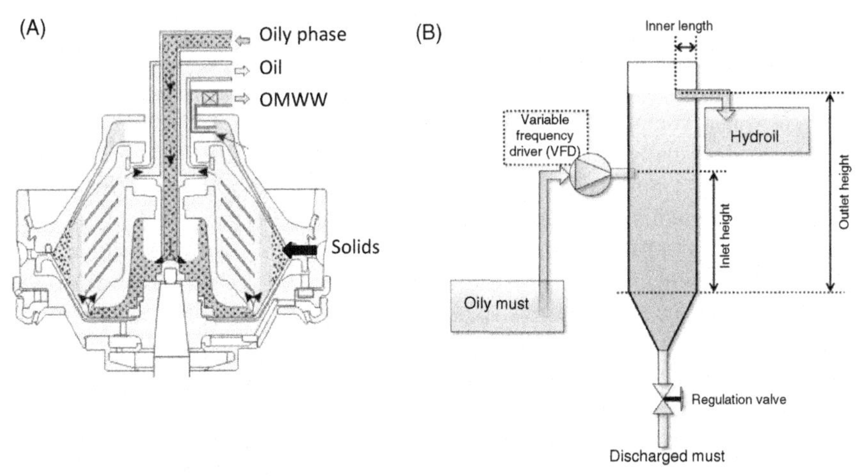

FIGURE 1.4 Systems Used for Oil Clarification Process

(A) Vertical centrifuge system and (B) Hydrocyclone prototype (Altieri et al., 2015) used for oil clarification.

feeder is rotating at differential speed (Fig. 1.4B) (Skaltsounis et al., 2015). Separation is conducted due to the centrifugal force developed in the drum. Olive paste could be treated either by two-phase centrifugal decanter or a three-phase centrifugal decanter. In the two-phase decanter, the product is separated into a liquid phase (oil) and a solid phase (kernel fragments, pulp, and vegetable water). In the three-phase centrifugal decanter the product is separated into a light liquid phase (oil), into a heavy liquid phase (water), and into a solid phase (kernel fragments and pulp). In both two- and three-phases decanter centrifuges, the oil phase is discharged by gravity. In three-phase decanter, the water phase is discharged using a centripetal or gravity pump. The solid phase is discharged in the drum conical terminal of the decanter after being pushed by the screw feeder. The two-phase process requires no dilution during the malaxation step. However, in the three-phase process, large amounts of water are added to the paste. After malaxation, the olive oil is either completely free or in the form of small droplets inside microgels, or emulsified in the aqueous phase (Clodoveo, 2012). Adding water during the centrifugation improved the release of the oil fraction locked in the microgels (Clodoveo, 2012).

Since the olive oil extraction by decanter centrifuge is greatly dependent on the paste rheological properties (water content, variety, and temperature) (Altieri et al., 2013), it is crucial to control the process in order to optimize the extraction yield while preserving a high quality of oil. Recent developments include the design of an automatic system for the decanter able to adjust the machine operating parameters according to the olive paste characteristics and guarantee a constant feed to the decanter centrifuge even in the presence of physical changes in the olive paste (Altieri et al., 2013, 2014).

1.2.7 OIL CLARIFICATION

Oil clarification is the final cleaning step of olive oil. It aims to separate the residual water and impurities existing in the extracted oil from the decanter. This operation is necessary to avoid fermentation, hydrolysis, and oxidation reactions causing alteration of the sensory properties of olive oil (Baiano et al., 2014). The final purification step is generally made by a filtration or vertical centrifugation

(Fig. 1.4A). Filtration is the most common technique used to clear the oil. A new processing arrangement was proposed and tested by Guerrini et al. (2015). It consists of the insertion of a steel prefilter into the system, which retains part of the suspension. Consequently, the plate filter-press retains only residual solids and water. The plate filter-press with the added prefilter was able to process about 1.8 times the amount of oil normally processed in a batch (Guerrini et al., 2015).

Vertical centrifugation contributes to a significant increase of dissolved oxygen concentration in oil and accelerates its oxidation. In order to reduce the oxygenation rate of olive oil during vertical centrifugation, Masella et al. (2012) suggested blanketing the vertical centrifuge with inert gas. Natural sedimentation is a good alternative to the vertical centrifuge but it is considered unsuitable for modern processes due to the extended time required to perform the operation (Gila et al., 2016). Altieri et al. (2014) compared the olive oil quality issued after improved process of natural sedimentation (made of a twin cylindrical columns) to the oil issued from the standard vertical centrifugation. The measurement of quality indices of oil, that is, peroxide, phenol, chlorophyll, carotenoid, turbidity, and K_{232} confirmed that sedimentation process reduces the oxygenation reaction and allows longer shelf-life of the oil. Recent developments suggest the use of bottom settling tanks for the purification step. These conic tanks have a working capacity between 400 and 10.000 L and can be used for both batch and continuous processes (Altieri et al., 2014). Altieri et al. (2015) introduced an innovative sedimentation plant for the separation of high quality VOO at industrial scale which is based on soft hydrocyclone action. The hydrocyclone system was introduced in order to define design requirements for a new olive oil clarification machine, aiming to improve oil quality, safety, and processing capacity (Altieri et al., 2015). Since the high variability of the oil issued from different extraction systems, the hydrocyclone should be carefully settled in order to optimize its use. The main parameters affecting this purification step are the density difference between liquid and solid particles, the particle size and the liquid viscosity, among others. These physical properties are dependent on the fatty acid composition of the oil and are strongly affected by temperature (Gila et al., 2015). Gila et al. (2016) proved that temperature is also an important parameter to monitor during static settling of olive oil. Therefore, the temperature of 30°C showed higher values of settling efficiency compared to lower temperatures (15 and 20°C) using both experimental and computational fluid dynamics procedures.

1.3 WATER CONSUMPTION FOR OLIVE OIL PRODUCTION

For the olive fruits production, the agricultural stage is responsible for an enormous consumption of fresh water including the water used in irrigation, fertilization, and pest control. A study was conducted by Avraamides and Fatta (2008) in Cyprus evaluating the consumption of water in various stages of olive oil production. The authors reported that a total of 3914 L of fresh water are consumed for the production of 1 L of olive oil. However, during olive processing, only 3.51 L of water are consumed for every liter of olive oil produced. This stage produces 4.34 kg of OMWW and 2.07 kg of solid waste (olive kernel) for every liter of olive oil produced (Avraamides and Fatta, 2008).

The water dilution of the olive paste affects the partition of hydrophilic phenols between oil and water and enhances their release in the water phase. The reduction of water dilution during the three phases process leads to an increase of the phenolic concentration in the olive oil, too (Amirante et al., 2002). Therefore, new generation of three-phase decanter centrifuges (water saving decanter) were designed for lower water consumption during the centrifugation process and consequently less generation of vegetation water. During vertical centrifugation, tap water is added to enhance the liquid–liquid separation (Masella et al., 2009). However, several researches have reported that the water addition in this step also reduces the hydrophilic phenolic content in olive oil (Masella et al., 2012).

1.4 WASTES AND WASTEWATERS GENERATED DURING OLIVE OIL PRODUCTION

Both, olive tree culture and olive oil industry produce large amounts of by-products, for example, pruning alone produces annually 25 kg of twigs and leaves/tree (Niaounakis and Halvadakis, 2006). However, OMW is the major environmental concern of olive oil industry, as leaves represent only 5% of olives' weight in oil extraction. Indeed, the annual world OMW production is estimated to be from 10 to over 30 million m^3 (McNamara et al. 2008).

The amount and the physicochemical characteristics of the generated wastes depend on the used oil extraction system, the processed fruits, and the operating conditions (added water, temperature, etc.). A limited amount of solid wastes (leaves and small twigs) is produced during the cleaning of the olives prior to milling. Nevertheless, these by-products do not present a management problem. OMWW is the main waste from three-phase extraction systems and traditional mills. It is constituted by vegetable water of the fruit and the water used in different stages of oil extraction. Besides, the three-phase extraction systems generate a solid waste (Table 1.1) that is used to extract olive kernel oil.

Table 1.1 Composition (in Percentage) of the Olive Mill Solid Wastes (Pastes/Cakes) (Vlyssides and Iaconidou, 2003)

Parameters	Press Process	Three-Phase Process	Two-Phase Process
Moisture	27.2 ± 1.048	50.23 ± 1.935	56.80 ± 2.188
Fats and oils	8.72 ± 3.254	3.89 ± 1.449	4.65 ± 1.736
Proteins	4.77 ± 0.024	3.43 ± 0.017	2.87 ± 0.014
Total sugars	1.38 ± 0.016	0.99 ± 0.012	0.83 ± 0.010
Cellulose	24.1 ± 0.283	17.37 ± 0.203	14.54 ± 0.170
Hemicellulose	11.0 ± 0.608	7.92 ± 0.438	6.63 ± 0.366
Ash	2.36 ± 0.145	1.70 ± 0.105	1.42 ± 0.088
Lignin	14.1 ± 0.291	10.21 ± 0.209	8.54 ± 0.175
Kjendahl nitrogen	0.71 ± 0.010	0.51 ± 0.007	0.43 ± 0.006
Phosphorous as P_2O_5	0.07 ± 0,005	0.05 ± 0.004	0.04 ± 0.003
Phenolic compounds	1.14 ± 0.060	0.326 ± 0.035	2.43 ± 0.150
Potassium as K_2O	0.54 ± 0.045	0.39 ± 0.033	0.32 ± 0.027
Calcium as CaO	0.61 ± 0.059	0.44 ± 0.043	0.37 ± 0.036
Total carbon	42.9 ± 3.424	29.03 ± 2.317	25.37 ± 2.025

Table 1.2 Physicochemical Characteristics of Olive Mill Wastewaters

Parameters	Unit	Paraskeva et al. (2007)	Asses et al. (2009)	Karpouzas et al. (2010)	El-Abbassi et al. (2013b)	Mekki et al. (2013)	Khoufi et al. (2015)	Range of Values
pH	—	5.2	5.1	5.7	5.3	5	4.8	4.8–5.7
Conductivity	mS/cm	5	—	11	24	81	17.5	5–81
COD	g/L	16.5	95	48	156	53	150	16.5–156
BOD	g/L	—	—	—	—	13.4	37.5	13.4–37.5
Dry residue	g/L	11.5	84.2	—	90	39.4	53.16	11.5–90
Lipids	g/L	—	—	—	7	—	—	7
Phenols	g/L	0.8	4.82	8.8	4.1	8.6	8.9	0.8–8.9
Sugar	g/L	1.3	—	—	4.3	—	—	1.3–4.3
Total nitrogen	g/L	0.06–0.3	—	0.9	—	0.5	—	0.06–0.9

On the other hand, wet olive pomace (from two-phase mills) is a solid waste with a strong odor and a doughy texture that makes its management and transport difficult. It is very humid and very difficult to treat (Table 1.1). It is generally subjected to a solar drying and subsequently used to extract oil with solvents. Table 1.2 shows the physicochemical characteristics of the OMWW according to different authors. The composition of OMWW is very variable and depends on olive variety, the fruit ripeness, the volume of added water, and the extraction process (press or centrifuge) (Ben Sassi et al., 2006). Typical OMWW composition by weight is: 83–94% water, 4–16% organic compounds, and 0.4–2.5% mineral salts (Davies et al., 2004). The organic fraction contains, among other components, 2–15% of phenolic compounds divided into low-molecular weight (tyrosol, hydroxytyrosol, *p*-coumaric acid, ferulic acid, syringic acid, protocatechuic acid etc.) and high molecular weight compounds (tannins, anthocyanins, etc.) as well as catechol-melaninic polymers (Obied et al., 2007). It is characterized by a dark-color (caused by lignin polymerization with phenolic compounds), increased acidity (pH about 5), and high electrical conductivity (Table 1.2). The latest parameter varies according to the salt content of OMWW which depends on the practices used for olive fruit conservation before milling.

The inorganic content of OMWW is mainly composed of metals. The mineral content of OMWW is shown in Table 1.3. Major elements in this wastewater are potassium (0.73–8.6 g/L), followed by calcium and sodium (0.03–1.1 and 0.05–0.8 g/L, respectively). Metals are important both from nutritional and toxicological viewpoints. Some metals, particularly iron, copper, and zinc, are essential to plant metabolism. It should be pointed out that OMWW had a high potential to be used as fertilizer if its polluting power was controlled.

1.5 ENVIRONMENTAL EFFECTS OF OMWW

OMWW are often disposed in evaporation ponds or various environmental receptors (Fig. 1.5) causing strong odor nuisance, soil contamination, plants growth inhibition, natural streams pollution as well as severe effects to the aquatic fauna and to the ecological status (Komnitsas et al., 2016). OMWW is one

Table 1.3 Mineral Composition of Olive Mill Wastewaters

Element	Unit	Mekki et al. (2006)	Karpouzas et al. (2010)	Moraetis et al. (2011)	Danellakis et al. (2011)	Piotrowska et al. (2011)	Range of Values
Pb	µg/L	—	—	6.7	10	—	6.7–10
Cd	µg/L	—	—	0.03	1	—	0.03–1
Fe	mg/L	23	6.5	—	8.88	20	6.5–23
Zn	mg/L	—	3.4	2.94	4.98	—	2.94–4.98
Cu	mg/L	—	2.4	—	2.96	—	2.4–2.96
Mn	mg/L	—	0.9	1.61	2.7	20	0.9–20
Mg	g/L	0.19	0.12	0.11	0.11	0.03	0.03–0.19
Ca	g/L	0.9	1.1	0.15	0.29	0.03	0.03–1.1
K	g/L	8.6	6.1	4.22	0.73	3.47	0.73–8.6
Na	g/L	0.8	0.07	—	0.15	0.05	0.05–0.8

of the most polluting effluents produced by the agrofood industries due to its high organic load and a wide range of contaminants, including organo-halogenated pollutants, fatty acids, phenolic compounds, and tannins (Karaouzas et al., 2011; Ntougias et al., 2013). The high phenolic nature of OMWW and its organic contents make it highly resistant to biodegradation (Zirehpour et al., 2014). OMWW shows poor biodegradability and high phytotoxicity due to the presence of phenolic compounds. Likewise, the presence of reduced sugars can stimulate microbial respiration and lower dissolved oxygen concentrations (McNamara et al., 2008). OMWW shows higher chemical oxygen demand and biological oxygen demand values of domestic sewage, ranging from 80 to 200 g/L and from 12 to 63 g/L, respectively

FIGURE 1.5

In most mediterranean countries, olive mill wastewaters are often disposed of in evaporation ponds (A), however, in some cases are thrown in the receiving environment without prior treatment (B) (Marrakech region, Morocco).

(Adhoum and Monser, 2004; Ehaliotis et al., 1999). These values are higher than the pollution indexes of domestic sewage.

1.5.1 EFFECTS ON ATMOSPHERE

Olive mills engender gas emissions ensuing in considerable odor grievances. Many of low-boiling organic compounds and volatile organic acids produce characteristic odors that can be smelled around the olive mills. OMWW is generally discharged into natural waters or on the land and/or stored in poorly engineered evaporation ponds. This effluent undergoes a natural fermentation and emanates pungent gases, such as phenols, sulfur dioxide, and hydrogen sulfide. This fact leads to a considerable odor pollution particularly during oil production season (Lagoudianaki et al., 2003).

Olive kernel with high moisture content is also a source of odor nuisance, in particular during spring and summer. During olive kernel drying, extremely pungent odors are released. An analysis of the condensates from the crude cake dry-distillation exhibited that fatty compounds, organic acids (i.e., butyric, caproic, valeric, and isobutyric acids) and low molecular weight compounds particularly 4-ethylphenol, are the main constituent of olive kernel odors (Ruiz-Mendez et al., 2013; Le Verge, 2004). The acidification of olive fruit during open air storage is known to reduce the formation of 4-ethylphenol and subsequently restrict odor environmental effects (Ruiz-Mendez et al., 2013).

1.5.2 EFFECTS ON AQUATIC LIFE AND WATER RESOURCES

OMWW disposal leads to significant effects on natural water bodies, too. These impacts are mainly related to concentration, composition, and seasonal production. The most visible effect is the discoloring of streams and rivers. This change in color is mainly due to the oxidation and subsequent polymerization of tannins giving darkly colored phenols (Niaounakis and Halvadakis, 2006; Hamdi, 1993). The high reduced sugars content in OMWW stimulates microorganisms growth, lowering dissolved oxygen concentrations in water, and thus decreases the share available for other living organisms. Other deteriorating effects can result from the high phosphorus concentration, which accelerates the growth of algae and lead to eutrophication (devastating the entire ecological balance in natural streams). Conversely to carbon and nitrogen (which flee after degradation as carbon dioxide and atmospheric nitrogen), phosphorus cannot be degraded. This means that phosphorus is taken up merely to a small level through the food chain, plant—invertebrates—fish—prehensile birds.

The presence of considerable amount of nutrients in OMWW supplies an appropriate medium for pathogens to multiply and infect streams, which have stern consequences to aquatic life, and humans, coming in contact with infected water. The impacts of OMWW discharge to freshwater and oligotrophic marine environments were assessed by Pavlidou et al. (2014). At this study, the enrichment of freshwater and the coastal zone of Messiniakos Gulf (Greece) in ammonia, nitrite, phenols, total organic carbon, copper, manganese, and nickel was directly correlated with OMWW disposal. Toxicity tests using 24 h LC50 *Palaemonidae* shrimp confirmed that OMWW possesses very high toxicity in the aquatic environment (Pavlidou et al., 2014).

A later study revealed the spatial and temporal structural worsening the aquatic system due to OMWW pollution with consequent reduction of the river ability for reducing the impacts of polluting compounds during self-purification. OMWW (even highly diluted) has important effects on the aquatic fauna and to the ecological status of fluvial ecosystems. The organic load of the effluent, substrate contamination, and distance from the mill outlet, are the main factors affecting macroinvertebrate assemblages. Besides, the typology of the stream site and volume of the effluent are the most important determinants of self-purification processes (Karaouzas et al., 2011).

1.5.3 EFFECT ON CROPS AND SOIL

OMWW contains large amount of organic substances, including phenolic compounds that may have negative impact when applied to soil. Because of the high concentrations of organic matter and mineral nutrients, OMWW could be a promising fertilizer for cropping systems. Ayoub et al. (2014) recommended the application of 10 L OMWW/m^2 to improve soil fertility and olive plant performance. Indeed, a significant increase in shoot growth, photosynthesis, fruit set, and fruit yield without any negative effects on oil quality parameters was observed. Besides, the concentrations of K, organic matter, phenolic compounds, and total microbial count were significantly increased in OMWW-treated soil (Ayoub et al., 2014). OMWW application shows short-term negative effects on soil chemical and biological properties, but it can be considered negligible after a suitable waiting period (Di Bene et al., 2013). In a recent study, the authors recommended soil amendment with OMWW 6 months before maize sowing for toxicity mitigation (Belaqziz et al., 2016). Along with the correct choice of convenient soils notably calcareous ones and tolerant crops, such as maize, this method could constitute an efficient approach for avoiding problems attributed to the uncontrolled disposal of OMWW and represents an economical alternative that provides a local fertilizer.

1.6 SUSTAINABLE MANAGEMENT OF OMWW: TREATMENT VERSUS VALORIZATION

Like all food processing wastes, OMWW have long been considered as a matter of treatment, minimization, and prevention due to the environmental effects induced by their disposal (Galanakis, 2012). Most of depolluting treatments of OMWW aim at the destruction of organic matter and phenolic compounds, hence the reduction of chemical oxygen demand and phytotoxicity, respectively (Mameri et al., 2000). However, the difficulties of OMWW treatment are mainly related to (Akdemir and Ozer, 2008):

1. its high organic loading,
2. seasonal operation,
3. high territorial scattering, and
4. the presence of nonbiodegradable organic compounds like long-chain fatty acids and phenols.

Many different processes have been suggested to treat OMWW, namely lagooning or direct watering on fields, cocomposting, physicochemical methods (flotation and settling, coagulation, oxidation by O_3 and Fenton reagent, flocculation, filtration, sedimentation, dilution open evaporating ponds, and incineration), ultrafiltration/reverse osmosis, chemical and electrochemical treatments, and manufacture into animal food (Rahmanian et al., 2014). Although the environmental problem is more intense for the disposal of OMWW, the earlier mentioned techniques are typically proposed to treat all kind of olive oil processing effluents. In an attempt to categorize the proposed methodologies of OMW management, three categories can be denoted:

1. Waste reduction via olive production systems conversion (i.e., two-phase instead of three-phase continuous systems).
2. Detoxification methods aiming at the reduction of impact of the pollution load to the recipient.
3. Recovery or recycling of components from OMW.

The conversion of olive oil production systems described earlier (Section 1.2). Table 1.4 presents a summary of detoxification technologies and their efficiency in spite of reducing pollution indexes. Physical processes are typically applied as a pretreatment step to remove the contained solids. Thermal processes are used to remove the contained water and condense the waste streams, however are ineffective due to the required operational cost. Despite their effectiveness, advanced oxidation processes are very expensive, too. Biological treatment requires a longer lag phase, whereas physicochemical methods, such as neutralization and precipitation are relatively cheaper, but they cannot diminish completely the pollution load of OMWW. Following these considerations, the researchers have proposed numerous combined treatments (chemical, physical, and biological). However, none of these treatments has found a widely accepted application since they have not been proved to be sustainable in a long-term base (Rahmanian et al., 2014).

As a consequence, researchers have reconsidered their point of view. More integrated methodologies should be developed combining treatment, recycling, valorization, and energy-producing processes, allowing the recovery of high added-value compounds and achieving the standard limits to reuse the purified effluent for irrigation or industrial purposes. This concept, typically considered within the biorefinery approach (de Jong and Jungmeier, 2015), makes the treatment process cost effective and results in an environmentally friendly olive oil production process. For instance, Zagklis et al. (2013) conducted a sustainability and benchmarking study of OMWW treatment methods, showing that the most effective processes in terms of organics reduction are membrane filtration, electrolysis, supercritical water oxidation, and photo-Fenton. Lower environmental impact was found with anaerobic digestion, coagulation, and lime processes, while the lowest-cost methods are composting and membrane filtration, owing to the added-value of composts and phenolic compounds, respectively (Zagklis et al., 2013).

Table 1.5 summarizes some of these valorization approaches. OMWW could be utilized either as a substrate for the growth of microorganisms and the production of fertilizers, bioproducts and animal feed, or as a cheap source for the recovery of compounds like phenols (e.g., hydroxytyrosol, oleuropein, phenolic acids, tannins, flavonols, anthocyanins, etc.) and dietary fiber (e.g., pectin) (Galanakis et al., 2010a,b,c,d,e; Galanakis, 2011). The latter are today used as additives in foodstuff due to their ability to provide advanced technological properties and health claims to the final product (Galanakis, 2012, 2013; Tsakona et al., 2012). Indeed, several studies have demonstrated that phenols are characterized by a wide array of biological activities like antioxidant, free radical scavenging, anti-inflammatory, anticarcinogenic, and antimicrobial activities (El-Abbassi et al., 2012b; Zbakh and El Abbassi, 2012). Kishikawa et al. (2015) assessed the bioactive characteristics of olive oil by-products, namely leaves, stems, flowers, OMW, fruit pulp, and seeds. According to their study, a large spectrum of biological activities was reported including the antibacterial activity of the leaves and flowers, the antiallergic activity of OMW, as well as the collagen-production-promoting activity of the leaves, stems, OMW, and fruit pulp. Hence, by-products of olive oil production have the potential to be further developed and used in the skin care industry (Kishikawa et al., 2015).

Table 1.4 Treatment Technologies of Olive Mill Wastewater

Category	Methodology	Description	Results	Notes	References
Physical	Sedimentation, filtration, flotation, centrifugation	Total solids removal	Total or partial removal of solids, 70% COD removal, 30% oil recovery	Common pretreatment methodology	Velioglu et al. (1987); Georgacakis and Dalis (1993)
	Micro-, ultra-, nanofiltration, reverse osmosis	Separation of compounds (existing in the same phase) according to their molecular weight	99% COD removal, but membrane fouling	They are applied in series, but the cost is high due to membrane fouling	Paraskeva et al. (2007); Russo (2007); Stoller and Angelo (2006); Turano et al. (2002)
	Evaporation				Masi et al. (2015)
	Sedimentation (settling)				De Martino et al. (2011)
	Centrifugation/ ultrafiltration				Turano et al. (2002)
	Adsorption/desorption	Nonionic XAD4, XAD16, and XAD7HP resins were used	A final concentration of 378 g/L in gallic acid equivalents was reached	The recovered phenolic compounds were concentrated through vacuum evaporation	Zagklis et al. (2015)
Thermal	Evaporation, distillation	Water removal and waste reduction	20–80% COD removal, needing additional treatment	High energy demands	Rozzi and Malpei (1996); Tsagaraki et al. (2007)
	Combustion, pyrolysis	Decomposition, waste elimination	Toxic gases production and high cost	Very high energy demands	Rozzi and Malpei (1996); Caputo et al. (2003)
	Solar distillation				Potoglou et al. (2004)
	Combined	Membrane distillation			El-Abbassi et al. (2012a,b, 2013a)
Physicochemical	Neutralization, precipitation, adsorption	Addition of $FeCl_3$, $Ca(OH)_2$/ MgO, Na_2SiO_3, adsorption on activated carbon	30–50% COD removal	80–95% COD removal by combining precipitation with adsorption	Sarika et al. (2005); Kestioglu et al. (2005); Adhoum and Monser (2004)

(*Continued*)

Table 1.4 Treatment Technologies of Olive Mill Wastewater (*cont.*)

Category	Methodology	Description	Results	Notes	References
	Oxidation and advanced oxidation processes	Ozonolysis, wet oxidation, O_3/H_2O_2 photolysis, photocatalysis	40–60% COD removal using simple oxidation practices	85% COD removal by combining methods	Javier Benitez et al. (2009); Chatzisymeon et al. (2009a); Chatzisymeon et al. (2009b)
	Chemical oxidation (Fenton reaction)	Ferric chloride catalyst was used for the activation of H_2O_2	86% COD removal	The produced water can be used for irrigation or can be directly discharged	Nieto et al. (2010)
	Wet hydrogen peroxide catalytic oxidation		System operating at 50°C reduced considerably the COD, color, and total phenolic contents		Azabou et al. (2010)
	Electro-Fenton				Kaplan et al. (2011); Kilic et al. (2013)
	Ozonation		Decrease of the total phenolic content (almost 50%) after 300 min of ozonation		Siorou et al. (2015)
	Lime treatment				Aktas et al. (2001)
	Electrocoagulation				Hanafi et al. (2010)
	Cloud point extraction (CPE)				Gortzi et al. (2008); El-Abbassi et al. (2014)
	Combined	Ferric chloride coagulation, lime precipitation, electrocoagulation and Fenton's reagent			Gursoy-Haksevenler et al. (2014)
Biological	Anaerobic processes	Dilution, nutrients addition, and alkalinity regulation	60–80% COD removal for 2–5 digestion days	90% COD removal for 25 digestion days	Dalis et al. (1996); Azbar et al. (2009); Azbar et al. (2008b)

	Aerobic processes	OMW cocomposting with sesame bark			Hachicha et al. (2009)
		Biofilm, activated sludge	5–75% COD removal for few days of digestion	80% COD for longer period	El Hajjouji et al. (2007); Velioglu et al. (1992)
	Mixing and digestion	Together with other agricultural wastes	75–90% COD removal	Nutrients and pH adjustment by combining wastes	Azbar et al. (2008a); Marques (2001)
	Enzymatic	Oxidative transformation of phenols by *Trametes trogii* laccases	laccases were capable of efficiently removing phenolic compounds		
Combined	Oxidation and biological processes	2–3 sequential methods	75% phenols 80–99 %COD removal	High cost due to processes combination	Bressan et al. (2004)
	Evaporation—condensation followed by biological treatment			The process generated a concentrate containing high concentrations of valuable chemicals, including polyphenols	Masi et al. (2015)

Table 1.5 Proposed Methods and Processes for the Valorization of OMWW

Process	Aim/Product	Statements	References
Continuous-flow adsorption/desorption process	Recovery of phenolic compounds	Amberlite XAD16 was used as adsorbent and acidified ethanol as desorption solvent. The adsorption yield was 20%.	Frascari et al. (2016)
Extraction using hydrophobic ionic liquids	Recovery of tyrosol	Hydrophobic ionic liquids were proposed to replace conventional volatile organic compounds as extraction solvents to recover tyrosol from OMWW.	Larriba et al. (2016)
Integrated process including fermentation, spray drying, and encapsulation technologies	Olive paste spread, olive powder, and encapsulated phenols	A number of valuable by-products were produced from OMWW to be included in food formulations.	Goula and Lazarides (2015)
Solar distillation	Drying of OMWW and the recovery of phenolic compounds	A solar distillator was used for the simultaneous solar drying of OMWW and the recovery of phenolic compounds with antioxidant properties in the distillate.	Sklavos et al. (2015)
Dark fermentation	Biohydrogen production	Dark fermentation of OMWW by pretreated thermophilic anaerobic digestate.	Ghimire et al. (2015)
Enzymatic hydrolysis/ microfiltration—ultrafiltration	Recovery of natural hydroxytyrosol	A chemical-free large-scale process consists of enzymatic pretreatment followed by two tangential flow membrane separation steps, allowing the production of a hydroxytyrosol rich extract.	Hamza and Sayadi (2015)
Impregnation on dry biomasses	Solid biofuels production	The addition of OMW leads to an increase of the biomass heating values with no negative effect on their firing quality.	Kraiem et al. (2014); Jeguirim et al. (2012)
Dyebath for dyeing wool	Natural dyes for textile dyeing	OMWW can represent a possible resource for the dyeing of textile materials. It contains a valuable source of abundant natural coloring substances.	Haddar et al. (2014); Meksi et al. (2012)
Fired clay brick production	Ceramic building materials	OMWW was used to replace fresh water in clay brick manufacture. The extrusion performance and technological properties were improved.	de la Casa et al. (2009)
Brick-making process	Building bricks	The incorporation of OMWW at a rate of 23% either maintained the physical and mechanical properties of bricks or improved them.	Mekki et al. (2008)

Several techniques (individually or in combination) have been proposed to recover phenols from OMW. These techniques include solvent extraction, chromatographic separations, centrifugation, membrane processes, (Rahmanian et al., 2014) and more recently high voltage electrical discharges, pulsed electric field, and ultrasounds (Roselló-Soto et al., 2015). Membrane separation processes are among the key physicochemical and nondestructive techniques applied for the separation of macro- and micromolecules in food waste streams (Galanakis, 2015; Galanakis et al., 2013, 2014). The treatment of OMWWs by membrane operations is generally focused on the development of integrated systems for producing effluents of acceptable quality for safe disposal into the environment (Paraskeva et al., 2007; Zirehpour et al., 2012).

Direct application of OMWW for agronomic purposes has been proposed by many researchers to take advantage of its high nutrients concentration and its potential for mobilizing soil ions, whereas, negative effects are associated with its high mineral salt content, low pH, and the presence of phenols (Kurtz et al., 2015; Belaqziz et al., 2016). To overcome the phytotoxic effects of OMWW spreading on soils, the cocomposting of OMWW with other agro-industrial wastes has been suggested (Hachicha et al., 2009; Abid and Sayadi, 2006). Composting is one of the main technologies for recycling OMW and transforming it into a fertilizer. This process allows the return of nutrients to cropland avoiding the negative effects often observed when applying these wastes directly to soil. The OMW could be absorbed in a solid substrate (lignocellulosic wastes or manures) before composting (Rigane and Medhioub, 2011). Cocomposting of OMW with poultry manure showed a significant phenols content decrease (99%) and a noticeable transformation of low to high molecular weight fraction during composting (Rigane et al., 2015). The application of OMW-compost showed to improve the chemical and physicochemical soil properties, mainly fertilizing elements, such as calcium, magnesium, nitrogen, potassium, and phosphorus (Rigane et al., 2015).

Soil amendment with OMWW is also known for its antimicrobial activity and suppressive effects against plant pathogens (Brenes et al., 2011). Several studies reported the potential of using OMWW as biopesticides against plant pathogens (Yangui et al., 2010, 2013; Debo et al., 2011). The OMWW can be used to suppress the growth of the main weed species (*Phelipanche ramosea*, *Amaranthus retroflexus Avena fatua*, *Alopecurus myosuroides*, among other species) without any negative effects on crop growth (Disciglio et al., 2015; Boz et al., 2010). Similarly, OMWW was effective against the main bacterial (i.e., *Clavibacter michiganensis*, *Erwinia toletana*, *Erwinia amylovora*) and fungal (i.e., *Aspergillus niger*, *Botrytis tulipae*, *Fusarium oxysporum*, *Penicillium* spp.) phytopathogens (Medina et al., 2011; Lykas et al., 2014). Besides, recent studies reported the biocide effect of OMWW on some pests, mollusks: *Isidorella newcombi* (Cayuela et al., 2008), nematodes: *Meloidogyne incognita* (Obied et al., 2007), and arthropods: *Aphis citricola*, *Euphyllura olivine* (Larif et al., 2013). Nevertheless, the pesticide effect of OMWW against mollusk, nematodes, and insects has been poorly investigated. Based on the available studies, OMWW can be utilized in agriculture for plant diseases control, in which case a by-product that is typically regarded as waste can be beneficial and useful. However, some measures should be respected at this case, especially in regards of dose and timing of use.

1.7 CONCLUSIONS AND FUTURE DIRECTIVES

Despite the overabundance of scientific researches aimed to increase the quality of olive oil and the efficiency of extraction systems, the industrial sector showed little predispositions toward the possibility of transforming the findings of the researchers into innovative plants. Indeed, the entire olive milling process has changed very little over the last decades. The introduction of the horizontal centrifuge (decanter), coupled with the malaxation process, has been the last big revolution in the technology field. However, the innovation transferred the problem to a new king of wastes without creating a real solution for the classical OMW.

Many solutions have been suggested for the valorization of OMW but many factors should be considered when selecting the best valorization method including total amount of effluents, investment costs, available land, industrial or agronomic environment, local laws, and most important needs. There is not a unique, sustainable solution. Sustainability depends on the specific needs of the local area and each olive oil industry separately. The majority of olive oil producer countries in the Mediterranean area suffer from desertification, so the water and organic matter reuse would be beneficial to improve soil fertility and control the erosion processes. Also in organic agriculture, the use of OMW as fertilizer and/or biopesticides could represent an important source of nutrients, protecting crops, and closing the cycle of residues–resources. However, phenols of OMW are too valuable to be diminished or discharged to the environment. Therefore, the recovery of phenols accompanied with their reutilization in different products and markets should come first. To conclude, the selection of the most suitable valorization strategy depends on the social, agricultural, or industrial environment of the olive oil production plants, but more importantly to the long-term sustainability of the provided solution. Solutions that promise the total treatment of OMW and at the same time collapse financially olive oil industries have been rejected in practice, not only because of sector's denial, but also because of society's tolerance that delays the enforcement of environmental legislations implementation.

REFERENCES

Abid, N., Sayadi, S., 2006. Detrimental effects of olive mill wastewater on the composting process of agricultural wastes. Waste Manag. 26 (10), 1099–1107.

Adhoum, N., Monser, L., 2004. Decolourization and removal of phenolic compounds from olive mill wastewater by electrocoagulation. Chem. Eng. Process. 43 (10), 1281–1287.

Akdemir, E.O., Ozer, A., 2008. Application of a statistical technique for olive oil mill wastewater treatment using ultrafiltration process. Sep. Purif. Technol. 62 (1), 222–227.

Aktas, E.S., Imre, S., Ersoy, L., 2001. Characterization and lime treatment of olive mill wastewater. Water Res. 35 (9), 2336–2340.

Aliakbarian, B., De Faveri, D., Converti, A., Perego, P., 2008. Optimisation of olive oil extraction by means of enzyme processing aids using response surface methodology. Biochem. Eng. J. 42 (1), 34–40.

Aliakbarian, B., Dehghani, F., Perego, P., 2009. The effect of citric acid on the phenolic contents of olive oil. Food Chem. 116 (3), 617–623.

Altieri, G., Di Renzo, G.C., Genovese, F., 2013. Horizontal centrifuge with screw conveyor (decanter): optimization of oil/water levels and differential speed during olive oil extraction. J. Food Eng. 119 (3), 561–572.

Altieri, G., Di Renzo, G.C., Genovese, F., Tauriello, A., D'Auria, M., Racioppi, R., Viggiani, L., 2014. Olive oil quality improvement using a natural sedimentation plant at industrial scale. Biosyst. Eng. 122, 99–114.

Altieri, G., Genovese, F., Tauriello, A., Di Renzo, G.C., 2015. Innovative plant for the separation of high quality virgin olive oil (VOO) at industrial scale. J. Food Eng. 166, 325–334.

Amirante, P., Catalano, P., Amirante, R., Clodoveo, M.L., Montel, G.L., Leone, A., Tamborrino, A., 2002. Prove sperimentali di estrazione di oli da olive snocciolate (experimental tests of olive oil extraction from depitted olives). OliVe Olio 6, 16–22.

Amirante, P., Clodoveo, M.L., Dugo, G., Leone, A., Tamborrino, A., 2006. Advance technology in virgin olive oil production from traditional and de-stoned pastes: influence of the introduction of a heat exchanger on oil quality. Food Chem. 98 (4), 797–805.

Asses, N., Ayed, L., Bouallagui, H., Ben Rejeb, I., Gargouri, M., Hamdi, M., 2009. Use of *Geotrichum candidum* for olive mill wastewater treatment in submerged and static culture. Bioresour. Technol. 100 (7), 2182–2188.

Avraamides, M., Fatta, D., 2008. Resource consumption and emissions from olive oil production: a life cycle inventory case study in Cyprus. J. Clean. Prod. 16 (7), 809–821.

Ayoub, S., Al-Absi, K., Al-Shdiefat, S., Al-Majali, D., Hijazean, D., 2014. Effect of olive mill wastewater land-spreading on soil properties, olive tree performance and oil quality. Sci. Hort. 175, 160–166.

Azabou, S., Najjar, W., Bouaziz, M., Ghorbel, A., Sayadi, S., 2010. A compact process for the treatment of olive mill wastewater by combining wet hydrogen peroxide catalytic oxidation and biological techniques. J. Hazard. Mater. 183 (1-3), 62–69.

Azbar, N., Keskin, T., Yuruyen, A., 2008a. Enhancement of biogas production from olive mill effluent (OME) by co-digestion. Biomass Bioenerg. 32 (12), 1195–1201.

Azbar, N., Keskin, T., Catalkaya, E.C., 2008b. Improvement in anaerobic degradation of olive mill effluent (OME) by chemical pretreatment using batch systems. Biochem. Eng. J. 38 (3), 379–383.

Azbar, N., Tutuk, F., Keskin, T., 2009. Biodegradation performance of an anaerobic hybrid reactor treating olive mill effluent under various organic loading rates. Int. Biodeter. Biodegr. 63 (6), 690–698.

Baiano, A., Terracone, C., Viggiani, I., Del Nobile, M.A., 2014. Changes produced in extra-virgin olive oils from cv. Coratina during a prolonged storage treatment. Czech J. Food Sci. 32 (1), 1–9.

Belaqziz, M., El-Abbassi, A., Lakhal, E.K., Agrafioti, E., Galanakis, C.M., 2016. Agronomic application of olive mill wastewater: effects on maize production and soil properties. J. Environ. Manag. 171, 158–165.

Ben Mansour, A., Flamini, G., Selma, Z.B., Dreau, Y.L., Artaud, J., Abdelhedi, R., Bouaziz, M., 2015. Comparative study on volatile compounds, fatty acids, squalene and quality parameters from whole fruit, pulp and seed oils of two tunisian olive cultivars using chemometrics. Eur. J. Lipid Sci. Technol. 117 (7), 976–987.

Ben Sassi, A., Boularbah, A., Jaouad, A., Walker, G., Boussaid, A., 2006. A comparison of olive oil mill wastewaters (OMW) from three different processes in Morocco. Process Biochem. 41 (1), 74–78.

Bordons, C., Nunez-Reyes, A., 2008. Model based predictive control of an olive oil mill. J. Food Eng. 84 (1), 1–11.

Boz, O., Ogut, D., Dogan, M.N., 2010. The phytotoxicity potential of olive processing waste on selected weeds and crop plants. Phytoparasitica 38 (3), 291–298.

Brenes, M., Garcia, A., de los Santos, B., Medina, E., Romero, C., de Castro, A., Romero, F., 2011. Olive glutaraldehyde-like compounds against plant pathogenic bacteria and fungi. Food Chem. 125 (4), 1262–1266.

Bressan, M., Liberatore, L., D'Alessandro, N., Tonucci, L., Belli, C., Ranalli, G., 2004. Improved combined chemical and biological treatments of olive oil mill wastewaters. J. Agric. Food Chem. 52 (5), 1228–1233.

Caponio, F., Squeo, G., Monteleone, J.I., Paradiso, V.M., Pasqualone, A., Summo, C., 2015. First and second centrifugation of olive paste: Influence of talc addition on yield, chemical composition and volatile compounds of the oils. LWT Food Sci. Technol. 64 (1), 439–445.

Caponio, F., Squeo, G., Difonzo, G., Pasqualone, A., Summo, C., Paradiso, V.M., 2016. Has the use of talc an effect on yield and extra virgin olive oil quality? J. Sci. Food Agric. 96 (10), 3292–3299.

Caputo, A.C., Scacchia, F., Pelagagge, P.M., 2003. Disposal of by-products in olive oil industry: waste-to-energy solutions. Appl. Therm. Eng. 23 (2), 197–214.

Catalano, P., Pipitone, F., Calafatello, A., Leone, A., 2003. Productive efficiency of decanters with short and variable dynamic pressure cones. Biosyst. Eng. 86 (4), 459–464.

Catania, P., Vallone, M., Pipitone, F., Inglese, P., Aiello, G., La Scalia, G., 2013. An oxygen monitoring and control system inside a malaxation machine to improve extra virgin olive oil quality. Biosyst. Eng. 114 (1), 1–8.

Catania, P., Vallone, M., Farid, A., De Pasquale, C., 2016. Effect of O_2 control and monitoring on the nutraceutical properties of extra virgin olive oils. J. Food Eng. 169, 179–188.

Cayuela, M.L., Millner, P.D., Meyer, S.L.F., Roig, A., 2008. Potential of olive mill waste and compost as biobased pesticides against weeds, fungi and nematodes. Science Total Environ. 399 (1–3), 11–18.

Chatzisymeon, E., Diamadopoulos, E., Mantzavinos, D., 2009a. Effect of key operating parameters on the non-catalytic wet oxidation of olive mill wastewaters. Water Sci. Technol. 59 (12), 2509–2518.

Chatzisymeon, E., Xekoukoulotakis, N.P., Mantzavinos, Dionissios, 2009b. Determination of key operating conditions for the photocatalytic treatment of olive mill wastewaters. Catal. Today 144 (1–2), 143–148.

Chiaiese, P., Francesca, P., Filippo, T., Carmine, L., Gabriele, P., Antonino, P., Edgardo, F., 2011. Engineered tobacco and microalgae secreting the fungal laccase POXA1b reduce phenol content in olive oil mill wastewater. Enzyme Microb. Technol. 49 (6–7), 540–546.

Clodoveo, M.L., 2012. Malaxation: influence on virgin olive oil quality. Past, present and future–an overview. Trends Food Sci. Technol. 25 (1), 13–23.

Clodoveo, M.L., Hbaieb, R.H., Kotti, F., Mugnozza, G.S., Gargouri, M., 2014. Mechanical strategies to increase nutritional and sensory quality of virgin olive oil by modulating the endogenous enzyme activities. Compr. Rev. Food Sci. Food Saf. 13 (2), 135–154.

Dalis, D., Anagnostidis, K., Lopez, A., Letsiou, I., Hartmann, L., 1996. Anaerobic digestion of total raw olive-oil wastewater in a two-stage pilot-plant (up-flow and fixed-bed bioreactors). Bioresour. Technol. 57 (3), 237–243.

Danellakis, D., Ntaikou, I., Kornaros, M., Dailianis, S., 2011. Olive oil mill wastewater toxicity in the marine environment: alterations of stress indices in tissues of mussel Mytilus galloprovincialis. Aquat. Toxicol. 101 (2), 358–366.

Davies, L.C., Vilhena, A.M., Novais, J.M., Martins-Dias, S., 2004. Olive mill wastewater characteristics: modelling and statistical analysis. Grasas Aceites 55 (3), 233–241.

De Faveri, D., Aliakbarian, B., Avogadro, M., Perego, P., Converti, A., 2008. Improvement of olive oil phenolics content by means of enzyme formulations: effect of different enzyme activities and levels. Biochem. Eng. J. 41 (2), 149–156.

de Jong, E., Jungmeier, G., 2015. Biorefinery concepts in comparison to petrochemical refineries. In: Ashok, P., Rainer, H., Christian, L., Mohammad, T., Madhavan, N. (Eds.), Industrial Biorefineries & White Biotechnology. Elsevier, Waltham, USA, pp. 3–33.

de la Casa, J.A., Lorite, M., Jimenez, J., Castro, E., 2009. Valorisation of wastewater from two-phase olive oil extraction in fired clay brick production. J. Hazard. Mater. 169 (1–3), 271–278.

De Martino, A., Arienzo, M., Iorio, M., Vinale, F., Lorito, M., Prenzler, P.D., Ryan, D., Obied, H.K., 2011. Detoxification of olive mill wastewaters by zinc-aluminium layered double hydroxides. Appl. Clay Sci. 53 (4), 737–744.

Debo, A., Yangui, T., Dhouib, A., Ksantini, M., Sayadi, S., 2011. Efficacy of a hydroxytyrosol-rich preparation from olive mill wastewater for control of olive psyllid, *Euphyllura olivina*, infestations. Crop Prot. 30 (12), 1529–1534.

Del Caro, A., Vacca, V., Poiana, M., Fenu, P., Piga, A., 2006. Influence of technology, storage and exposure on components of extra virgin olive oil (Bosana cv) from whole and de-stoned fruits. Food Chem. 98 (2), 311–316.

Demicheli, M.C., Bontoux, L., 1997. Novel technologies for olive oil manufacturing and their incidence on the environment. Fresen. Environ. Bull. 6 (5), 240–247.

Di Bene, C., Pellegrino, E., Debolini, M., Silvestri, N., Bonari, E.., 2013. Short- and long-term effects of olive mill wastewater land spreading on soil chemical and biological properties. Soil Biol. Biochem. 56, 21–30.

Disciglio, G., Lops, F., Carlucci, A., Gatta, G., Tarantino, A., Frabboni, L., Tarantino, E., 2015. Effects of different methods to control the parasitic weed *Phelipanche ramosa* (L.) Pomel in processing tomato crops. Ital. J. Agron. 11 (1), 39–46.

EC, European Union Commission, 2001. Council Regulation (EC) No. 1513/2001 of 23 July 2001 amending regulation (EC) 136/66/EEC and No. 1638/98 as regards the extension of the period of validity of the aid scheme and the quality strategy for olive oil. Off. J. Eur. Commun. L 201, 4–7.

Ehaliotis, C., Papadopoulou, K., Kotsou, M., Mari, I., Balis, C., 1999. Adaptation and population dynamics of Azotobacter vinelandii during aerobic biological treatment of olive-mill wastewater. FEMS Microbiol. Ecol. 30 (4), 301–311.

El Hajjouji, H., Fakharedine, N., Ait Baddi, G., Winterton, P., Bailly, J.R., Revel, J.C., Hafidi, M., 2007. Treatment of olive mill waste-water by aerobic biodegradation: an analytical study using gel permeation chromatography, ultraviolet-visible and Fourier transform infrared spectroscopy. Bioresour. Technol. 98 (18), 3513–3520.

El-Abbassi, A., Kiai, H., Hafidi, A., Garcia-Payo, M.C., Khayet, M., 2012a. Treatment of olive mill wastewater by membrane distillation using polytetrafluoroethylene membranes. Sep. Purif. Technol. 98, 55–61.

El-Abbassi, A., Kiai, H., Hafidi, A., 2012b. Phenolic profile and antioxidant activities of olive mill wastewater. Food Chem. 132 (1), 406–412.

El-Abbassi, A., Hafidi, A., Khayet, M., Garcia-Payo, M.C., 2013a. Integrated direct contact membrane distillation for olive mill wastewater treatment. Desalination 323, 31–38.

El-Abbassi, A., Khayet, M., Kiai, H., Hafidi, A., Garcia-Payo, M.C., 2013b. Treatment of crude olive mill wastewaters by osmotic distillation and osmotic membrane distillation. Sep. Purif. Technol. 104, 327–332.

El-Abbassi, A., Kiai, H., Raiti, J., Hafidi, A., 2014. Cloud point extraction of phenolic compounds from pretreated olive mill wastewater. J. Environ. Chem. Eng. 2 (2), 1480–1486.

Espinola, F., Moya, M., de Torres, A., Castro, E., 2015. Comparative study of coadjuvants for extraction of olive oil. Eur. Food Res. Technol. 241 (6), 759–768.

Frascari, D., Molina Bacca, A.E., Zama, F., Bertin, L., Fava, F., Pinelli, D., 2016. Olive mill wastewater valorisation through phenolic compounds adsorption in a continuous flow column. Chem. Eng. J. 283, 293–303.

Galanakis, C.M., 2011. Olive fruit dietary fiber: components, recovery and applications. Trends Food Sci. Technol. 22 (4), 175–184.

Galanakis, C.M., 2012. Recovery of high added-value components from food wastes: conventional, emerging technologies and commercialized applications. Trends Food Sci. Technol. 26 (2), 68–87.

Galanakis, C.M., 2013. Emerging technologies for the production of nutraceuticals from agricultural by-products: a viewpoint of opportunities and challenges. Food Bioprod. Process. 91 (4), 575–579.

Galanakis, C.M., 2015. Separation of functional macromolecules and micromolecules: from ultrafiltration to the border of nanofiltration. Trends Food Sci. Technol. 42 (1), 44–63.

Galanakis, C.M., Tornberg, E., Gekas, V., 2010a. A study of the recovery of the dietary fibres from olive mill wastewater and the gelling ability of the soluble fibre fraction. LWT Food Sci. Technol. 43 (7), 1009–1017.

Galanakis, C.M., Tornberg, E., Gekas, V., 2010b. Clarification of high-added value products from olive mill wastewater. J. Food Eng. 99 (2), 190–197.

Galanakis, C.M., Tornberg, E., Gekas, V., 2010c. Dietary fiber suspensions from olive mill wastewater as potential fat replacements in meatballs. LWT Food Sci. Technol. 43 (7), 1018–1025.

Galanakis, C.M., Tornberg, E., Gekas, V., 2010d. Recovery and preservation of phenols from olive waste in ethanolic extracts. J. Chem. Technol. Biotechnol. 85 (8), 1148–1155.

Galanakis, C.M., Tornberg, E., Gekas, V., 2010e. The effect of heat processing on the functional properties of pectin contained in olive mill wastewater. LWT Food Sci. Technol. 43 (7), 1001–1008.

Galanakis, C.M., Markouli, E., Gekas, V., 2013. Recovery and fractionation of different phenolic classes from winery sludge using ultrafiltration. Sep. Purif. Technol. 107, 245–251.

Galanakis, C.M., Chasiotis, S., Botsaris, G., Gekas, V., 2014. Separation and recovery of proteins and sugars from Halloumi cheese whey. Food Res. Int. 65 (Pt. C), 477–483.

Georgacakis, D., Dalis, D., 1993. Controlled anaerobic digestion of settled olive-oil waste-water. Bioresour. Technol. 46 (3), 221–226.

Ghimire, A., Frunzo, L., Pontoni, L., d'Antonio, G., Lens, P.N.L., Esposito, G., Pirozzi, F., 2015. Dark fermentation of complex waste biomass for biohydrogen production by pretreated thermophilic anaerobic digestate. J. Environ. Manag. 152, 43–48.

Gila, A., Jiménez, A., Beltrán, G., Romero, A., 2015. Correlation of fatty acid composition of virgin olive oil with thermal and physical properties. Eur. J. Lipid Sci. Technol. 117 (3), 366–376.

Gila, A., Beltrán, G., Bejaoui, M.A., Sánchez, S., Nopens, Ingmar, Jiménez, A., 2016. Modeling the settling behavior in virgin olive oil from a horizontal screw solid bowl. J. Food Eng. 168, 148–153.

Gimeno, E., Castellote, A.I., Lamuela-Raventós, R.M., De la Torre, M.C., Lopez-Sabater, M.C., 2002. The effects of harvest and extraction methods on the antioxidant content (phenolics, α-tocopherol, and β-carotene) in virgin olive oil. Food Chem. 78 (2), 207–211.

Gortzi, O., Lalas, S., Chatzilazarou, A., Katsoyannos, E., Papaconstandinou, S., Dourtoglou, E., 2008. Recovery of natural antioxidants from olive mill wastewater using Genapol-X080. J. Am. Oil Chem. Soc. 85, 8.

Goula, A.M., Lazarides, H.N., 2015. Integrated processes can turn industrial food waste into valuable food by-products and/or ingredients: the cases of olive mill and pomegranate wastes. J. Food Eng. 167, 45–50.

Guermazi, Z., Gharsallaoui, M., Perri, E., Gabsi, S., Benincasa, C., 2015. Characterization of extra virgin olive oil obtained from whole and destoned fruits and optimization of oil extraction with a physical coadjuvant (Talc) using surface methodology. J. Anal. Bioanal. Tech. 6 (6), 1.

Guerrini, L., Masella, P., Migliorini, M., Cherubini, C., Parenti, A., 2015. Addition of a steel pre-filter to improve plate filter-press performance in olive oil filtration. J. Food Eng. 157, 84–87.

Gursoy-Haksevenler, Hande, B., Dogruel, S., Arslan-Alaton, I., 2014. Effect of ferric chloride coagulation, lime precipitation, electrocoagulation and the Fenton's reagent on the particle size distribution of olive mill wastewater. Int. J. Global Warm. 6 (2−3), 194–211.

Hachicha, S., Cegarra, J., Sellami, F., Hachicha, R., Drira, N., Medhioub, K., Ammar, E., 2009. Elimination of polyphenols toxicity from olive mill wastewater sludge by its co-composting with sesame bark. J. Hazard. Mater. 161 (2−3), 1131–1139.

Haddar, W., Baaka, N., Meksi, N., Elksibi, I., Mhenni, M.F., 2014. Optimization of an ecofriendly dyeing process using the wastewater of the olive oil industry as natural dyes for acrylic fibres. J. Clean. Prod. 66, 546–554.

Hamdi, M., 1993. Thermoacidic precipitation of darkly colored polyphenols of olive mill wastewaters. Environ. Technol. 14 (5), 495–500.

Hamza, M., Sayadi, S., 2015. Valorisation of olive mill wastewater by enhancement of natural hydroxytyrosol recovery. Int. J. Food Sci. Technol. 50 (3), 826–833.

Hanafi, F., Assobhei, O., Mountadar, M., 2010. Detoxification and discoloration of Moroccan olive mill wastewater by electrocoagulation. J. Hazard. Mater. 174 (1-3), 807–812.

Javier Benitez, F., Acero, J.L., Leal, A.I., Gonzalez, Manuel, 2009. The use of ultrafiltration and nanofiltration membranes for the purification of cork processing wastewater. J. Hazard. Mater. 162 (2–3), 1438–1445.

Jeguirim, M., Chouchene, A., Reguillon, A.F., Trouve, G., Buzit, G.L., 2012. A new valorisation strategy of olive mill wastewater: impregnation on sawdust and combustion. Resour. Conserv. Recy. 59, 4–8.

Jiménez, B., Sánchez-Ortiz, A., Rivas, Ana, 2014. Influence of the malaxation time and olive ripening stage on oil quality and phenolic compounds of virgin olive oils. Int. J. Food Sci. Technol. 49 (11), 2521–2527.

Kapellakis, I.E., Tsagarakis, K.P., Crowther, J.C., 2008. Olive oil history, production and by-product management. Rev. Environ. Sci. Biotechnol. 7 (1), 1–26.

Kaplan, F., Hesenov, A., Gozmen, B., Erbatur, O., 2011. Degradations of model compounds representing some phenolics in olive mill wastewater via electro-Fenton and photoelectro-Fenton treatments. Environ. Technol. 32 (7), 685–692.

Karaouzas, I., Skoulikidis, N.T., Giannakou, U., Albanis, T.A., 2011. Spatial and temporal effects of olive mill wastewaters to stream macroinvertebrates and aquatic ecosystems status. Water Res. 45 (19), 6334–6346.

Karpouzas, D.G., Ntougias, S., Iskidou, E., Rousidou, C., Papadopoulou, K.K., Zervakis, G.I., Ehaliotis, C., 2010. Olive mill wastewater affects the structure of soil bacterial communities. Appl. Soil Ecol. 45 (2), 101–111.

Kestioglu, K., Yonar, T., Azbar, N., 2005. Feasibility of physico-chemical treatment and advanced oxidation processes (AOPs) as a means of pretreatment of olive mill effluent (OME). Process Biochem. 40 (7), 2409–2416.

Khoufi, S., Louhichi, A., Sayadi, S., 2015. Optimization of anaerobic co-digestion of olive mill wastewater and liquid poultry manure in batch condition and semi-continuous jet-loop reactor. Bioresour. Technol. 182, 67–74.

Kilic, M.Y., Yonar, T., Kestioglu, K., 2013. Pilot-scale treatment of olive oil mill wastewater by physicochemical and advanced oxidation processes. Environ. Technol. 34 (12), 1521–1531.

Kiritsakis, A., Nanos, G.D., Polymenopulos, Z., Thomai, T., Sfakiotakis, E.M., 1998. Effect of fruit storage conditions on olive oil quality. J. Am. Oil Chem. Soc. 75 (6), 721–724.

Kishikawa, A., Ashour, A., Zhu, Q.C., Yasuda, M., Ishikawa, H., Shimizu, K., 2015. Multiple biological effects of olive oil by-products such as leaves, stems, flowers, olive milled waste, fruit pulp, and seeds of the olive plant on skin. Phytother. Res. 29 (6), 877–886.

Komnitsas, K., Modis, K., Doula, M., Kavvadias, V., Sideri, D., Zaharaki, D., 2016. Geostatistical estimation of risk for soil and water in the vicinity of olive mill wastewater disposal sites. Desalin. Water Treat. 57 (7), 2982–2995.

Kraiem, N., Jeguirim, M., Limousy, L., Lajili, M., Dorge, S., Michelin, L., Said, R., 2014. Impregnation of olive mill wastewater on dry biomasses: impact on chemical properties and combustion performances. Energy 78, 479–489.

Kurtz, M.P., Peikert, B., Bruehl, C., Dag, A., Zipori, I., Shoqeir, J.H., Schaumann, G.E., 2015. Effects of olive mill wastewater on soil microarthropods and soil chemistry in two different cultivation scenarios in Israel and Palestinian territories. Agriculture 5 (3), 857–878.

Lagoudianaki, E., Manios, T., Geniatakis, M., Frantzeskaki, N., Manios, V., 2003. Odor control in evaporation ponds treating olive mill wastewater through the use of Ca(OH)(2). J. Environ. Sci. Health A Tox. Hazard. Subst. Environ. Eng. 38 (11), 2537–2547.

Larif, M., Zarrouk, A., Soulaymani, A., Elmidaoui, A., 2013. New innovation in order to recover the polyphenols of olive mill wastewater extracts for use as a biopesticide against the *Euphyllura olivina* and *Aphis citricola*. Res. Chem. Intermediat. 39 (9), 4303–4313.

Larriba, M., Omar, S., Navarro, P., Garcia, J., Rodriguez, F., Gonzalez-Miquel, M., 2016. Recovery of tyrosol from aqueous streams using hydrophobic ionic liquids: a first step towards developing sustainable processes for olive mill wastewater (OMW) management. RSC Adv. 6 (23), 18751–18762.

Le Verge, S., 2004. La fertilization a partir des residues de la trituration des olives. Le Nouvel Olivier (OCL) 42, 5–21.

Leone, A., Romaniello, R., Zagaria, R., Tamborrino, A., 2014. Development of a prototype malaxer to investigate the influence of oxygen on extra-virgin olive oil quality and yield, to define a new design of machine. Biosys. Eng. 118, 95–104.

Leone, A., Romaniello, R., Zagaria, R., Sabella, E., De Bellis, L., Tamborrino, A., 2015. Machining effects of different mechanical crushers on pit particle size and oil drop distribution in olive paste. Eur. J. Lipid Sci. Technol. 117 (8), 1271–1279.

Leone, A., Esposto, S., Tamborrino, A., Romaniello, R., Taticchi, A., Urbani, S., Servili, M., 2016. Using a tubular heat exchanger to improve the conditioning process of the olive paste: evaluation of yield and olive oil quality. Eur. J. Lipid Sci. Technol. 118 (2), 308–317.

Lykas, C., Vagelas, I., Gougoulias, N., 2014. Effect of olive mill wastewater on growth and bulb production of tulip plants infected by bulb diseases. Span. J. Agric. Res. 12 (1), 233–243.

Mameri, N., Halet, F., Drouiche, M., Grib, H., Lounici, H., Belhocine, D., Pauss, A., Piron, D., 2000. Treatment of olive mill washing water by ultrafiltration. Can. J. Chem. Eng. 78 (3), 590–595.

Marques, I.P., 2001. Anaerobic digestion treatment of olive mill wastewater for effluent re-use in irrigation. Desalination 137 (1–3), 233–239.

Masella, P., Parenti, A., Spugnoli, P., Calamai, L., 2009. Influence of vertical centrifugation on extra virgin olive oil quality. J. Am. Oil Chem. Soc. 86 (11), 1137–1140.

Masella, P., Parenti, A., Spugnoli, P., Calamai, L., 2012. Vertical centrifugation of virgin olive oil under inert gas. Eur. J. Lipid Sci. Technol. 114 (9), 1094–1096.

Masi, F., Bresciani, R., Munz, G., Lubello, C., 2015. Evaporation-condensation of olive mill wastewater: Evaluation of condensate treatability through SBR and constructed Wetlands. Ecol. Eng. 80, 156–161.

McNamara, C., Anastasiou, C., Oflaherty, V., Mitchell, R., 2008. Bioremediation of olive mill wastewater. Int. Biodeter. Biodegr. 61 (2), 127–134.

Medina, E., Romero, C., de los Santos, B., de Castro, A., Garcia, A., Romero, F., Brenes, M., 2011. Antimicrobial activity of olive solutions from stored Alpeorujo against plant pathogenic microorganisms. J. Agric. Food Chem. 59 (13), 6927–6932.

Mekki, A., Dhouib, A., Sayadi, S., 2006. Changes in microbial and soil properties following amendment with treated and untreated olive mill wastewater. Microbiol. Res. 161 (2), 93–101.

Mekki, H, Anderson, M., Benzina, M, Ammar, E., 2008. Valorization of olive mill wastewater by its incorporation in building bricks. J. Hazard. Mater. 158 (2−3), 308–315.

Mekki, A., Dhouib, A., Sayadi, S., 2013. Review: effects of olive mill wastewater application on soil properties and plants growth. Int. J. Recyc. Org. Waste Agric. 2 (1), 1–7.

Meksi, N., Haddar, W., Hammami, S., Mhenni, M.F., 2012. Olive mill wastewater: a potential source of natural dyes for textile dyeing. Ind. Crop. Prod. 40, 103–109.

Moraetis, D., Stamati, F.E., Nikolaidis, N.P., Kalogerakis, N., 2011. Olive mill wastewater irrigation of maize: impacts on soil and groundwater. Agric. Water Manag. 98 (7), 1125–1132.

Niaounakis, M., Halvadakis, C.P., 2006. Olive Processing Waste Management Literature Review and Patent Survey. vol. 5, Waste management series: Elsevier, Amsterdam.

Nieto, L.M., Hodaifa, G., Vives, S.R., Casares, J.A.G., 2010. Industrial plant for olive mill wastewater from two-phase treatment by chemical oxidation. J. Environ. Eng. 136 (11), 1309–1313.

Ntougias, S, Gaitis, F., Katsaris, P., Skoulika, S., Iliopoulos, N., Zervakis, G.I., 2013. The effects of olives harvest period and production year on olive mill wastewater properties—evaluation of Pleurotus strains as bioindicators of the effluent's toxicity. Chemosphere 92 (4), 399–405.

Obied, H.K., Bedgood, Jr., D.R., Prenzler, P.D., Robards, K., 2007. Bioscreening of Australian olive mill waste extracts: biophenol content, antioxidant, antimicrobial and molluscicidal activities. Food Chem. Toxicol. 45 (7), 1238–1248.

Paraskeva, P., Diamadopoulos, E., 2006. Technologies for olive mill wastewater (OMW) treatment: a review. J. Chem. Technol. Biotechnol. 81, 1475–1485.

Paraskeva, C.A., Papadakis, V.G., Tsarouchi, E., Kanellopoulou, D.G., Koutsoukos, P.G., 2007. Membrane processing for olive mill wastewater fractionation. Desalination 213 (1–3), 218–229.

Pavlidou, A., Anastasopoulou, E., Dassenakis, M., Hatzianestis, I., Paraskevopoulou, V., Simboura, N., Rousselaki, E., Drakopoulou, P., 2014. Effects of olive oil wastes on river basins and an oligotrophic coastal marine ecosystem: a case study in Greece. Sci. Total Environ. 497, 38–49.

Piotrowska, A., Rao, M.A., Scotti, R., Gianfreda, L., 2011. Changes in soil chemical and biochemical properties following amendment with crude and dephenolized olive mill waste water (OMW). Geoderma 161 (1−2), 8–17.

Potoglou, D., Kouzeli-Katsiri, A., Haralambopoulos, D., 2004. Solar distillation of olive mill wastewater. Renew. Energy 29 (4), 569–579.

Rahmanian, N., Jafari, S.M., Galanakis, C.M., 2014. Recovery and removal of phenolic compounds from olive mill wastewater. J. Am. Oil Chem. Soc. 91 (1), 1–18.

Ranalli, A., Marchegiani, D., Pardi, D., Contento, S., Girardi, F., Kotti, F., 2009. Evaluation of functional phytochemicals in destoned virgin olive oil. Food Bioprocess Technol 2 (3), 322–327.

Reboredo-Rodriguez, P., Gonzalez-Barreiro, C., Cancho-Grande, B., Fregapane, G., Salvador, M.D., Simal-Gandara, J., 2015. Characterisation of extra virgin olive oils from Galician autochthonous varieties and their co-crushings with Arbequina and Picual cv. Food Chem. 176, 493–503.

Reboredo-Rodríguez, P., González-Barreiro, C., Cancho-Grande, B., Simal-Gándara, J., 2014. Improvements in the malaxation process to enhance the aroma quality of extra virgin olive oils. Food Chem. 158, 534–545.

Rigane, H., Medhioub, K., 2011. Cocomposting of olive mill wastewater with manure and agro-industrial wastes. Compost Sci. Util. 19 (2), 129–134.

Rigane, H., Chtourou, M., Mahmoud, I.B., Medhioub, K., Ammar, E., 2015. Polyphenolic compounds progress during olive mill wastewater sludge and poultry manure co-composting, and humic substances building (Southeastern Tunisia). Waste Manag. Res. 33 (1), 73–80.

Rodis, P.S., Karathanos, V.T., Mantzavinou, A., 2002. Partitioning of olive oil antioxidants between oil and water phases. J. Agric. Food Chem. 50 (3), 596–601.

Roselló-Soto, E., Galanakis, C.M., Brnčić, M., Orlien, V., Trujillo, F.J., Mawson, R., Knoerzer, K., Tiwari, B.K., Barba, F.J., 2015. Clean recovery of antioxidant compounds from plant foods, by-products and algae assisted by ultrasounds processing. Modeling approaches to optimize processing conditions. Trends Food Sci. Technol. 42 (2), 134–149.

Rozzi, A., Malpei, F., 1996. Treatment and disposal of olive mill effluents. Int. Biodeter. Biodegr. 38 (3–4), 135–144.

Ruiz-Mendez, M.V., Romero, C., Medina, E., Garcia, A., de Castro, A., Brenes, M., 2013. Acidification of Alperujo paste prevents off-odors during their storage in open air. J. Am. Oil Chem. Soc. 90 (3), 401–406.

Russo, C., 2007. A new membrane process for the selective fractionation and total recovery of polyphenols, water and organic substances from vegetation waters (VW). J. Membr. Sci. 288 (1–2), 239–246.

Sadkaoui, A., Jiménez, A., Pacheco, R., Beltrán, G., 2015. Micronized natural talc with a low particle size and a high carbonate rate is more effective at breaking down oil in water emulsion. Eur. J. Lipid Sci. Technol. 118 (4), 545–552.

Sarika, R., Kalogerakis, N., Mantzavinos, D., 2005. Treatment of olive mill effluents. Part II. Complete removal of solids by direct flocculation with poly-electrolytes. Environ. Int. 31 (2), 297–304.

Servili, M., 2014. The phenolic compounds: a commercial argument in the economic war to come on the quality of olive oil? OCL 21 (5), D509.

Servili, M., Selvaggini, R., Esposto, S., Taticchi, A., Montedoro, G.F., Morozzi, G., 2004. Health and sensory properties of virgin olive oil hydrophilic phenols: agronomic and technological aspects of production that affect their occurrence in the oil. J. Chromatogr. A 1054 (1–2), 113–127.

Servili, M., Taticchi, A., Esposto, S., Urbani, S., Selvaggini, R., Montedoro, G.F., 2007. Effect of olive stoning on the volatile and phenolic composition of virgin olive oil. J. Agric. Food Chem. 55 (17), 7028–7035.

Servili, M., Esposto, S., Taticchi, A., Urbani, S., Di Maio, I., Veneziani, G., Selvaggini, R., 2015. New approaches to virgin olive oil quality, technology, and by products valorization. Eur. J. Lipid Sci. Technol. 117 (11), 1882–1892.

Siorou, S., Vgenis, T.T., Dareioti, M.A., Vidali, M.S., Efthimiou, I., Kornaros, M., Vlastos, D., Dailianis, S., 2015. Investigation of olive mill wastewater (OMW) ozonation efficiency with the use of a battery of selected ecotoxicity and human toxicity assays. Aquat. Toxicol. 164, 135–144.

Skaltsounis, A.L., Argyropoulou, A., Aligiannis, N., Xynos, N., 2015. Recovery of high added value compounds from olive tree products and olive processing byproducts. In: Boskou, D. (Ed.), Olive, Olive Oil Bioactive Constituents. AOCS Press, Urbana, Illinois USA, pp. 333–356.

Sklavos, S., Gatidou, G., Stasinakis, A.S., Haralambopoulos, D., 2015. Use of solar distillation for olive mill wastewater drying and recovery of polyphenolic compounds. J. Environ. Manag. 162, 46–52.

Stoller, M., Angelo, C., 2006. Technical optimization of a batch olive wash wastewater treatment membrane plant. Desalination 200 (1-3), 734–736.

Tamborrino, A., Romaniello, R., Zagaria, R., Leone, A., 2014. Microwave-assisted treatment for continuous olive paste conditioning: Impact on olive oil quality and yield. Biosyst. Eng. 127, 92–102.

Tsagaraki, E., Lazarides, H.N., Petrotos, K.B., 2007. Olive mill wastewater treatment. In: Oreopoulou, V., Russ, W. (Eds.), Utilization of By-Products and Treatment of Waste in the Food Industry. Springer, New York, NY, USA, pp. 133–157.

Tsakona, S., Galanakis, C.M., Gekas, V., 2012. Hydro-ethanolic mixtures for the recovery of phenols from Mediterranean plant materials. Food Bioprocess Technol. 5 (4), 1384–1393.

Turano, E., Curcio, S., De Paola, M.G., Calabro, V., Iorio, G., 2002. An integrated centrifugation-ultrafiltration system in the treatment of olive mill wastewater. J. Membr. Sci. 209 (2), 519–531.

Velioglu, S.G., Curi, K., Camlilar, S.R., 1987. Laboratory experiments on the physical treatment of olive oil wastewater. Int. J. Dev. Technol. 5 (1), 49–57.

Velioglu, S.G., Curi, K., Camlilar, S.R., 1992. Activated-sludge treatability of olive oil-bearing waste-water. Water Res. 26 (10), 1415–1420.

Vichi, S., Boynuegri, P., Caixach, J., Romero, A., 2015. Quality losses in virgin olive oil due to washing and short-term storage before olive milling. Eur. J. Lipid Sci. Technol. 117 (12), 2015–2022.

Vlyssides, A.G, Iaconidou, K., 2003. Olive Oil Production in Greece. EU IMPELS Olive Oil Workshop, Cordoba, Spain.

Vlyssides, A.G., Loizides, M., Karlis, P.K., 2004. Integrated strategic approach for reusing olive oil extraction by-products. J. Clean. Prod. 12 (6), 603–611.

Yangui, T., Sayadi, S., Gargoubi, A., Dhouib, A., 2010. Fungicidal effect of hydroxytyrosol-rich preparations from olive mill wastewater against *Verticillium dahliae*. Crop Prot. 29 (10), 1208–1213.

Yangui, T., Sayadi, S., Dhouib, A., 2013. Sensitivity of *Pectobacterium carotovorum* to hydroxytyrosol-rich extracts and their effect on the development of soft rot in potato tubers during storage. Crop Prot. 53, 52–57.

Yorulmaz, A., Tekin, A., Turan, S., 2011. Improving olive oil quality with double protection: destoning and malaxation in nitrogen atmosphere. Eur. J. Lipid Sci. Technol. 113 (5), 637–643.

Zagklis, D.P., Arvaniti, E.C., Papadakis, V.P., Paraskeva, C.A., 2013. Sustainability analysis and benchmarking of olive mill wastewater treatment methods. J. Chem. Technol. Biotechnol. 88 (5), 742–750.

Zagklis, D.P., Vavouraki, A.I., Kornaros, M.E., Paraskeva, C.A., 2015. Purification of olive mill wastewater phenols through membrane filtration and resin adsorption/desorption. J. Hazard. Mater. 285, 69–76.

Zbakh, H., El Abbassi, A., 2012. Potential use of olive mill wastewater in the preparation of functional beverages: a review. J. Func. Foods 4 (1), 53–65.

Zinnai, A., Venturi, F., Sanmartin, C., Taglieri, I., Andrich, G., 2015. The utilization of solid carbon dioxide in the extraction of extra-virgin olive oil VOO/EVOO yield and quality as a function of extraction conditions adopted. Agro Food Ind. Hi. Tec. 26 (3), 24–26.

Zirehpour, A., Jahanshahi, M., Rahimpour, A., 2012. Unique membrane process integration for olive oil mill wastewater purification. Sep. Purif. Technol. 96, 124–131.

Zirehpour, A., Rahimpour, A., Jahanshahi, M., Peyravi, M., 2014. Mixed matrix membrane application for olive oil wastewater treatment: process optimization based on Taguchi design method. J. Environ. Manag. 132, 113–120.

CHAPTER

OLIVE MILL WASTE: RECENT ADVANCES FOR THE SUSTAINABLE DEVELOPMENT OF OLIVE OIL INDUSTRY

2

Maria K. Doula*, Jose Luis Moreno-Ortego, Federico Tinivella[†], Vassilis J. Inglezakis[‡], Apostolos Sarris[§], Konstantinos Komnitsas[¶]**

**Benaki Phytopathological Institute, Department of Phytopathology, Kifissia, Greece; **CEBAS-CSIC, Department of Soil and Water Conservation and Organic Resources Management, Espinardo, Murcia, Spain; [†]Center for Agricultural Experimentation and Assistance, Albenga (SV), Italy; [‡]School of Engineering, Department of Chemical Engineering, Environmental Science & Technology Group (ESTg), Nazarbayev University, Astana, Republic of Kazakhstan; [§]Foundation for Research and Technology, Hellas, Institute for Mediterranean Studies, Laboratory of Geophysical-Satellite Remote Sensing and Archaeo-environment, Nik. Foka & Melisinou, Rethymnon, Crete, Greece; [¶]School of Mineral Resources Eng, Technical University of Crete, Chania, Crete, Greece*

2.1 INTRODUCTION

2.1.1 EFFECTS OF OLIVE MILL WASTES DISPOSAL ON SOIL AND WATER

For Mediterranean countries, disposal of olive mill wastes (OMW) is considered a major environmental problem. The uncontrolled disposal of OMW on soil causes strong phytotoxic and antimicrobial effects, increases soil hydrophobicity, decreases water retention and infiltration rate, as well as affects acidity, salinity, nitrogen immobilization, microbial activity, nutrient leaching, lipids concentration, organic acids, and naturally occurred phenols (Sierra et al., 2007).

In 2009, the European Commission funded a demonstration LIFE project, namely "*Strategies to improve and protect soil quality from the disposal of olive oil mill wastes in the Mediterranean region*" (PROSODOL), which was also awarded "Best" project status in 2014 (http://ec.europa.eu/environment/life/bestprojects/bestenv2014/index.htm). The project aimed to develop strategies for the protection of Mediterranean soils and water bodies against uncontrolled disposal of OMW, demonstrate soil remediation methodologies as well as develop and propose a legislative framework that could be adopted by the EU and Member States' national frameworks.

The experience gained through the PROSODOL project indicates that uncontrolled disposal of OMW increases substantially the risk of soil degradation. In the framework of the project, a number of OMW disposal areas in Greece were monitored for a period of 2 years. It was revealed that almost all soil physical and chemical parameters are affected, some of them permanently.

Olive Mill Waste. http://dx.doi.org/10.1016/B978-0-12-805314-0.00002-9

Table 2.1 Approximate Input-Output Data for the Three Types of Olive Oil Production Processes (Azbar et al., 2004; Caputo et al., 2003)

Production Process	Input	Output
Traditional	Olives = 1 ton Wash water = 0.1–0.12 m^3 Energy = 40–60 kWh	Oil ≈200 kg Solid waste = 200–400 kg (25% water + 6% oil) Wastewater ≈ 400–600 kg (88% water + solids and oil)
Three-phase	Olives = 1 ton Wash water = 0.1–0.12 m^3 Fresh water for decanter =0.5–1 m^3 Water to polish the impure oil ≈ 10 L Energy = 90–117 kWh	Oil = 200 kg Solid waste = 500–600 kg (50% water + 4% oil) Wastewater = 1,000–1,200 kg (94% water + 1% solids and oil)
Two-phase	Olives = 1 ton Wash water = 0.1–0.12 m^3 Energy = 90–117 kWh	Oil = 200 kg Solid waste = 800–950 kg (60% water + 3% oil) Wastewater = 85–110 kg

As regards deterioration of water quality, monitoring results indicated that:

- The risk for groundwater depends mainly on the soil type, the depth of groundwater table, and on the presence of limestones in low depth. The presence of clay in soil also affects pollutants movement by reducing substantially the toxic load during infiltration.
- In general, the risk for humans is low. Higher risk is anticipated if humans drink water from public wells where high concentration of phenols has been determined in specific periods.

Therefore, it is anticipated that potential impacts will affect mainly recipients at local scale. In case of more intense activities, larger affected areas, and different soil qualities (e.g., sandy soil) risk for humans and ecosystems will be much higher.

2.1.2 OLIVE MILL WASTES GENERATION AND TREATMENT TECHNOLOGIES

Three types of olive processing technologies are currently used worldwide, producing different type and amount of wastes, that is, the traditional olive-pressing process, the three-phase, and the two-phase extraction processes.

The approximate input-output data for these three types of technologies are presented in Table 2.1.

Fig. 2.1 illustrates the most common OMW management scheme (OH: olive husks) while the most common treatment methods are denoted in Table 2.2 (Caputo et al., 2003; Azbar et al., 2004; Roig et al., 2006).

2.2 ANALYSIS OF EUROPEAN AND MEDITERRANEAN COUNTRIES LEGISLATIVE FRAMEWORKS

At the moment there is no European Union legislation regulating OMW management. Preconditions for safe disposal, threshold values for receptors, and wastes to be disposed as well as physicochemical parameters for treated wastes are left to be set by the Member States (MS), individually. Among

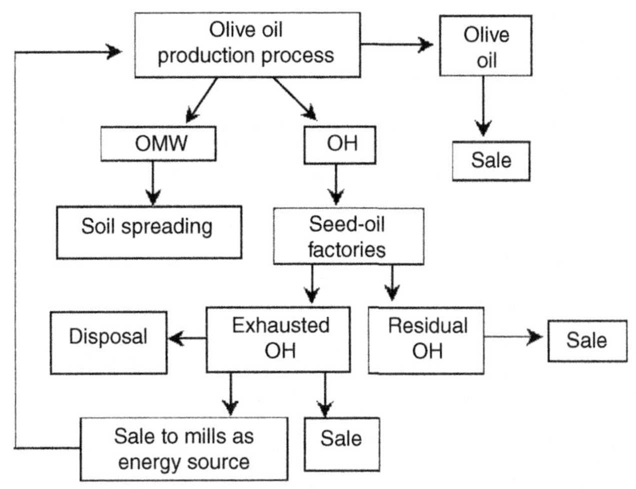

FIGURE 2.1 Most Common Management Scheme for Olive Mill Wastes (Caputo et al., 2003)

European MS, Italy, Portugal, and recently Spain have put in force specific legislation for disposal/application of OMW on agricultural soil (Ouzounidou et al., 2010). Furthermore, the issue of whether pomace should be considered as "waste" has been addressed in Italy and Portugal and has been resolved; it is not considered as hazardous waste. In the other olive oil producing countries, no specific legislation exists (IMPEL, 2003). In the following, brief analysis is provided on the specific issue of OMW regulations in Italy, Greece, and Spain as well as in Portugal and Cyprus. The latter cases have some interest as they are olive oil producing countries and have special provisions on this kind of waste.

2.2.1 ITALY

The Legislative Decree no. 152 of 1999, transposition of the European Directives 91/271/EEC and 91/676/EEC, regulates water protection from pollution (EEC, 1991a,b). The article 38 of this act makes reference to the Italian Law no.574 of 1996 with regards to the agronomic use of sewage sludge and other wastes such as OMWW. With Law no. 574, the agronomic use of these by-products is allowed on the ground of their composition and the characteristics of soils. Such use has to be authorized each time by the competent public authority on the ground of simple documentation but subordinate to limitations, verifications, and possible sanctions in order to avoid any fraudulent activity that can pollute water tables (MORE, 2008). This law allows the direct application of OMWW without previous treatment (Kapellakis et al., 2008).

Some technical aspects according to the Italian law 574/96 are (MORE, 2008; RES-HUI, 2006):

1. maximum tolerance limit for soils: 50 m^3/ha/year for OMWW deriving from traditional mills (discontinuous extraction systems) and 80 m^3/ha/year for vegetable water deriving from centrifugal extraction (continuous extraction systems);
2. possibility for the Mayor of any municipality of modifying those limits or suspending fertigation in case of environmental risk;

Table 2.2 Treatment Methods of Olive Mill Wastewater (OMWW), Solid and Semisolid Olive Mill Waste (OMW)

Method	Notes	Advantages	Disadvantages
Wastewater			
Biological treatment	Reduction of BOD and COD. Enables the removal of organic and inorganic suspended solids. • Anaerobic process is used to remove organic matter in high concentration streams by converting organic compounds to methane and carbon dioxide • Aerobic process is used on lower concentration streams to further remove residual organic matter and nutrients.	Anaerobic technology has clear advantages in comparison to aerobic processes including higher removal efficiency, lower excess sludge production (20-fold lower than the aerobic process), low space requirements. It produces biogas, a valuable by-product that can be used to cover the mill energy demand.	Pretreatment of wastewater is needed, for example, sedimentation—filtration. The presence of oil and antimicrobial phenolic compounds create a challenge for biological treatment. Due to high investment costs and complex process management, this technology is suited for industrial-scale olive mills or as a centralized treatment facility serving several olive mills.
Composting	Composting is one of the main technologies used for recycling wastewater and transforming it into a fertilizer. Wastewater is added repeatedly during the thermophilic phase.	Valorization of waste and production of natural fertilizers. This process allows the return of nutrients to cropland.	Wastewater has to be removed before composting (e.g., using simple precipitation units, by adsorption on a solid substrate).
Membrane filtration	Microfiltration, ultrafiltration, nanofiltration, and reverse osmosis can be applied on olive mill wastewaters. The concentrate can be sent to incineration or landfilled.	Optimal COD reduction. Since conventional biological treatment is not always sufficient, other processes, such as membrane filtration (e.g., reverse osmosis or ultrafiltration) are required.	Hardly suited for treatment of OMW in small-scale oil mills due to the very high investment and operational costs, high energy demand, and rather complex process control that requires highly qualified personnel.
Chemical–physical treatment	Removal of nonbiodegradable dissolved organic pollutants by adsorption, precipitation, and flocculation, characterized by the addition of specific chemicals: activated carbon, polyelectrolytes, flocculants/coagulants such as ferric chloride, aluminum chloride, ferrous sulfate, and calcium hydroxide.	High removal efficiency. Low investment and operational costs, better suited for small-scale olive mills.	Large quantities of produced sludge and subsequent disposal problems.
Natural evaporation in open storage ponds	The most common management method.	Low investment is required; the approach is favored by the climatic conditions in Mediterranean countries.	It needs large areas and causes several problems such as bad odor, infiltration, and insect proliferation. Evaporation of OMW results in the production of sludge which is typically disposed of in landfill sites. Most studies targeting revalorisation of OMW sludge focus on composting.

Forced evaporation	Water separation by using a multistage evaporation system enables heat recovery. The concentrate has to be treated, prior to its disposal, generally by biological treatment, such as aerobic digestion and activated sludge process.	Heat recovery possible.	High energy demand, considerable air emissions, and complex control process that requires qualified personnel. High operational costs are associated. It is a solution suited for industrial-scale oil mills only.
Solid and semi-solid waste (olive kernel, olive husks/cake and wet olive pomace)			
Drying	Heat is transferred to waste by means of hot gases (drum dryers, belt dryers, and fluidized-bed dryers).	The resulting dried waste may be incinerated for energy production, reused in agriculture, or landfilled, while air emissions must be treated appropriately.	High-energy demand, however, the resulting final product can be reused for the production of the required energy. High investment and operating costs are associated to drying plants; moreover, trained and qualified personnel is required to ensure trouble-free operation. The high moisture content of the wet olive pomace causes technical problems during drying in traditional husk-extraction mills. These problems initiated research for new drying methods.
Thermal treatment	The treatment provides the possibility of energy recovery while mixtures of OMW with other organic wastes can be incinerated as well. After the second extraction of the wet olive pomace, the exhausted olive cake is usually used as fuel in the husk mill to obtain thermal or electric energy through combustion.	Possible energy recovery	Very expensive, considering the high cost of combustors and the associated air pollution control systems. Most of the energy obtained by combustion is used for drying of the fresh wet olive pomace and therefore the total energy recovery is low.
Biological treatment	Both aerobic (composting) and anaerobic (fermentation) processes can be used. Bulking material such as wood shavings has to be added during waste composting to achieve proper moisture level and good aeration of the pulp. Unlike wastewater, the wet olive pomace is added at once at the beginning of the composting process.	Fairly low air emissions and insignificant consumption of energy and resources. As far as the fermentation process is concerned, energy and space requirements are very low, while biogas production enables energy recovery.	Compared to composting, problems often arise in process control, hence qualified personnel is required. Moreover, this technology is affected by higher investment costs. The toxicity of phenolic compounds present in wet olive pomace affects anaerobic digestion. Due to its semi-solid state, pomace needs to be mixed with bulking agents, such as straw, cotton-waste, poplar sawdust, and bark chips, before composting. The physical characteristics of wet olive pomace cause difficulties during composting by forced aeration systems and mechanical turning is often preferred.

3. submission of the agronomic report to the Municipality at least 30 days before spreading. The report has to be written by an expert technician and has to cover topics such as characteristics of the soil and the means of spreading;
4. uniform spreading is required in order to avoid surface runoff;
5. it is forbidden to spread vegetable water on:
   a. soils which are less than 300 m from the preservation areas for water collection destined to human consumption;
   b. soils which are less than 200 m from inhabited areas;
   c. soils which are cultivated with vegetable crops;
   d. soils where water tables are in the depth of less than 10 m;
   e. soils which are frozen, covered by snow, awashed, or saturated with water.
6. waste storage in the olive mill should be less than 30 days (limit protracted to 3 months—D Lgs 22/1997).

The July 6, 2005 Decree, "*Criteria and technical rules regarding regional regulation of the agronomic use of olive mill waste waters and other mill wastes*" makes a reference to article no. 38 of Decree no. 152, whereas more exclusions of lands are added (RES-HUI, 2006) as follows:

- distance <10 m from river banks
- distance <10 m from sandy shore or lake water
- lands with slope >15% and lack of hydraulic and agricultural setting
- woods
- gardens and public areas
- quarries

Furthermore, the same Decree prohibits mixing of OMWW with other wastewater streams (e.g., animal slurry) or wastes. Finally, olive mill water with stone fragments and the fibrous part of the fruit can be used in agriculture and are not subject to Fertilizer Law No. 748.

According to Law no. 574/1996, wet olive husks can be used as soil amendment notwithstanding to the indications given in the Italian Law no. 748/1984 on fertilizers and subsequent modifications such as the legislative Decree no. 217/2006 "*Revision of regulations on fertilizer use*." According to the later, wet olive husks can be considered as a "*simple not-composted plant amendment*" and therefore they can be applied on soil without any specific limitations. This can occur if they comply with the thresholds of specific parameters, that is, humidity, pH, organic carbon, organic nitrogen, total Cu referred to dry matter, total Zn referred to dry matter, total peat content, and other heavy metals contents.

Finally, the disposal of wastewaters of any kind (e.g., OMWW) in sewage systems or in superficial water bodies (rivers, lakes etc.) is regulated by the Decree 152/2006. The specific thresholds set by the Decree for all parameters to be taken in consideration for disposal are listed in Table 2.3 of Annexure 5 of the Decree.

2.2.2 SPAIN

Following the introduction of the three-phase process in the 1970s, the production of large quantities of OMWW caused major surface water pollution problems. In 1981, the Spanish government adopted a legal framework to prohibit the discharge of untreated OMWW into rivers and subsidized the

Table 2.3 Wastewater Limit Values For Food Industries in Greece (Law 1180/1981)

Pollutant	Limit Value-Daily Maximum (kg/ton of Product)	Limit Value-Monthly Average (kg/ton of Product)
BOD	4.0	2.0
COD	6.0	3.0
Suspended solids	5.0	2.0
Oils	1.0	0.5

construction of about 1000 ponds for its storage during the milling period, aiming at the evaporation of its water during the warm Andalusian summer (Kapellakis et al., 2008). The result of these initiatives was the improvement of water quality in the nearby rivers and streams. During 1991–92, a shift from the three-phase to two-phase system was implemented resulting in lower wastewater generation. Nowadays, more than 90% of the olive mills operate with the two-phase system that generates wet olive pomace instead of OMWW (Kapellakis et al., 2008).

In Spain, there is a Ministerial Decree about the required operations for valorisation and disposal of wastes, based on the European list of Wastes (O.M. MIMAM 304/2002, February 19, 2002, activities of valorization and disposal of wastes). Wet olive pomace or OMWW are not generally considered dangerous wastes by the Spanish legislation. These wastes are considered secondary products, which can be valorized in order to prevent soil or water contamination. However, olive mills generate other minor waste streams that are considered dangerous waste such as motor oils, used lubricants, boiler parts, rejected chemicals, fluorescents tubes, and other wastes that contain mercury.

The recently issued Decree 4/2011 of the Regional Government of Andalusia allows the use of wastewater produced by olive mills as soil amendment in agriculture. In particular, Art. 7 specifies that:

- The volume of the effluent to be applied to agricultural land should in no case exceed 50 m^3/ha/year.
- Applications should be designed so as not to generate surface runoff, leaching, or invasive lesions of the soil water table.
- The field application of the effluent shall be subjected to the following areas of exclusion:
  - Those areas located within 500 m from urban areas.
  - The landside stripe of 100 m wide on each side, counted from the line delimiting the riverbed, (the riverbed is defined in Article 6.2.b of Regulation for Public Water), approved by Royal Decree 849/1986.
  - The easement area of protection is100 m, which includes the Maritime Terrestrial Public Domain, as defined in Article 23.1 of Law 22/1988.

In general, wastewater is sent to ponds where it is evaporated or used for irrigation of soil. However, wastewater can affect the hydraulic public network. The related Spanish law considering this aspect is Law 46/1999, which modified Law 29/1985. Therefore, according to Law 46/199, it is compulsory that olive mill owners have a specific authorization for every catchment of continental water above or below ground. Also an authorization of wastewater application on soil is required for all these activities that may cause pollution or degradation of the hydraulic public network, while a project aiming at processing, as well as hydrogeological and environmental studies or any documentation that will be considered necessary may be requested from the mill owners. There are some wastewater parameters that must be considered to select the best wastewater depuration treatment. Actually the discharge of OMWW to hydraulic network is forbidden in Spain. The Royal Decree 849/86, entitled "*Reglamento del Dominio Público Hydráulico*" (Regulations of the Hydraulic Public Domain) specifies in the "*Anexo al titulo IV*" (Annexure of the Title 4th) certain discharge limits, which depend upon the extent and efficiency of the applied wastewater treatment. The discharge limit values (DLVs) have been set at national level (IPPC BREF, 2006b). The legislation concerning wastewater discharges into sea waters, either directly or via continental waters, is laid down by the *Real Decreto* 258/1989 (Royal Decree 258/1989). According to this decree the discharge limits are fixed for each specific industrial plant.

2.2.3 GREECE

In Greece there are no specific regulations regarding the discharge of OMW. The main principles for OMW management are based on the Law 1650/86 "*For the Protection of the Environment*" according to which, olive mill owners are obliged to provide an environmental impact assessment study. The updated circular letter YM/5784/1992 (No 4419/1992) refers to the problems encountered due to OMW disposal, the need for an efficient pretreatment and the care required in order to avoid disposal to various water resources. The present legislative status (Laws 1650/86 and 3010/2002) does not allow application of untreated OMW to soil surface.

Each Greek Prefecture is responsible for adopting proper OMW management practices encouraging different waste management approaches. For example, OMW management in the Prefecture of Messinia (Peloponnese) is based on the modification of a three-phase decanter system into a two-phase, in the Prefecture of Lesvos, OMW was discharged until recently untreated onto aquatic ecosystems, while in the Prefecture of Iraklio, Crete, the disposal of OMW in the aquatic environment is forbidden (Kapellakis et al., 2008).

Wastewater limit values have not been addressed at national level, but they have been issued for many prefectures at regional level (IPPC BREF, 2006a). The limits are applied for any kind of discharges, including other streams of the food industry (IPPC BREF, 2006a; ECOIL, 2005). However, the values shown in Table 2.3 are considered as guidance values, specifically for the production and treatment discharges of oil (industrial sector: "Production and processing of vegetable/animal fats and oils"). It is noted that the limit values for discharges to water are issued by each prefectural authority according to the location of the specific water recipient (ECOIL, 2005).

Finally, the recent Joint Ministerial Decision (KYA) 145116/2011 regulates the reuse of treated wastewater for several purposes, including irrigation in arable lands. The minimum requirement is the use of biological treatment and disinfection units.

2.2.4 PORTUGAL

Law No. 626/2000 provides a special permission that allows OMWW spreading on land, similar to the Italian practice (Kapellakis et al., 2008). The maximum tolerance limit for soils is 80 m^3/ha/year.

General water discharges are regulated by Decreto Lei 236/98. Wastewater discharge limits are defined in its Annexure XVIII. The ELVs are set for national level (IPPC BREF, 2006b).

2.2.5 CYPRUS

OMW management should be conducted according to the Cyprus Ordinance No. 254/2003 on Water Pollution Control (Waste Disposal Permit) Ordinance of 2003, Official Newspaper of the Cyprus Government No. 3649. The waste streams generated by olive mills differ according to the process used for oil extraction; for example, two-phase or three-phase centrifuge. The types and amounts of OMW allowed to be deposited are shown in Tables 2.4 and 2.5 (ECOIL, 2005; Anastasiou et al., 2011).

Regardless the type of process used (two-phase or three-phase) (ECOIL, 2005):

- liquid wastes [wastes types (a) and (b) in the Tables earlier] should be temporarily stored in waterproof sealed tanks. Whether the streams will be mixed or separated depends on the disposal method.
- sludge [olive dregs—type (c) in the earlier Tables] should be temporarily stored in a covered area with concrete base (platform). Liquids originating from leakages or run-offs from the temporary solid waste or sludge storage areas should be collected and transferred to the liquid wastes tanks, via open-air waterproof pipes.

Table 2.4 Maximum Annual OMW Volumes Allowed to be Disposed for Two-Phase Centrifuge Olive Mills in Cyprus

Waste Stream Generated	Maximum Annual Waste Quantities Allowed (m^3)
a. Liquid waste from the washing of olives.	180
b. Liquid waste (water and minimal olive oil mill wastewater) originating from the centrifuging decanters, where separation of the plant liquids of the fruit from oil takes place.	400
c. Sludge (mixture of olive dregs and olive oil mill wastewater) originating from the decanter.	750
d. Sludge settling at the liquid wastes evaporation tanks.	—
e. Leaves from the defoliation.	—

Table 2.5 Maximum Annual Waste Quantities Allowed for Three-Phase Centrifuge Olive Mills in Cyprus

Waste Stream Generated	Maximum Annual Waste Quantities Allowed (m^3)
a. Liquid waste from the washing of the olives.	1600
b. Liquid waste (water and minimal olive oil mill wastewater) originating from the centrifuging decanters, where separation of the plant liquids of the fruit from oil takes place.	1400
c. Solid waste (olive dregs) originating from the horizontal centrifuging decanter.	750
d. Sludge settling at the liquid wastes evaporation tanks.	—
e. Leaves from the defoliation.	—

Table 2.6 Quality of Liquid Waste Entering Evaporation Tanks in Cyprus

Parameter	Maximum Value Allowed
pH	5.0–7.0
Electric conductivity	10,000 μS/cm
Suspended solids	5,000 mg/L
BOD5	10,000 mg/L
Fat	6,000 mg/L
Phenols	1,000 mg/L

Waste stream (a) of Table 2.5 can be used for irrigation of cultivations (trees, forest-trees, etc.) around the olive mill. In cases that the effluents are mixed with liquid waste originating from the centrifuging decanters (b), the liquid wastes should be transferred in evaporation tanks for final disposal. Evaporation tanks should be open, waterproof, earthen, and shallow with a maximum depth of 1.2 m. Liquid wastes should be transferred in the evaporation tank with closed pipes or with a tanker. The required quality of the liquid wastes to be disposed in the evaporation tank is shown in Table 2.6 (maximum allowance) (ECOIL, 2005; Anastasiou et al., 2011).

Sludge produced by the decanter (c) of a two-phase mill, should be collected and transferred by a tanker to the appropriate facilities for incineration or composting. At the end of olive production period, no sludge should be present at the temporary storage area. Solid wastes (c) produced by a three-phase mill should be collected and used as animal stocking, fertilizer, or sent to a seed-oil production facility for further treatment. The institution exploiting the waste, should maintain a database for the quantities and the ways of waste disposal. In cases that the olive drugs are used as soil improver (fertilizer), the application should be at least take place 300 m away from residential areas, with maximum disposal rate of 3.5 ton/ha/y. At the end of the operating period, no sludge or solid waste should be present in the temporary storage area for both of the above cases. Sludge disposed at the bottom of evaporation tanks (d) should be collected when needed, after the liquid present in the tank has been dried, and transferred for disposal to an approved public area or used as soil improver (under the conditions stated earlier) (ECOIL, 2005; Anastasiou et al., 2011).

2.3 POTENTIAL USE OF OMWW FOR OLIVE TREES IRRIGATION AND FERTILIZATION

An alternative, environmentally friendly, and economically feasible approach to manage OMWW would take advantage of its valuable ingredients (e.g., organic matter, nutrients, and water). In this respect, OMWW could be considered as soil amendment, and natural fertilizer when applied under specific preconditions for trees irrigation (LIFE Prosodol, 2016a). Italy is the leading country in using OMWW for these purposes. Capitalizing the Italian experience in reusing OMWW into olive tress orchards, an irrigation system was installed in Albenga, Italy, during the implementation of PROSODOL project. For this, a hanging dripline system at 60 cm above soil surface per each olive row (Fig. 2.2) was

FIGURE 2.2 Dripline Used for OMWW Distribution

FIGURE 2.3 A Detail of the Hanging Pipeline Used for OMWW Distribution

installed and pressure compensated tubes normally used in the agricultural sector for the distribution of water or the application of liquid fumigants were adopted. Such driplines have the following properties:

- flow: 2 L/h
- distance between drips: 30 cm
- pressure: 1.5 bar

Each plant was provided with 2 drips aligned with plant row roughly at 15 cm apart from olive stem. Driplines are fastened to iron wires (Fig. 2.3). Two orders of iron wires were tightened between concrete poles, (1) one to provide support to dripline, and (2) one to provide support to the plants. The distance between poles was 10 m. A plastic tube which was connected to the pump for OMWW distribution to feed the driplines was also used (Fig. 2.4).

FIGURE 2.4 Valve and Connection Pipes Connected to Distribution Tubes and Driplines

2.3.1 IRRIGATION SYSTEMS AND CRITICAL ASPECTS

It is clear that irrigation requires extra costs in terms of design, set up, and system management in comparison to an olive orchard without any irrigation system. It is also crucial to evaluate costs due to the exploitation of water in areas where it is often lacking.

Different watering systems can be set up:

1. dripline hanging on the soil surface,
2. dripper plugged into the watering tube,
3. sprinklers for microaspersion (e.g., on centuries-old plants),
4. buried dripline at 20–30 cm depth (subirrigation using antisiphon technology).

When installing an irrigation system, the following critical aspects should be taken into consideration:

1. *Filtration system*: Each installation should be provided with a suitable filtration system in order to retain organic and inorganic particles for the overall proper operation of the entire system. Such an aspect is even more crucial when OMWW are distributed because of their lipidic nature and the presence of suspended particles.
2. *Maintenance of the irrigation system*: In the case of hanging driplines, sprinklers should lay on the soil so that troubles and obstructions could be easily detectable. With regard to subirrigation, it is necessary to verify the correct operation and flow of the system through water counters along with a precautionary maintenance.

2.3.2 ECONOMICAL ASPECTS RELATED TO WATERING SYSTEMS

The selection of a viable technical solution to be installed in an olive orchard is linked with the amount of the investment the user is ready to undertake. Costs related to the materials depend on technical, technological, and agronomic factors affecting the process, for example, spacing between rows affects the quantity of dripline used, the quality of the water affects the filtration system needed, etc. Due to the numerous affecting factors, it is not easy to define general cost assessment rules. Indicatively, dripline is cheaper than sprinklers and both of them are cheaper than a subirrigation system characterized by high laying costs. Thus, there are many choices for the irrigation system that will be adopted and for this reason the final investment cost is highly site specific.

2.3.3 FILTRATION

Filtration represents the key aspect during OMWW distribution through dripline in consideration of the waste composition. Distribution ease can vary depending on the thickness and the presence of suspended solids in OMWW. Such parameters may present a very high variability during the olive season related to the characteristics of olives milled. A significant reduction in the workability is related to the setup of a filtration system between the feeding tank and the dripline. It is therefore highly advisable to adopt a filtration system before OMWW collection in the plastic container (tank) in order to have OMWW feeding the dripline with the lowest content of suspended solids. Such filtration system should be composed by superimposed filtering elements made of steel, which are fed from the top part and provide output in the lower part.

Sediments and solids that accumulate within the filter can be then disposed on the soil together with well mixed olive husks. This is conducted to avoid too concentrated hot spots that could be harmful when kept in contact with the soil for a prolonged time.

If OMWW is characterized by a significant presence of suspended solids, it is also advisable to apply a different distribution system, based on dripline provided with "sip" drippers, in order to improve workability.

2.4 SOIL REMEDIATION TECHNIQUES

So far, no specific remedial methodology for OMW disposal areas has been developed or implemented. Some of the already existing soil remedial methodologies/technologies could be conformed to this specific waste type and should be considered as remedial targets to reduce the values of the proposed soil indicators (Section 2.5.2.5) to an acceptable level.

Therefore, a soil remediation and protection plan suitable for OMW disposal areas, should include methodologies for polyphenols reduction and retention or immobilization of inorganic constituents. In situ bioremediation may be applied to reduce polyphenols concentration in soil, since it targets biodegradation of organic pollutants in soil by taking full advantages of the natural biodegradation process of organic molecules by soil microorganisms (Cookson, 1995). On the other hand, the use of natural zeolite, namely clinoptilolite as soil amendment was shown, during PROSODOL project, to be the most suitable solution to reduce inorganic soil constituents because of the well-known properties of clinoptilolite to attract, retain, and slowly release inorganic cations. Moreover, this method is of very low cost and very easy to be implemented, even by not qualified personnel (Doula et al., 2011, 2012). Both these methods were applied in a pilot area with very satisfactory results (LIFE Prosodol, 2016a).

Prior to the selection and implementation of a remedial methodology, it is important to set up target values for all soil parameters that exceed thresholds.

2.4.1 IN SITU BIOREMEDIATION (BIOPILING) AT OMW DISPOSAL AREAS

Bioremediation is a relatively low-cost and simple process, which generally has high public acceptance and can often be carried out on site. However, it is not suitable for all situations. Indeed, a detailed study of local soil conditions is required in order to identify if the organic contaminants could be biodegraded by soil microorganisms and if the residual contaminant levels are acceptable (Vidali, 2001). Under favorable conditions, microorganisms can completely metabolize organic contaminants and convert them into nontoxic by-products, such as carbon dioxide and water or organic acids and methane (US-EPA, 1991). Generally, bioremediation can be used in any soil type with adequate moisture content, although it is difficult to supply oxygen and nutrients into low permeability soils. Very high concentrations of contaminants may be toxic to microorganisms and thus may inhibit their activity. Thus for heavily contaminated sites, bioremediation may not be the best remediation option. Therefore, prior to implementation, a feasibility study is required to determine if biodegradation is a viable option for the specific site, soil type, and contamination level (Aggarwal et al., 1990).

For the determination of the bioremediation potential of a site contaminated with organic wastes, treatability studies are required to provide specific information, such as the potential rate and extent of bioremediation, the fate and behavior of each organic pollutant in surface soil, and deeper vadose zone. Treatability studies include both laboratory and field tests. A flowchart for the determination of the bioremediation potential of an OMW contaminated site is presented in Fig. 2.5.

The application of bioremediation foresees many stages, which, in general, include:

1. feasibility studies to identify the bioremediation potential of the site of interest;
2. implementation of the selected bioremediation technology; and
3. effective monitoring.

During the first stage, laboratory studies should be performed in order to determine the optimum conditions for microbial activity and if these exist in the area of interest. By this way, the existence of microorganisms capable to biodegrade polyphenols could be determined. If such microorganisms do not exist, they should be inoculated in soil (bioaugmentation). Thereafter, the conditions, that is, soil aeration, moisture, nutrients' concentration that ensure the optimum microbial activity should be determined and ensured.

The microbial activity and some soil properties under various conditions of aeration, moisture, and nutrients' levels should be tested in soil samples collected from the targeted area.

The factors that need to be determined during feasibility studies are:

- separation, identification, and quantification of polyphenols (using HPLC);
- water soluble polyphenols;
- microbial biomass carbon;
- water soluble carbon fraction;
- soil pH, electrical conductivity, and all other soil parameters;
- phytotoxicity tests by conducting germination tests;
- ecotoxicity tests;

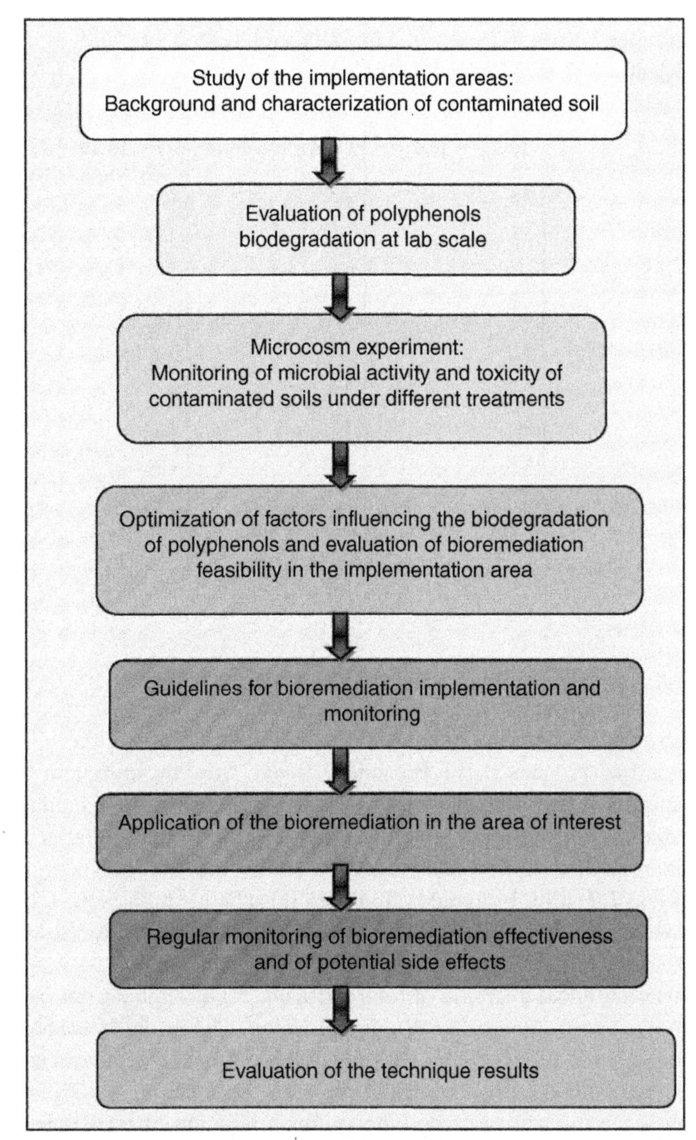

FIGURE 2.5 Flow Chart for Determining the Bioremediation Potential of OMW Disposal Sites

- soil respiration; and
- soil enzymatic activities, that is, phenoloxidase and dehydrogonase activity.

Soil samples should be analyzed for the above parameters before and after the application of the different conditions (scenarios) of aeration, moisture, and nutrients' level at laboratory scale. Thereafter, a series of laboratory experiments should be carried out in order to identify the optimum treatment methodology to be applied at the area of interest.

Table 2.7 Soil parameter Values Before and After Bioremediation Implementation in Relation to the Remedial Objectives (LIFE Prosodol, 2016a)

Soil Parameter	Target Value	Control Sample	Value Before Treatment	Value After Treatment
Organic matter (%)	5.0	4.3	6.4	6.02
Total polyphenols (mg/kg)	57	57	117	32
Electrical conductivity (μS/cm)	4.0	0.67	1.89	0.67
Total nitrogen (mg/g)	3.0	2.3	4.4	3.0
Exchangeable potassium (cmol/kg)	1.2	0.6	7.8	4.1
Exchangeable magnesium (cmol/kg)	2.2	2.9	4.0	3.6
Available iron (mg/kg)	50	46	106	67
Available copper (mg/kg)	3.0	2.6	4.6	4.4
Available phosphorus (mg/kg)	28	16	113	77
Available boron (mg/kg)	1.5	0.2	0.6	0.3

After the completion of the feasibility studies, the optimum conditions for the implementation of bioremediation at the area of interest and a set of instructions on how to implement the technology can be provided. Such a process was implemented in a pilot field in Crete, Greece, which accepts non-treated OMW for more than 12 years (LIFE Prosodol, 2016a). After the completion of bioremediation implementation, results indicated that the method could be effective in reducing soil polyphenols, total nitrogen, available iron (DTPA-extractable), and boron, although the latter did not exceed the threshold value before remediation. Soil analyses revealed that the total polyphenols content was significantly reduced (Table 2.7) after bioremediation. The initial very high value (i.e., 117 mg/kg) was reduced by 72.6%, while the final polyphenols concentration was very low (lower than the control sample of the area).

Moreover, there was a gradual decrease of total nitrogen (N) throughout the sampling period. Soil bioremediation procedures seem to enhance N mineralization. The initial N content was significantly high, considering the threshold of 3.0 mg/g (above which a soil is characterized as very rich in nitrogen) and the mean value of the control samples of the pilot area (2.3 mg/g). The final N values were considered acceptable since the soil contained the sufficient amount of total nitrogen. Available iron (DTPA-extractable) was significantly reduced, however the final value was still higher than the threshold of 50 mg/kg for soils. However, this result is accepted and the method is considered as effective in reducing available Fe.

On the contrary, the method was not effective to reduce soil organic matter, exchangeable potassium and magnesium, available copper and phosphorus at acceptable values. In situ bioremediation seemed to have no effect on available Cu concentration and on the exchangeable Mg. The reduction in the concentration of exchangeable K and available P was substantial, but the values of these two parameters were still higher than the accepted thresholds.

2.4.2 CLINOPTILOLITE AS SOIL AMENDMENT AT OMW DISPOSAL AREAS

Zeolites are highly porous aluminosilicates with different cavity structures. They consist of a three-dimensional framework, having a negatively charged lattice which is balanced by exchangeable cations. The high ion-exchange capacity, the relatively high specific surface area, and more importantly the relatively cheap price allow zeolite to be considered as attractive adsorbents. Zeolites consist of a wide variety of species (more than 40), however, the most abundant and frequently studied zeolite is clinoptilolite, a mineral of the heulandite group. Clinoptilolite is known to have high selectivity for certain pollutants. The characteristics and environmental applications of zeolites have been extensively studied (Inglezakis et al., 2003; Doula et al., 2011, 2012). Clinoptilolite price ranges from 120 to 150 €/ton, depending on the quality of the mineral. Zeolites are becoming widely used as alternative materials in areas where adsorptive applications are required. Recently, they have been intensively studied because of their applicability in removing trace quantities of pollutants such as heavy metal ions and phenols.

The pilot application of clinoptilolite took place at the OMW disposal area of Section 2.4.1 during PROSODOL project. It should be highlighted that the pilot area accepted OMW disposal during the period of zeolite treatment.

The methodology included three stages:

1. Complete physicochemical characterization of the area of interest.
2. Area configuration and application of zeolite on soil.
3. Effective monitoring.

Clinoptilolite was added at two rates (5% and 10%) as fine powder with particles diameter <0.8 mm, and of larger size (particles diameter of 0.8–2.5 mm). After clinoptilolite application, the area was tilled until 25 cm depth with a small tilling machine.

Table 2.8 presents the values of soil parameters measured during the treatment in relation to the remedial objectives. Since the area continued to accept OMW during treatment, a range of values

Table 2.8 Soil Parameters Values Before and After Zeolite Application on Soil

Soil Parameter	Target Value	Control Sample	Value Before Treatment	Value After Treatment
Organic matter (%)	5.0	4.3	6.0–29.0	5.6–7.1
Total polyphenols (mg/kg)	57	57	54–118	32–94
Electrical conductivity (μS/cm)	4.0	0.67	1.40–6.93	1.08–2.91
Total nitrogen (mg/g)	3.0	2.3	4.0–17	4.0–7.0
Exchangeable potassium (cmol/kg)	1.2	0.6	7.0–12	13–20
Exchangeable magnesium (cmol/kg)	2.2	2.9	7.1–11	5.1–5.4
Available iron (mg/kg)	50	46	119–202	123–333
Available copper (mg/kg)	3.0	2.6	3.9–8.5	4.6–8.5
Available phosphorus (mg/kg)	28	16	162–591	171–387
Available boron (mg/kg)	1.5	0.2	1.0–3.0	0.9–3.0

for all the parameters (and not the absolute values) is given. The treatment resulted in stabilization of soil organic matter values, although still at values higher than the target ones. This is considered, however, satisfactory considering that before treatment organic matter values ranged between 6.0% and 29%. Even though high organic matter values could be beneficial, thus, very high values of organic matter enhance anaerobic phenomena in soils. This is due to the improvement of soil aeration (a known effect of zeolite addition on soil) and thus to the enhancement of soil microorganism activity to biodegrade soil organic matter. The effect of zeolite on total nitrogen content is similar to that on organic matter content, due to the same reasons, however, the final values of total nitrogen are considered unacceptable.

Exchangeable K and available Fe were significantly increased in soil due to the retention of these elements by clinoptilolite. However, the increase is not attributed to the increase of these elements in soil particles but in zeolite framework. Consequently, K and Fe are not excessively leached but they are released slowly to soil solution. On the contrary, no significant effect was recorded on soil available Cu.

Soil electrical conductivity was decreased due to the retention of ions within the zeolite framework. The recorded values for electrical conductivity were lower than the target value of 4 μS/cm. Thus, despite an increase in exchangeable K and available metals contents in soil (Table 2.8) after treatment, soil electrical conductivity was decreased since ions were not found in soil solution but they were retained in/on the zeolite framework.

The content of total polyphenols was reduced, but not as much to meet the target value. The concentration of available phosphorus was decreased, however, the final values were unacceptable. Exchangeable Mg was also significantly decreased and stabilized at lower concentrations compared to the initial ones. Nevertheless, these values were almost doubled compared to the target ones. No significant effect was recorded on the concentration of soil available B, which remained higher than the target value after treatment. Finally, no significant difference was shown when different ratios and grain sizes of clinoptilolite were used. Therefore, the use of zeolite on soil is proposed to be up to 5%.

2.5 INTEGRATED STRATEGY, MEASURES, AND MEANS SUITABLE FOR MEDITERRANEAN COUNTRIES

2.5.1 LEGISLATIVE PROPOSALS

In many countries, olive mills are family businesses and small-scale enterprises, scattered around olive production areas, making thus individual onsite treatment options unaffordable (Paraskeva and Diamadopoulos, 2006). Since olive mills have a low benefit/cost ratio, forcing them to meet environmental regulations could lead them to close down. For example, it is assumed that small olive mills cannot build and operate anaerobic treatment plants because of their high cost. Due to the seasonal olive oil production (the harvest period ranges from September to February) and the variable character of the residue, anaerobic plants would not be advised since they need a long running in phase of several weeks or months. However, this system could be applicable to cooperatives that work also with other agricultural products generated in different seasons. Furthermore, the quantity of wastewater generated during olive oil processing is small in comparison to other commercial industrial operations and thus the purification cost cannot be justified in terms of volume and production duration.

Considering the current financial crisis and the above parameters, implementation of detailed regulatory waste rules would put the sustainability of these enterprises at risk. However, although this situation is understandable, it should not lead to the conclusion that the law is "bending" to meet the needs of a specific sector. Therefore, an acceptable approach could be to start from an objective revision of the legal framework on waste management and accommodate the particularities of OMW (Taccogna, 2010).

The following recommendations (concerning the waste legislative part) were proposed to the European Commission for the European olive oil productive Member States (LIFE Prosodol, 2016a).

Statutory legislative proposals should be:

- Untreated waste/wastewater disposal into the environment should be strictly banned.
- Irrespectively if waste/wastewater is hazardous or not, it should be treated before any disposal to land/surface waters takes place. Specific discharge limit values should be defined, especially in the case of land spreading, for which no statutory standards exist but only indicative application rates are proposed by national legislation in some countries.
- As OMW is potentially hazardous, the legislation should provide statutory limits (especially on phenols) under which the waste is characterized as nonhazardous. OMWW should be categorized as H14 (ecotoxic), whereas limits, tests and monitoring measures should be provided depending on the receiving media, that is, soil or surface waters.
- The legislative act should clearly specify that the waste should be analyzed for its physicochemical characteristics, for example, vegetative trials, germination tests, phytotoxicity tests, growing-on test, etc., testing its potential toxicity regarding plant growth and the environment in general. Standard sampling and analytical procedures, harmonized at EU level, could be introduced.
- There should be a categorization of olive mills according to their production capacity and/or volume of generated waste in order to draw specific measures for waste management, that is, for mills with high production capacity, waste/wastewater management could be performed in situ while small facilities could establish collective schemes to manage and treat their wastes.
- In case evaporation ponds are used, the minimum requirement should be the use of protective liners in engineered evaporation ponds.
- As landspreading is a common and low-cost practice, especially for small production units, specific regulations should be developed.
- In case of landspreading and under the condition that the olive mill waste/wastewater fulfills the requirements of the existing legislation, OMW could be considered as fertilizer. Thus, the annual dose estimation should follow the general rules of soil fertilization depending on the properties and usage purpose.
- When reusing treated wastewater for agricultural land irrigation, application guidelines should be developed in order to provide a common level of environmental and public health protection.
- If OMW is considered as waste, national law should allow it to be treated as municipal waste when produced by smaller olive mills.
- The European Commission should provide technical specifications, pursuant to art. 5 of Directive 2008/98, on the conditions for using OMW as by-product regardless of their economic value and the possible need of a drying phase.
- National laws should be brought in line with this new concept of by-product, mainly, the part that still provides for the economic value of by-products as a requirement (as in the case of the Italian law).
- The regulations should take into account (1) the land-use (e.g., agriculture, food products, nonfood products, residential/parkland, commercial or industrial), (2) the soil type, and (3) the period of reuse.

- OMW is usually discharged in small stream catchments (<10 km^2), which are not considered within the Water Framework Directive 2000/60/EC. Therefore, there is a need to include small streams into monitoring and assessment schemes as small streams contribute to the pollution load of the river basin.
- Environmental quality standards should be set in an EU Directive in the same way that is done for water bodies, at least as minimum requirements per soil type. The threshold for pollutant (as phenols) concentrations in soil could be set in such values as to reflect existing soil maximum background concentrations in natural undisturbed soils.
- Emission limit values (ELVs) should be provided in national legislation as in the case of Italy and Spain, however local soil conditions should be taken into account for estimating ELVs.
- More favorable national laws should be introduced for obtaining permits for facilities producing energy from biomass (especially when they are small).

The optional legislation proposals could be:

- Support of technology change to two-phase process for minimization of waste/wastewater volume. When two-phase systems are used, the fresh water consumption is reduced and the wastewater streams are eliminated.
- Introduction of laws that expressly facilitate initiatives for municipalities to build installations and cooperate through regional agreements with olive mill owners and with other parties that would significantly contribute in providing biomass for energy production and other uses.
- National law should clearly state that, in the absence of adequate private initiative, municipalities are able to build such facilities and operate them within the scope of their local public services.
- Promotion of the establishment of collective/centralized treatment systems.

2.5.2 BUILDING A STRATEGY TO PROTECT SOIL AT OMW DISPOSAL AREAS

Protecting soil at OMW disposal areas requires the design and implementation of a strategy that should be adopted by all involved stakeholders (mill owners, disposal areas' owners, authorities, and the general public). Considering the specific climatic conditions of the Mediterranean countries, it is recommended that a monitoring system fully suited to OMW disposal areas should include a series of steps, as these are explained in the following sections.

2.5.2.1 Identification of Potential and Current Waste Disposal Areas and Recording Them in a GIS Geo-Database

Authorities must identify the current disposal areas in their territory and record them in an inventory. The inventory will contain all licensed disposal areas and as many as possible nonlicensed ones. Local inventories should be created as a first step under the responsibility of local or regional authorities, which afterwards will be integrated into a national inventory under the responsibility of governmental agencies. GIS mapping of the disposal areas and the establishment of a digital database is strongly recommended.

2.5.2.2 Characterization of Disposal Areas—Risk Assessment

Recorded disposal areas should be characterized considering location, hydrogeology, physiography, geomorphology, land use, soil structure, texture, water permeability, coefficient of hydraulic conductivity (saturated or unsaturated), porosity, presence, and depth of impermeable soil layers. Additionally, the collected data may include:

1. history of the site;
2. extent and types of contaminants that may exist, hydrogeological and hydrological regime for the broader area;
3. known/anticipated presence and behavior of receptors;
4. sampling of soil and groundwater: comparison with generic guideline values or quality standards;
5. sampling of soil and groundwater: site-specific modeling of fate, transport, and exposure and comparison with toxicological values; and
6. other parameters which may be considered necessary for the complete characterization of the area.

Such a characterization will allow the implementation of the risk assessment study of the area and the identification of the sites posing higher risk to human health and the environment.

Indicatively, a risk assessment study could be comprised of the following steps:

1. Preliminary investigation (desk study, site reconnaissance, and sometimes limited exploratory investigation). The goal of this preliminary stage is to assess whether potentially contaminating activities have taken place on the site, whether soil and/or water pollution is suspected, and in some cases to confirm the existence of pollution. In short, this phase should focus on hazard identification.
2. Detailed investigation. The aims during the main site investigation stage are:
   a. to define the extent and degree of contamination,
   b. to assess the risks associated with identified hazards and receptors, and
   c. to determine the need for remediation in order to reduce or eliminate the risks to polluted or actual receptors.
3. Supplementary or feasibility investigations to better define the need and the type of remedial action or monitoring. The aim may be to assess the feasibility of various remediation techniques. This may include more detailed physical and chemical characterization of soils and laboratory studies on groundwater treatability. Supplementary investigations may also be designed to improve understanding of the nature, extent, and behavior of contaminants.

The risk assessment, however, should not be limited to toxic constituents (e.g., polyphenols for the case of OMW) but to consider also the potential progressive soil degradation due to the presence of other less hazardous or nonhazardous (e.g., nutrients, other inorganic compounds, and so on) constituents in wastes. This factor is often underestimated and the majority of risk assessment studies focus on the exposed toxicity to soil and humans. Therefore, specific care should also be taken for inorganic constituents (e.g., K, Cl^-, NO_3^-, SO_4^{2-}, P, Mg, Fe, Zn, and others), since the very high concentrations disposed on soil may severely affect its quality, even many years after the last disposal. Therefore, the performance of a complete soil physicochemical analysis and identification of organic and inorganic soil constituents are strongly recommended. Assessment of phytotoxicity potential is also recommended.

2.5.2.3 Evaluation of Risk Level

The fourth step is the evaluation of risk level of the potentially degraded areas by excluding further future disposal areas under *high risk*. Indeed, a remediation plan should be developed and implemented immediately. For the areas under *medium risk*, further assessment of the threat type and potential extent is strongly recommended in order to decide the conditions of waste disposal or the design and implementation of remediation actions. Decisions should be taken considering data collected during the risk assessment study. For those areas under *low* or *near zero risk*, a management plan for the safe disposal of wastes should be developed and implemented under the supervision of local authorities and the responsible governmental agencies.

2.5.2.4 Adoption of Soil Quality Indicators and Thresholds

Through continuous monitoring of many OMW disposal areas, it was revealed that not all soil parameters are affected by the disposal. In particular, eight soil parameters were identified to be appropriate to describe changes in soil quality due to OMW disposal (Doula et al., 2013), namely electrical conductivity, organic matter, total nitrogen, total polyphenols, available phosphorus, exchangeable potassium, available iron, and pH (mainly for acidic soil types).

Specific care should also be taken for Ni, Cr, and Mo. These metals may be also slowly released as a result of corrosion of poor quality steel parts, when they come in contact with the acidic OMW in olive mill facilities. These steel alloying elements may be finally transported to deeper soil horizons and also contaminate ground water. Contamination of soils with recalcitrant heavy metals is an issue that needs to be seriously considered in OMW disposal sites.

Monitoring of soil indicators requires the design of sampling strategies allowing assessment of changes in soil quality. In general, changes in soil quality can be assessed by monitoring soil indicators and comparing them with target values (critical limits or threshold levels) at different time intervals, for a specific use in a selected area-system (Arshad and Martin, 2002).

In general, when a set of indicators is proposed, thresholds levels should be also established in order to assist evaluation of collected data and chemical analysis results. Thresholds could be identified from EU directives and national laws, but also from international literature. The particularity of the indicators proposed in the case of OMW disposal is that they mainly correspond to soil properties associated with fertility and not to pollutants (such as heavy metals) in the classical sense, and therefore they are not included in national laws or EU directives. Nevertheless, international literature can provide general limits as these properties have been extensively studied for many years. Given the complexity of setting limits and the uniqueness of each targeted area/region, it may be more efficient to develop guidelines that can help in setting up limits under certain land and environmental conditions. Therefore, the definition of indicators' thresholds would be more effective and representative for each target area, if they would be determined after evaluation of data collected from the areas of interest and by taking into account local characteristics. In other words, it is recommended to define thresholds for indicators by collecting and analyzing representative control (clean) samples (Doula et al., 2013).

For instance, the assessment of polyphenols in soil is considered difficult and entails high degree of uncertainty due to the lack of generally accepted threshold. Thereby, it is recommended to use local and site specific thresholds as guidelines/normal values (Kavvadias et al., 2010).

2.5.2.5 Physical, Chemical and Biological Characterization of the Wastes

Wastes must be fully characterized considering the variability of their composition. Therefore, a well-designed sampling strategy must be developed and performed by experts.

The appropriateness for landspreading must be proved through appropriate chemical, physical, and biological characterization, whereas waste must be characterized for its hazardous potential. According to the results of this assessment and the national or international legislative restrictions, the competent local/regional/governmental authority may permit (or not) landspreading. Given that the wastes could be distributed on land (with or without preconditions), the authorities must proceed to the next steps and define the preconditions and the monitoring measures.

2.5.2.6 Land Suitability Maps—Defining the Conditions of Landspreading

The development of land suitability maps for OMW distribution is strongly recommended. Such maps may assist authorities in decision-making regarding the level of land suitability to accept wastes. An example of land suitability maps can be seen for the case of pistachio waste landspreading on the web site of another LIFE project, AgroStrat (LIFE-AgroStrat, 2016, http://www.agrostrat.gr). The maps have been developed after evaluating and rating a number of soil and area parameters (i.e., drainage, slope, soil depth, infiltration rate, erosion level, texture, salinity, exchangeable sodium percentage, cation exchange capacity, soil indicators, and waste properties). According to the United Nations Food and Agriculture Organization—FAO (1976):

- *Land evaluation* is the process of estimating the potential of land for alternative kinds of land use so that the consequences of change can be predicted.
- *Land suitability* is the fitness of a given area for a land utilization type (or land use), or the degree to which it satisfies the land user. It is generally presented as a class or rating.

Where a landscape characteristic does not meet the requirements for a particular land use, it constitutes a potential limitation or "constraint." Constraints are commonly rated to express the degree of severity to which they may impair land use. Internationally recognized suitability classes outlined by FAO can be adapted and applied at both regional and local scale. Adapted FAO (1976) suitability classes, as cited in Van Gool et al. (2008), are presented in Table 2.9.

Table 2.9 Land Suitability Classes (FAO, 1976)

Suitability Classes	Description
S1, Highly suitable	Land having no significant limitations to sustained application for a given land use or only minor limitations. Nil to minor negative economic, environmental, health, and/or social outcomes.
S2, Moderately suitable	Land having limitations which in aggregate are moderately severe for sustained application of a given land use. Appreciably inferior to S1 land. Potential negative economic, environmental, health, and/or social outcomes if not adequately managed.
S3, Marginally suitable	Land having limitations which in aggregate are severe for sustained application of a given use. Moderate to high risk of negative economic, environmental, health, and/or social outcomes if not adequately managed.
N1, Not suitable	Land having limitations, which may be insurmountable. Limitations are so severe as to preclude successful sustained use of the land. Very high risk of negative economic, environmental, and/or social outcomes if not managed.
N2, Not suitable	Land having limitations, which appear so severe as to, preclude any possibilities of successful sustained use of the land in the given manner. Almost certain risk of significant negative economic, environmental, and/or social outcomes.

In order to implement the methodology, FAO principles and guidelines should be conformed to the specific area characteristics and to OMW composition, as well as to regional action plans. Legislative and other generally accepted restrictions for waste landspreading should be taken into consideration, too (MAFF, 1989; Soil Science Society of America, 1986; LIFE-AgroStrat, 2016).

Considering land suitability classes of Table 2.9, OMW can be applied without limitations in areas characterized as S1. At this case, a management plan should be developed and implemented under the supervision of local authorities and the responsible governmental agencies. For areas characterized as S2 and S3, authorities should define the degree of limitations as well as the restricted factors and decide if these areas can be included in the landspreading plans. Further assessment of the threat type and its potential extent is strongly recommended in order to decide the conditions of landspreading or the design and implementation of improvement/remediation actions. N1 and N2 areas must be excluded from landspreading plans due to the high risk of degradation (environmental, economic, and societal), whereas an improvement or remediation plan should be developed and implemented (Doula et al., 2016).

2.5.2.7 Quantification of Landspreading—Doses Estimation

Having identifying the suitable areas and ensuring that no legislative restrictions exist for OMW landspreading, this step aims to define the optimum amount of OMW to be distributed on soil. First, the *maximum permitted OMW amount* should be calculated. This is the maximum amount of waste that a soil can afford based on its physicochemical properties. To estimate this amount one should consider the concentration of each element/substance in soil and waste, as well as to define the one that is the restrictive factor for the application. In order to have an efficient evaluation conformed to the specific waste type, the soil indicators could be considered as potential restrictive parameters. In addition, the legislative framework in many countries defines specific amounts of heavy metals that can be added on soils annually, as well as the upper limits of their concentration in soils and waste. This aspect is recommended to be considered, although heavy metals are rarely found in OMW (except for some steel alloying elements such as Ni, Cr, and Mo, as explained earlier). After the identification of the most restrictive element/substance, its concentration can be used for the estimation of the maximum amount of waste to be distributed. For instance, this element/substance could be the one with the highest concentration or the lowest threshold in soil or in waste (Doula et al., 2016). Moreover, the estimation of the waste amount should take into account the thresholds of soil indicators, which should not be exceeded after waste spreading. Therefore, in order to calculate the correct waste amount, the concentration of each indicator in soil and waste should be known. For nitrogen, the maximum amount that is allowed by the relevant legislation for NO_3^- addition on soils (e.g., the "Nitrates Directive" 91/676/EEC which defines 170 kg N/ha/y or other amounts as defined by national laws) should be considered as the upper limit (Doula et al., 2016).

For each soil indicator, the following procedure should be followed (Doula et al., 2016):

1. Definition of the amount of the element that could be added on soil, by subtracting the concentration in soil from the upper threshold value,
2. Following calculation (1) and the concentration of the element in waste, the amount of waste that could be added on soil can be calculated.

After the completion of the estimations for all soil indicators, the derived amounts of wastes are compared with each other. The smallest one is the one proposed for landspreading. The final comparison that should be done is between the amount derived from the above calculation (2) and the *maximum permitted amount.*

If the *maximum permitted waste amount* is higher than the estimated one, then the latter can be applied on land. Otherwise, the amount to be distributed is the *maximum* one.

2.5.2.8 Soil Monitoring

Given that safe OMW landspreading has been ensured through the implementation of all previous steps, authorities should design a monitoring strategy to ensure environmental sustainability over time.

All information should be available to the authorities at any time so as to be able to implement monitoring strategies and assess the collected data. Therefore, mill and disposal areas owners must inform the responsible authorities and provide them with detailed plans and data collected during the previous steps. Period and duration of landspreading should be taken into account and a time plan should be submitted, so that the authorities are able to design the appropriate monitoring plan.

An effective monitoring system has to consider the geomorphology, the hydrology, and soil types of the application area, the peculiarities and the characteristics of OMW as well as the local meteorological conditions. Authorities should design an effective monitoring strategy and implement it in cooperation with scientists and local waste users. Monitoring of soil quality indicators should be performed once a year and preferably before waste landspreading. This requires annual soil sampling and chemical analysis in the framework of a defined monitoring strategy that land users or polluters must follow. The results of the chemical analysis should be evaluated by experts (e.g., agronomists, soil scientists, environmentalists) and a technical report should be submitted to the responsible authorities. The report should also include a detailed description of the wastes' landspreading plan (amount, timing, equipment used). Depending on the evaluation results, the responsible authorities may permit wastes disposal or not.

A specific inventory (a database) for each disposal site should be established and updated annually. This will facilitate the immediate identification of risky areas and provide data regarding the history of the site, specific local geomorphological characteristics, amounts of disposed waste, and soil composition as well as any other data that are considered useful and necessary for the effective protection of soil quality and function. If, it is found that a disposal area is under soil deterioration risk, then the risk level should be evaluated and changes in land distribution plans should be proposed. The authorities may also require the development and implementation of a remedial strategy.

Soil samples should be collected annually from hot spots and analyzed only for soil quality indicators. A geo-referenced soil-sampling scheme should be planned for the annual monitoring of the area and implemented by authorities or through them by areas owners (LIFE Prosodol, 2016b; LIFE-AgroStrat, 2016). It is strongly recommended to record and monitor areas of waste reuse using GIS web tools that allow easy and visualized evaluation by authorities, scientists, and other stakeholders (Chliaoutakis et al., 2012).

Software application tools for soil monitoring could facilitate the adoption of the monitoring system by authorities and individuals as well as enhance continuous and effective monitoring. The LIFE-Prosodol project developed a web based GIS tool, which presents soil constituents' distribution versus time, and depth. Through this tool, local and regional authorities have the opportunity to map and screen disposal areas rapidly, identify potential risky conditions, carry out systematic monitoring of the areas of interest, and facilitate decision making with the appropriate measures to be taken at field or regional scale. The tool monitors the eight soil indicators selected for this waste type (i.e., soil pH, electrical conductivity, organic matter, total nitrogen, polyphenols, exchangeable potassium, available phosphorus, and available iron) and integrates the continuous monitoring of the OMW disposal areas into the regular activities of local/regional authorities. Thus, it allows the proper and continuous monitoring of such areas. Nevertheless, the tool can be adapted to any type of agricultural waste and conformed easily to any local/regional peculiarities.

However, the adoption and application of such a tool requires the cooperation of the disposal areas' owners, since repeated soil sampling at various sites is necessary for maps creation and update. The proposed application tool uses interpolation surfaces that indicate the variation of the different physical and chemical parameters in the area of interest, so the user can rapidly screen the degree of risk in the vicinity of the waste disposal areas. Potentially, this allows the establishment of an Operational Centre in cooperation with the Environmental Protection Office of the Local Government. This joint management center which could be located, for instance, at the premises of a Municipality or a Regional Authority and can undertake the continuous monitoring of areas under risk as well as the scientific and consulting support of the owners.

The design of the particular software package needs to monitor a number of fields that are spread around and make queries based on various spatial and chemical attributes. It is proposed that, for each disposal area, one initial mapping should be carried out by performing soil sampling at various sites (e.g., at least 4 times every 2 months). The sampling sites will be selected according to the generally accepted soil sampling rules, whereas a qualified person should be present and undertake the overall control. The maps that will be created should be used for no more than 5–8 years. After this period, the maps must be updated by repeating the sampling procedure.

2.5.2.9 Code of Good Practices for Soil Management

A Code of Good Practice should be developed. This code would give advice in relation to soil management practices, adopted by all sectors having the potential to impact on soil quality. The Code should be a practical guide that will assist owners of the OMW disposal areas, farmers who may use OMW for soil fertilization or land managers to protect the environment in which they operate. The Code should describe key actions that the main actors can take to protect and further improve soil, water, and air quality to meet legal obligations. The Code should not be just a manual guide on how to manage such areas but should assist on selecting the appropriate actions for the specific conditions (environmental, climatic, social, financial). Such a Code entitled "Good Practices for the agronomic use of olive oil mills wastes" has been developed within the framework of the PROSODOL project and is available on the web site of the project (LIFE Prosodol, 2016c). It includes guidelines for soil sampling, use of OMW for crops irrigation (technical and financial aspects), OMW disposal on soil, periodical soil monitoring, soil remediation techniques, composting, and existing legislative framework.

REFERENCES

Aggarwal, P.K., Means, J.L., Hinchee, R.E., Headington, G.L., Gavaskar, A.R., 1990. Methods to Select Chemicals for In-Situ Biodegradation of Fuel Hydrocarbons. Air Force Engineering and Services Center, Tyndall AFB, FL.

Anastasiou, C.C., Christou, P., Michael, A., Nicolaides, D., Lambrou, T., 2011. Approaches to olive mill wastewater treatment and disposal in Cyprus. Environ. Res. J. 5 (2), 49–58.

Arshad, M.A., Martin, S., 2002. Identifying critical limits for soil quality indicators in agro-systems. Agric. Ecosys. Environ. 88, 153–160.
Azbar, N., Bayram, A., Filibeli, A., Muezzinoglu, A., Sengul, F., Ozer, A., 2004. A review of waste management options in olive oil production. Crit. Rev. Env. Sci. Tec. 34 (3), 209–247.
Caputo, A.C., Scacchia, F., Pelagagge, P.M., 2003. Disposal of by-products in olive oil industry: waste-to-energy solutions. Appl. Therm. Eng. 23 (2), 197–214.
Chliaoutakis, A., Kydonakis, A., Doula, M.K., Kavvadias, V., Sarris, A., Papadopoulos, N., 2012. Geospatial tools for olive oil mills' wastes (OOMW) disposal areas management. Adv. Geosci., EARSeL, Mykonos island, Greece.
Cookson, J.T., 1995. Bioremediation Engineering Design and Application. McGraw-Hill, New York.
EEC, 1991a. Council Directive 91/271/EEC concerning urban wastewater treatment was adopted on 21 May 1991 to protect the water environment from the adverse effects of discharges of urban waste water and from certain industrial discharges.
EEC, 1991b. Council Directive 91/676/EEC of 12 December 1991 concerning the protection of waters against pollution caused by nitrates from agricultural sources.
FAO, 1976. A framework for land evaluation. Soil resources management and conservation service land and water development division. FAO Soil Bulletin No. 32. FAO-UNO, Rome.
Doula, M.K., Kavvadias, V.A., Elaiopoulos, K., 2011. Zeolites in soil-remediation processes. In: Inglezakis, V., Zorpas, A. (Eds.), Natural Zeolites Handbook. Bentham Publisher, China, pp. 330–355, (Chapter 4.3).
Doula, M.K., Elaiopoulos, K., Kavvadias, V.A., Mavraganis, V., 2012. Use of clinoptilolite to improve and protect soil quality from the disposal of olive oil mills wastes. J. Hazard. Mater. 207–208, 103–110.
Doula, M.K., Kavvadias, V., Elaiopoulos, K., 2013. Proposed soil indicators for olive mill waste (OMW) disposal areas. Water Air Soil Pollut. 224, 1621–1632.
Doula, M.K., Sarris, A., Hliaoutakis, A., Kydonakis, A., Argyriou, L., 2016. Building a strategy for soil protection at local and regional scale-the case of agricultural waste landspreading. Environ. Monit. Assess. 188 (3), 1–14.
ECOIL, 2005. LIFE+ project, Life Cycle Assessment (LCA) as a Decision Support Tool (DST) for the ecoproduction of olive oil, Deliverable 2, TASK 1, Recording and assessment of the existing situation, Chania, Greece.
IMPEL, 2003, Olive Oil Project, European Union Network for the Implementation and Enforcement of Environmental Law (IMPEL), EU.
Inglezakis, V., Zorpas, A.A., Loizidou, M.D., Grigoropoulou, H., 2003. Simultaneous removal of metals Cu^{2+}, Fe^{3+} and Cr^{3+} with anions SO_4^{2-} and HPO_4^{2-} using clinoptilolite. Microp. Mesop. Mater. 61, 167–171.
IPPC BREF, 2006a. Integrated pollution prevention and control, reference document on best available techniques in the, food, drink and milk industries, European Commission
IPPC BREF, 2006b. Integrated pollution prevention and control, member states' national legislation and standards, additional information provided by the TWG of the food, drink and milk industries BREF, European Commission.
Kapellakis, I.E., Tsagarakis, K.P., Crowther, J.C., 2008. Olive oil history, production and by-product management. Rev. Environ. Sci. Biotechnol. 7, 1–26.
Kavvadias, V., Doula, M.K., Komnitsas, K., Liakopoulou, N., 2010. Disposal of olive oil mill wastes in evaporation ponds: effects on soil properties. J. Hazard. Mater. 182, 144–155.
Life-AgroStrat, 2016. Sustainable strategies for the improvement of seriously degraded agricultural areas: The example of Pistachia vera L. (2016b). Land Suitability maps. Available from: http://www.agrostrat.gr/?q=en/node/546
LIFE Prosodol, 2016a. Strategies to improve and protect soil quality from the disposal of Olive Oil Mill Wastes in the Mediterranean region (2012). Results and Achievements of a 4-year demonstration project—What to consider; What to do. Available from: http://ec.europa.eu/environment/life/project/Projects/index.cfm?fuseaction=home.showFile&rep=file&fil=PRODOSOL_Results_Achievements.pdf

LIFE Prosodol, 2016b. Strategies to improve and protect soil quality from the disposal of Olive Oil Mill Wastes in the Mediterranean region (2012). Surface Interpolation. Available from: http://www.prosodol.gr/?q=node/3445

LIFE Prosodol, 2016c. Strategies to improve and protect soil quality from the disposal of Olive Oil Mill Wastes in the Mediterranean region (2012). Good practices for the agronomic use of olive mill wastes. Available from: http://www.prosodol.gr/sites/prosodol.gr/files/DGr_7.pdf

MAFF, Department of Environment, 1989. Code of Practice for agricultural use of sewage sludge. London.

MORE, 2008. Market of olive residues for energy, Intelligent Energy for Europe (IEE) project, Deliverable 3.1, One joint report for the 5 Regional "state of the art" reports from each involved area describing the current olive-milling residues market with a focus on energy uses.

Ouzounidou, G., Zervakis, G.I., Gaitis, F., 2010. Raw and microbiologically detoxified olive mill waste and their impact on plant growth. Terr. Aquat. Environ. Toxicol. 4, 21–38.

Paraskeva, P., Diamadopoulos, E., 2006. Technologies for olive mill wastewater (OMW) treatment: a review. J. Chem. Technol. Biotechnol. 81, 1475–1485.

RES-HUI, 2006. Integrated Management of Olive Oil-Mill residues and wastewater/RES-HUI, Project num. C3-06, INTEREG IIIC project, Dipartamento di Energetica "Sergion Stecco", Universita degli Studi di Firenze.

Roig, A., Cayuela, M.L., Sanchez-Monedero, M.A., 2006. An overview on olive mill wastes and their valorisation methods. Waste Manage. 26, 960–969.

Sierra, J., Marti, E., Garau, A.M., Cruãnas, R., 2007. Effects of the agronomic use of olive oil mill wastewater: field experiment. Sci. Total Environ. 378, 90–94.

Soil Science Society of America, 1986. Utilization, treatment and disposal of waste on land. Proceedings of a Workshop held in Chicago 6–7 Dec., U.S.A.

Taccogna, G., 2010. The legal regime of olive pomace deriving from olive oil extraction at olive mills, waste, by-products and biomass, Department of Public and Procedural Law, University of Genoa, on behalf of ARE s.p.a.—Agenzia regionale per l'energia della Liguria, member of the community project: "MORE: Market of Olive Residues for Energy".

US-EPA, 1991. Understanding Bioremediation: A Guidebook for Citizens. Office of Research and Development, Washington, D.C.

Van Gool, D., Maschmedt, D., McKenzie, N., 2008. Conventional land evaluation. In: McKenzie, N.J., Grundy, M.J., Webster, R., Ringrose-Voase, A.J. (Eds.), Guidelines for Surveying Soil and Land Resources. second ed. CSIRO Publishing, Australia.

Vidali, M., 2001. Bioremediation. An overview. Pure Appl. Chem. 73, 1163–1172.

CHAPTER

THE BIOREFINERY CONCEPT FOR THE INDUSTRIAL VALORIZATION OF RESIDUES FROM OLIVE OIL INDUSTRY

3

María José Negro*, Paloma Manzanares*, Encarnación Ruiz, Eulogio Castro**, Mercedes Ballesteros***

**Biofuels Unit, Energy Department-CIEMAT, Madrid, Spain; **Department of Chemical, Environmental and Materials Engineering, Agrifood Campus of International Excellence, ceiA3, University of Jaén, Jaén, Spain*

3.1 INTRODUCTION

Nowadays, biorefineries include a wide range of technologies to separate main biomass components (carbohydrates, lignin, protein, etc.) and convert them into biofuels and chemicals. This approach has been applied to several biomasses and, currently, a broad spectrum of different large scale biorefineries, using a single feedstock, are under development (EBTP, 2016). On the other hand, the use of mixed biomass feedstocks in a single facility allows enlarging the view to a multi-feedstock biorefinery concept, wherein several materials are converted into a series of added-value products. In this context, agroindustrial and agrofood wastes, due to their high diversity of biomass derived products, offer a high potential of feedstocks to be processed in such a multifeedstock biorefinery plant.

This approach would be consistent with the development of small scale biorefineries better adapted to the rural areas where the wastes are generated, in contrast to the above mentioned large scale biorefineries that require high capital costs and face barriers for sustainable biomass supply and distribution. Conversely, small-scale biorefineries will require a significantly lower CAPEX, and solve several challenges facing their larger competitors. Nevertheless, despite their advantages, numerous technological and strategic challenges hamper commercial development. One of them is the inherent characteristic of heterogeneous biomasses that require complex technological transformation processes. A novel integrated self-sustainable biorefinery would be capable of transforming feedstocks by means of different process units and unit operations to produce a vast array of biomaterials and bioproducts, while maximizing the resources and energy efficiency.

Among main agroindustries worldwide, olive oil production is an economically important industry, especially in the Mediterranean countries. According to recent estimations from FAOSTAT (2015), close to 10.31 Mha of olive crop were cultivated in 2013 worldwide, with close to 50% corresponding to cultivation in nine countries of the EU, leaded by Spain (24.2%), Italy (11.1%), and Greece (9.0%). Total olive fruit production in the EU accounted for 13.24 Mt, the major producer being Spain with

Olive Mill Waste. http://dx.doi.org/10.1016/B978-0-12-805314-0.00003-0

7.87 Mt, close to 60% of total EU yield. Regarding olive oil production in 2013, 1.11 Mt oil were produced in Spain, which accounted for close to 57 % of total EU yield (1.96 Mt).

Although olive tree is cultivated as a source of olives and olive oil, several by-products are generated from the cleaning operations of olive trees and the different steps of the olive processing to obtain olive oil. A detailed description of these residues, which can be in principle classified into the categories of low/medium-moisture (olive tree biomass residues, olive stones, and pomaces) and high-moisture residues (wastewaters) is given below. This chapter examines the utilization of the residues produced from start to finish in olive oil industry within an integrated biorefinery to obtain value-added products, such as antioxidants, biofuels, energy, etc. The technological processes to convert the solid and wet residues generated into different by-products are described in detail.

3.2 BIOMASS RESIDUES FROM OLIVE CROP AND OLIVE-OIL INDUSTRY

From olive tree cultivation and olive oil industry, a number of by-products are generated together with olive oil, that is, biomass residues from olive tree pruning operations (OTP), olive leaves, olive stones and pomace residues from olive oil extraction, and olive mill wastewaters (OMWW), which refers to all wastewater streams generated in washing steps and in the different stages of oil production (Fig. 3.1).

Tree pruning is an essential operation performed every two days after the fruit harvesting by cutting less productive branches for tree regeneration, thus improving fruit production. OTP includes a mixture of leaves, thin branches and wood, in variable proportions depending on culture conditions, tree age, and/or local pruning practice. OTP biomass contains a significant amount of cellulose, hemicelluloses, and lignin, as well as a significant amount of soluble compounds (extractives) and ash. The overall chemical composition of OTP biomass may differ slightly depending on tree age, soil makeup, and climate conditions, but as average carbohydrates may account from 46.1% to 61.6% (dwb) (Table 3.1). Extractives found in OTP, range from 14.1% to 31.4% (dwb), while an important part of them consists of nonstructural glucans. This composition allows considering OTP as an important energy and chemicals source that can be processed in a multiproduct industry.

Olive leaves come from the cleaning, and wastewater originate from the washing of the olives. Olive leaves are usually removed using a blower machine. They show a quite low carbohydrate content (close to 10%), but a high extractives content of around 40% (Table 3.1), that could be a source of polyphenols (Aydinoglu and Sargin, 2013) or other antioxidant compounds.

Regarding the process to extract oil from olives previously crushed, it can follow the two-phase and three-phase separation modes. In the three-phase system water is added to the paste, which is next separated into a solid (olive pomace), an oily phase and a water phase. The two-step mode is essentially the same process, but without adding water to the initial olive mash. In both separation systems, the resultant olive oil is washed to remove impurities and the wash water separated by centrifugation generating the "olive oil washing wastewater", which is normally processed together with other residual water streams generated. Olive pomace is the main residue after both types of separation systems and it is constituted by crushed olive stones, process water and all material coming from the olive fruits except the olive oil. It represents the main residue of the olive oil extraction process by weight and differs in composition depending on the production process (two- or three-phase). An example of a two-phase pomace composition is included in Table 3.1, showing a high content of extractives around 48% (dwb) and a carbohydrate content close to 20% (dwb). This residue contains some residual olive oil and also

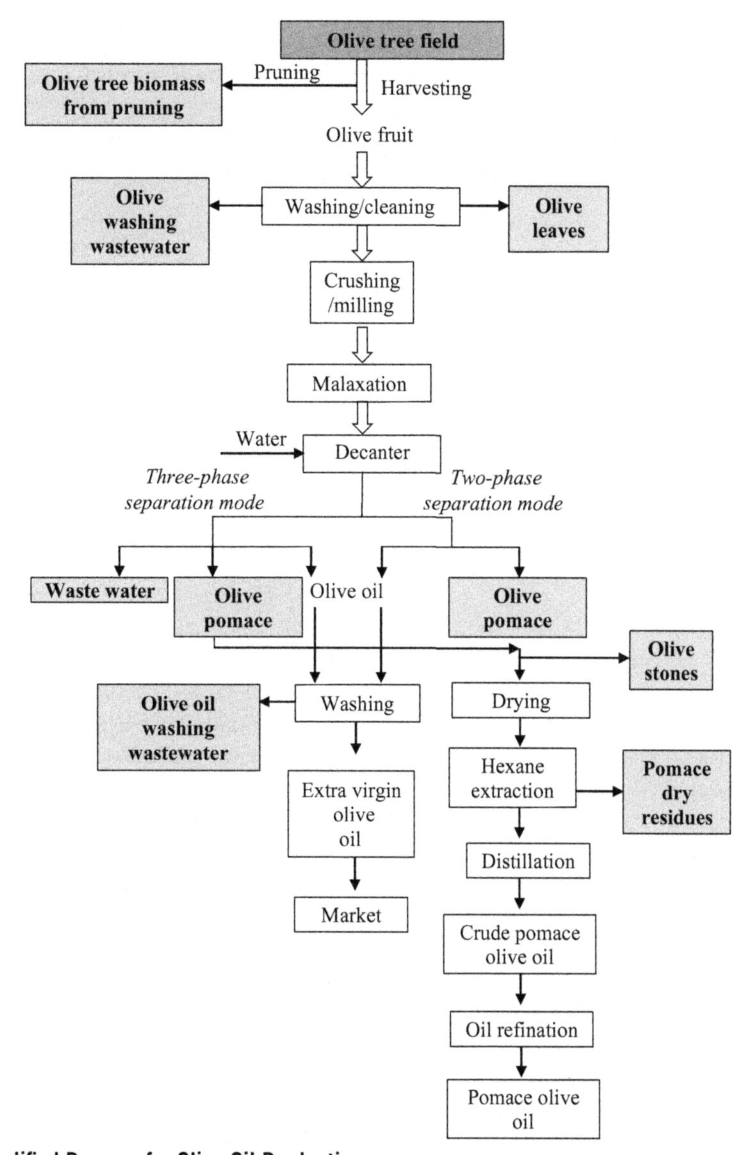

FIGURE 3.1 Simplified Process for Olive Oil Production

Reprinted from Romero-García, J.M., Niño, L., Martínez-Patiño, C., Álvarez, C., Castro, E., Negro, M.J., 2014. Biorefinery based on olive biomass. State of the art and future trends, Bioresour. Technol. 159, 421–432. Copyright (2016), with permission from Elsevier.

some polyphenols. It can be further processed to extract the oil and obtain the so-called "pomace oil" and a final solid residue is generated called " extracted dry pomace residue," which can be integrated in the biorefinery for energy production or other applications. It is characterized by a similar composition in dry weight basis than olive pomace; high extractives content that can overcome 30% (dwb) and carbohydrates accounting from 20% to 30% (dwb) (Table 3.1).

Table 3.1 Reported Chemical Composition Values of Main Lignocellulosic Biomass Residues From Olive Crop and Olive-Oil Industry (When Available, Value Range is Given)

	Main Component (wt.% dry basis)					
Raw Material	Cellulose	Hemicellulose	Lignin	Ash	Extractives	References
Olive tree pruning	26.1–36.6	16.4–25.0	16.6–27.7	3.3	14.1–31.4	Romero-García et al. (2014); Ballesteros et al. (2011)
Olive stones	28.1–40.4	18.5–32.2	25.3–27.2	0.7	19.2	Lama-Muñoz et al. (2014); Cuevas et al. (2015b); Ballesteros et al. (2001)
Olive oil pomace	12.8	7.5	28.4	2.2	47.8	Ballesteros et al. (2001)
Extracted dry pomace	11.1–15.8	9.0–15.6	24.2–40.4	7.7–13.9	20.0–33.7	Unpublished data from the authors
Olive leaves	5.7	3.8	39.6		36.5	García-Maraver et al. (2013)

Olive stones can be recovered from the olive pomace after oil separation. Cellulose, hemicellulose, and lignin content from olive stone are in the range 28.1–40.4%; 18.5–32.2%; and 25.3–27.2% (dwb), respectively. Other components are proteins and phenolic compounds, such as tyrosol, hydroxytyrosol, oleuropein, and others, which make this residue an important by-product source of valuable compounds (Rodríguez et al., 2008).

Regarding olive mill wastewater (OMWW), several water streams are generated at three different points: during the cleaning of the olive fruits; from the horizontal centrifuge (decanter) during the three-phase separation step; and during the washing process from the secondary centrifuge of virgin olive oil. Chemical composition of OMWW is very variable depending of many factors, such as the system used for oil extraction, the variety of olive trees, and the degree of maturity and fruit storage time. In general, apart from water (83–92%), the average content of organic matter (polysaccharides, proteins, organic acids, and polyphenols) is in the range of 4–16% and also contains 0.4–2.5% of salts (Dermeche et al., 2013). The high chemical organic demand (COD) and the high concentrations of polyphenols, up to 12 g/L, (Daâssi et al., 2013) makes the treatment of OMWW difficult and costly.

3.3 REVALORIZATION TECHNOLOGIES TO CONVERT OLIVE RESIDUES INTO BY-PRODUCTS AND BIOFUELS

As mentioned earlier, residues produced in olive oil industry can be categorized as either low/medium- or high-moisture materials. The conversion of low/medium biomasses (in general, lignocellulose) can be directed to heat and power production by thermochemical processes, or be upgraded into added-value products by biochemical/chemical routes (Fig. 3.2). Biochemical processes generally involve fractionation and hydrolysis of polymers (carbohydrates and lignin) into oligosaccharides or

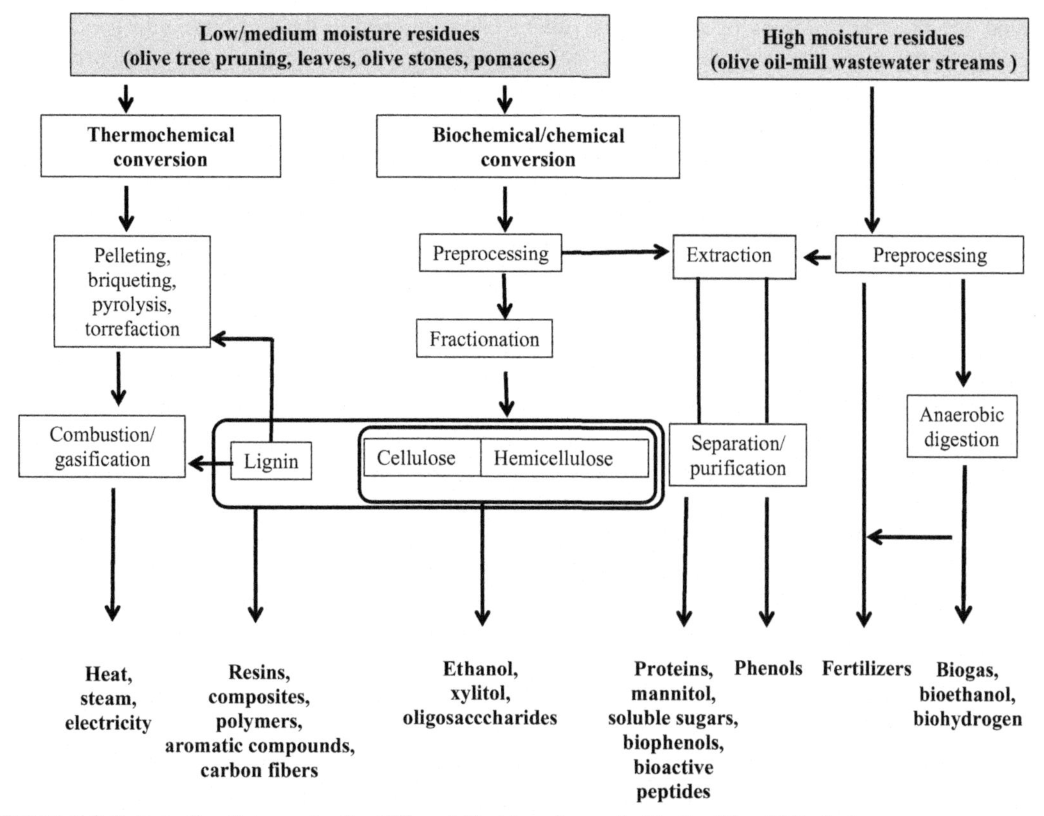

FIGURE 3.2 Valorization Scheme for the Different Residues Generated in the Olive Oil Industry

monomers that can further be converted into a wide variety of products. The conversion of wet biomass (wastewater streams) generally involves biofertilization and/or anaerobic digestion.

3.3.1 PROCESSES FOR LOW/MEDIUM-MOISTURE RESIDUES

3.3.1.1 Direct Conversion to Heat, Steam, and Electricity

The conversion of wastes generated from the olive mill industry into thermal and electrical power is an interesting option in the case of low-moisture residues, as the olive stones or the extracted dry pomace residue produced in the olive pomace oil industry (Galanakis, 2011). Currently, the use of olive stones as a biofuel for thermal applications is quite widespread in olive oil productive regions, especially in agroindustries, livestock farms, greenhouses, and domestic heating systems. Extracted dry pomace residue is used as a fuel in some local industries too, like the ceramic industry. However, this residue is less used than olive stones in domestic heating systems because of the higher generation of pollutant emissions. In the pomace olive oil extraction industry, extracted dry pomace residue is usually used to dry the wet olive pomace whereas olive stones are often employed in the industrial boilers to obtain process steam because it causes less corrosion problems. In recent years, the use of olive stones as a fuel in olive mills boilers instead of extracted dry pomace residue has been progressively increased.

Indeed, the practice of removing most of the olive stone contained in the wet olive mill pomace has been becoming more usual. Consequently, the extracted dry pomace residue composition resulted poorer in olive stones leading to worse combustion properties.

On the other hand, olive stones and extracted dry pomace residue are also currently applied for the generation of electrical energy or cogeneration (simultaneous steam and electricity production) through combustion. OTP can be also used in plants of electric production, in a lesser extent. Concerning medium-moisture wastes, such as the two-phase olive mill pomace, there are also some experiences about its energetic use, but it is more difficult due to its moisture content (Alburquerque et al., 2004).

Besides combustion, other thermochemical technologies have been investigated in recent years for this kind of residues, such as pyrolysis or gasification (Christoforou and Fokaides, 2016). Pyrolysis implies the biomass heating without air. Depending on the temperature and the rate of heating different products can be obtained as solid, liquids, or gaseous biofuels. Concerning gasification, it refers to the biomass conversion into syngas by partial oxidation at temperatures generally in the range 800–900°C. This technology has been emerging as a promising way for the production of energy from olive tree and olive mill residues. For example, Campoy et al. (2014) found that extracted dry pomace residue is a suitable fuel in a pilot fluidized bed gasifier, while Vera et al. (2014) investigated the use of olive stones and OTP in a downdraft gasifier.

Chemical composition is an important factor that affects the performance of the thermochemical processes. The main structural components of the different residues coming from the olive cultivation and the olive mill industry are generally cellulose, hemicellulose, and lignin. The thermal degradation of these components is different. In thermogravimetric analysis (TGA) with olive oil pomace, Ozveren and Ozdogan (2013) reported hemicellulose, cellulose, and lignin degradation temperature ranges of 150–295°C, 300–400°C, and 150–480°C, respectively. Apart from the moisture content and chemical composition, other important parameters involved in the process effectiveness are the reaction temperature, the particle size, and the heating rate (Christoforou and Fokaides, 2016).

Logistic challenges should be overcome for the energetic valorization of the solid olive derived residues, especially in the case of the wastes generated from the olive tree cultivation (OTP) because of this low density, dispersion, and the short period of production. Preparation methods as pelleting, briquetting, and pyrolysis could be applied to increase the energy density (Hanandeh, 2015). Concerning medium-moisture wastes, such as the two-phase olive mill pomace, some treatments as the torrefaction combined with briquetting have been investigated to assess the viability of the energy production from these residues (Benavente and Fullana, 2015). Besides that, in order to explore the feasibility of the potential use of the different residues derived from the olive oil industry as solid biofuels, economic, and sustainability studies should be performed (Lanfranchi et al., 2016). In this way, thermochemical biorefineries based on multiproduction (fuels, chemical, and services) could be a promising alternative in order to diminish the risk of investment (Haro et al., 2014).

3.3.1.2 Biochemical/Chemical Conversion Routes

Apart from the revalorization of low/medium biomasses by thermochemical processes, a variety of valuable products can be obtained by biochemical and chemical conversion technologies based on lignocellulosic nature of these residues, as explained later.

3.3.1.2.1 Preprocessing Strategies

Lignocellulosic-type residues from olive crop and oil mill, apart from the structural carbohydrates (hemicellulose and cellulose) and lignin, contains other nonstructural minority compounds that could be extracted before the process conversion of structural components, contributing to the economic viability of a possible biorefinery based on these materials. This is the case of phenolic compounds with antioxidant properties that have a great interest for the food industry as substitutes of synthetic antioxidants. Therefore, a preprocessing strategy, such as an extraction step, could be applied to the residues in order to recover these bioactive components, while improving the yields of the biochemical conversion route of structural components. For example, with OTP, it has been proved that a water extraction step in autoclave at 120°C during 60 min improves the yields of structural sugars that can be obtained in further steps (Ballesteros et al., 2011; Negro et al., 2014). The liquid obtained from this aqueous extraction contains soluble sugars, mannitol, and phenolic compounds with antioxidant capacity, whose removal has also the positive effect of increasing the fermentability of hemicelulosic hydrolysates obtained in subsequent steps (Martínez-Patino et al., 2015). This strategy has also been applied to olive stones (Lama-Muñoz et al., 2014), aimed at the optimization of antioxidant recovery by water extraction at 130°C during 90 min, followed by a dilute acid hydrolysis step for the recovery of fermentable sugars from structural carbohydrates. Protein extraction through a continuous water flow using a high-pressure system (Kazan et al., 2015) has also been proposed as a previous step of a biochemical process to obtain sugars from olive pomace. Another thermal preprocessing method has been applied to improve the recovery of bioactive compounds from olive pomace (Lama-Muñoz et al., 2011). To recover different products from olive cake, such as antioxidant compounds (Rubio-Senent et al., 2012), short-chain oligosaccharides, phenolic glycosides, secoiridoids (Fernández-Bolaños et al., 2014), and pectins (Rubio-Senent et al., 2015), steam-explosion treatment at mild temperatures (150–170°C) has been applied. Bioactive peptides extraction from olive seeds by enzymatic hydrolysis with different proteases followed by ultrafiltration fractionation has also been recently investigated (Esteve et al., 2015).

Olive leaves have been widely studied for the production of different bioactive components with therapeutic and functional applications, including phenolic compounds, secoiridoids, and flavonoids. Besides conventional extraction methods as maceration and Soxhlet extraction, other promising extraction techniques have being developed recently, such as pressurized liquid extraction, subcritical or supercritical extraction, microwave, ultrasound, etc. Different solvents can be used for this purpose apart from water, as short-chain alcohols, ethyl acetate, hexane, and acidic steam (Rahmanian et al., 2015).

3.3.1.2.2 Carbohydrate Conversion

A key issue for the biorefineries is the cost-effective conversion of cellulose and hemicellulose contained in lignocellulose biomass into fermentable sugars, as a first step in the production of high-added value molecules. Particularly for fuel ethanol production from lignocellulose biomass, process technology generally includes a pretreatment step followed by enzymatic hydrolysis and the subsequent fermentation to ethanol of the sugars released, by fermenting microorganisms. Pretreatment is an important first step for efficient carbohydrates (cellulose and hemicellulose) conversion processes, and it is required to alter the structure of lignocellulose biomass making polysaccharides more accessible to the enzymes that hydrolyse them into fermentable sugars. Moreover, an ideal pretreatment should result in a complete fractionation of the biomass into its key constituents, cellulose, hemicelluloses, and lignin, which will facilitate a subsequent conversion of main components into fuels and high value products at high yield. Many pretreatment methods have been evaluated for ethanol production (Alvira et al., 2010). Specifically for OTP biomass, pretreatments tested include mechanical, chemical methods, and various combinations thereof (Romero-García et al., 2014). After pretreatment, lignocellulose biomass is submitted to enzymatic hydrolysis (EH) process, in which the polysaccharides contained in the pretreated lignocellulose biomass are depolymerized. Due to the complex structure of lignocellulose biomass, enzymatic hydrolysis requires the combined action of different hydrolytic enzymes (cellulases, hemicellulases, esterases, arabinofurosidases, etc.) that must be used in an appropriate proportion to achieve a complete hydrolysis.

Steam explosion is one the most suitable method for pretreatment of herbaceous and hardwood biomass in terms of low reaction time, high solid loading, and minimum use of chemicals. It is a physicochemical pretreatment that combines mechanical forces with the chemical effects of acetyl groups easily released from hemicellulose (autohydrolysis). In this pretreatment, biomass is rapidly heated by high pressure steam for a period of time and then the pressure is suddenly reduced, which makes the material undergo an explosive decompression (Talebnia et al., 2010). Temperatures in the range of 190–240°C for 5 min have been tested in olive tree wood (Cara et al., 2006) and OTP (Cara et al., 2008a). However, relatively low enzymatic hydrolysis yields (about 40% and 60% for wood and OTP, respectively) after steam explosion pretreatment were found. It has been reported that an alkaline peroxide delignification step after steam explosion pretreatment of olive tree wood can improve enzymatic hydrolysis yields up to about 61%. (Cara et al., 2006). Also, the addition of sulfuric acid during steam explosion pretreatment of OTP was found to improve the rate and extension of hemicellulose solubilization and the enzymatic hydrolysis yield, reaching a maximum value of about 70% of theoretical. On the other hand, extractive removal previous to steam explosion has been shown to result in 20% more total sugars recovery in comparison to OTP biomass without water extraction stage (Ballesteros et al., 2011).

Liquid hot water pretreatment has also been investigated on OTP for sugar production (Cara et al., 2007; Requejo et al., 2012). In liquid hot water the biomass is pretreated with pressurized water at high temperature. Temperature and time revealed the most significant effect in the hemicellulosic-sugars recovery and the yield of subsequent enzymatic hydrolysis of pretreated OTP biomass. This pretreatment produces high hemicellulose solubilization releasing oligosaccharides, which remain soluble in the liquid fraction. However, the allowable solid load is much less than in other hydrothermal pretreatments, that is, steam explosion, which is frequently greater than 40%.

Inorganic acids, such as sulfuric and phosphoric acid have been explored for pretreatment of OTP (Cara et al., 2008b; Díaz-Villanueva et al., 2012; Martínez-Patino et al., 2015). Diluted acid pretreatment results in almost 84% of the hemicellulosic-sugars recovery, and a yield of glucose after enzymatic hydrolysis up to 66% at optimal conditions (180°C, 1% H_2SO_4).

Other chemical agents, such as inorganic salts ($FeCl_3$) have been tested to pretreat olive tree biomass (López-Linares et al., 2013). Results show improved results in comparison to untreated material, reaching a maximum glucose yield of 39 g glucose/100 g glucose in untreated olive tree biomass, at 160°C and 0.275M $FeCl_3$ for 30 min. This value corresponded to an efficiency value of 88.7% of theoretical.

All these pretreatments mentioned earlier have in common the production of a pretreated material in which soluble fraction is mainly composed of the hemicellulose sugars, while cellulose and lignin remain in insoluble solid fraction, while technologies described below aim, in general, at solubilizing lignin fraction from the material, the cellulose, and hemicellulose components remaining in the pretreated solid.

Organosolv pretreatment, which uses organic or aqueous solvents (ethanol, methanol, ethylene glycol, acetone, glycerol, etc.) to extract lignin, has been tested on OTP by Díaz et al. (2011). They found pretreatment severity (in terms of ethanol content and temperature) was positively correlated with delignification (up to 64% at 210°C for 60 min and 66% w/w aqueous ethanol). On the contrary, xylan hydrolysis was promoted by low severity conditions in term of ethanol content, reaching a maximum of 92%. Organosolvent pretreatment of OTP was also addressed by Toledano et al. (2011), who concluded that reaction temperature and ethanol concentration were the most important variables, while reaction time was less significant. In a further step in the research work on organosolv pretreatment of OTP, the authors (Toledano et al., 2013) have proven that it results in a highly digestible cellulose substrate (87.2% EH yield), and that lignin can be successfully recovered after pretreatment.

Apart from hydrothermal and chemical pretreatment technologies, thermo-mechanical processes, such as extrusion have been applied to OTP and other olive derived residues. Extrusion is a process of simultaneous heating, mixing, and shearing that results in physical and chemical changes along the extruder barrel (Duque et al., 2013). Recently, fractionation of OTP by one-step alkaline extrusion has been investigated by Negro et al. (2015). The pretreatment resulted in a cellulose and hemicellulose enriched-solid, which contained 92–100% and 94% of the glucan and xylan, respectively, of untreated OTP. A high value of hemicellulose-sugars recovery in the pretreated solid as that obtained in alkaline extrusion of OTP, is interesting to increase the total fermentable sugars production by enzymatic hydrolysis.

The generation of fermentable sugars from olive stones has also been investigated by using different pretreatments: steam explosion (Felizón et al., 2000; Ballesteros et al., 2001); diluted acid (Saleh et al., 2014; Lama-Muñoz et al., 2014), liquid hot water (Cuevas et al., 2009), and organosolv (Cuevas et al., 2015a). Diluted acid and liquid hot water pretreatments were reported to be suitable technologies for hemicellulose solubilization from olive stones at high temperatures. Oligosaccharides were the main carbohydrates obtained in liquid hot water prehydrolysates (16.9 kg per 100 kg of dry olives stones at 210°C), while diluted acid pretreatment provides the maximum xylose recovery (21 kg per 100 kg of dry olive stones) at 200°C. The ethanol-based organosolv pretreatment with addition of sulphuric acid achieved high delignification (>88%) and provided a solid with cellulose content around 83.3% (dwb) (Cuevas et al., 2015a).

The next step in the conversion of lignocellulose biomass to ethanol is the fermentation of sugars, either from enzymatic hydrolysis of cellulose (C6 sugars) or hemicellulose (C6 and C5 sugars), which is commonly carried out by using yeasts. For C6 fermentation, strains of *Saccharomyces* are generally used, while for C5 sugars microorganisms that are able to ferment them, such as *Pachysolen tannophilus, Candida shehatae, and Pichia stipitis* are utilized.

Regarding the process configuration used to convert OTP sugars from cellulose and hemicellulose fractions into ethanol by enzymatic hydrolysis and fermentation, there are several options (Fig. 3.3); a separated enzymatic hydrolysis and fermentation (SHF) process, a simultaneous saccharification and fermentation (SSF) process and a variation of the last one called liquefaction or presaccharification plus SSF (PSSF). In SHF, in which both step are carried out in different reactors, hydrolysis of the pretreated substrate is carried out at optima enzyme temperature and pH (namely 50°C and pH around 5) and once completed; the resulting sugars are fermented at optima pH and temperature for the fermenting yeast (30–37°C). In SSF, a compromise between the optimal temperature for the hydrolytic enzymes and the yeast is needed (37°C) and the process is carried out in a single step. The SSF scheme minimizes the inhibition by end-product on the enzyme activity, resulting in improved yields, shorter residence time, and allows a reduction in the enzymes doses, contributing to decrease the final process cost. In PSSF, the substrate is incubated with the enzymes at 50°C during a relatively short period of time (8–24 h) and next the media is inoculated with the microorganism to proceed with the SSF.

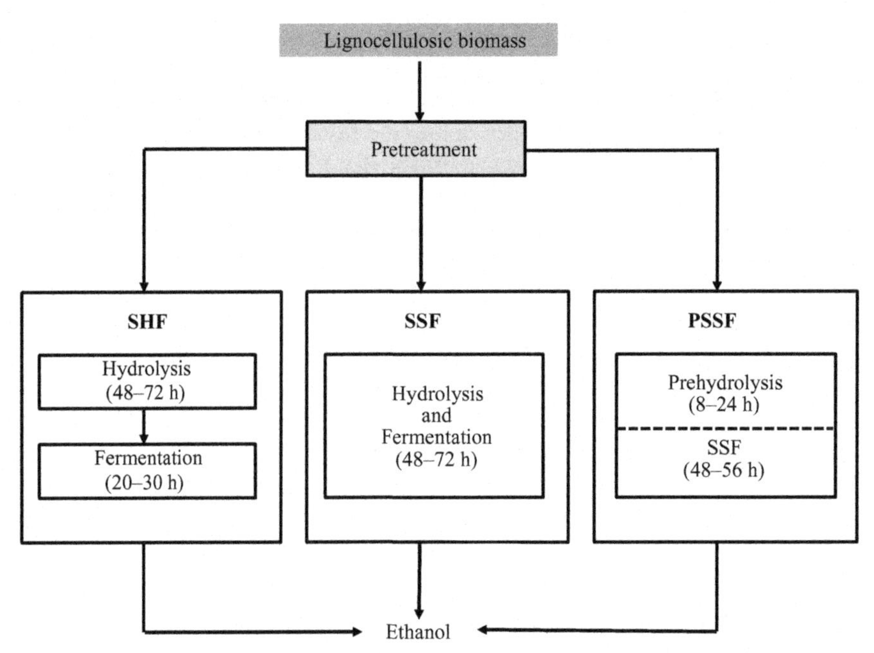

FIGURE 3.3 Different Process Configurations for Lignocellulosic Biomass Conversion to Ethanol

SHF, separated hydrolysis and fermentation; *SSF*, simultaneous saccharification and fermentation, and *PSSF*, prehydrolysis and simultaneous saccharification and fermentation. In *brackets* standard time for each process step.

Manzanares et al. (2011) have reported on the effect of pretreated (liquid hot water and diluted-acid) OTP at different substrate loads on ethanol production in above-mentioned process configurations; SHF, SSF, and PSSF. When comparing both pretreatments, better results were found with liquid hot water-pretreated OTP, reaching an overall ethanol yield (based on the amount of cellulose in the raw material) of 38.7 g ethanol per 100 g cellulose (8.8 g ethanol per 100 g raw material) by the SSF process strategy. In both cases, increasing substrate loading led to a raise in ethanol concentration, which peaked in at about 3.7% (v/v) in experiments performed in liquid hot water-pretreated olive pruning at 23% (w/w) substrate loading by SHF or PSSF. Toledano et al. (2013) performed SHF trials in organosolv pretreated OTP and found similar ethanol concentrations around 36 g/L.

In general, prior to the fermentation step a detoxification treatment of the hydrolysates is required to reduce the inhibitory effect of compounds released from acid pretreated-OTP (Díaz et al., 2009; Negro et al., 2014). Díaz-Villanueva et al. (2012) have studied the ethanolic fermentation of liquid fraction from pretreated olive tree biomass after detoxification (overliming) utilizing *P. stipitis and P. tannophilus,* and compared the performance of both yeasts. The best results were obtained with *P. tannophilus* with values as high as 0.44 g ethanol/g sugar, and conditioning by overliming was proven to improve fermentability of hydrolysates. The fermentation with *P. tannophilus* of sulfuric acid hydrolysed-OTP has also been successfully carried out by Romero et al. (2007), too, who reported maximum ethanol yield of 0.38 g/g glucose. Ethanol production yield from OTP has been assessed in hydrolysates from steam explosion impregnated with phosphoric acid pretreatment and enzymatic hydrolysis, using *S. cerevisiae* and *S. stipitis* (Negro et al., 2014). Again, a detoxification treatment was required before fermentation to alleviate the inhibitory effect of compounds present in the hydrolysate. High ethanol yields about 80% of theoretical were attained in SSF experiments at 15% (w/w) solid loading. Recently, Martínez-Patino et al. (2015) have used an acetate-tolerant ethanologenic strain of *E. coli* (MS04) generated by metabolic engineering (Fernández-Sandoval et al., 2012), to ferment all sugars present in the prehydrolysates from water-extracted OTP pretreated at 170°C and 0.5% phosphoric acid concentration. A total ethanol yield of 13.2 g of ethanol/100 g of original OTP was obtained (16.3 g of ethanol/100 g of extracted OTP), both from liquid and solid fraction after enzymatic hydrolysis.

Despite the fact that most of the research work on ethanol production has been focused on OTP residue, other olive mill-wastes have been studied, too. For instance, olive cake (the solid residue from the traditional oil extraction system) has been explored as substrate for ethanol production by a recombinant *E. coli* strain. Soluble sugars obtained after diluted acid pretreatment (18.1 g/L) and detoxification steps were fermented by *E.coli* FRB5 strain to produce ethanol at a concentration of 8.1 g/L (El Asli and Qatibi, 2009). Besides, the pomace and fragmented olive stones have been tested for ethanol production by the SSF process using a fed-batch strategy. Experiments with fed-batch pretreated olive stones provided SSF yields significantly lower than those obtained at standard SSF procedure (Ballesteros et al., 2001). Recently, Cuevas et al. (2015a) have reported an overall ethanol yield up to 13.1 kg ethanol per 100 kg of olive stones (ethanol concentration 47 g/L) using a thermotolerant yeast (*S. cerevisiae* IR2-9a), after ethanol-based organosolv pretreatment of stones with addition of sulfuric acid.

In addition to ethanol, other compounds can be obtained from sugars contained in olive-mill residues following alternative conversion routes. Particularly from xylose, which is the main pentose in hemicelluloses, xylitol can be produced. Xylitol is a value-added product by its sweetening properties, similar to that of sucrose but with 40% less calories, universally used as sweetener for diabetics. Currently, it is produced in a complex, labor- and energy-intensive process based on the catalytic hydrogenation of xylose that is obtained from hydrolysis of woody hemicellulose or from extracted corn cobs, which are the remains of corn ears after the kernels have been extracted (IEA, 2015). But xylitol may be also produced by biological conversion of xylose fermentation present in hemicellulosic hydrolysates with yeast species, such as *Candida, Pichia, and Pachysolen.*

An example of this application is the study of xylitol production from the liquid fraction obtained after diluted sulfuric acid pretreatment of olive stones carried out by Saleh et al. (2014). The fermentation of the prehydrolysates obtained after optimal pretreatment conditions (195°C, 5 min, and 0.025 M sulfuric acid) and after a detoxification step, 9.2 g of xylitol per 100 g of olive stones, was attained using *Pachysolen tannophiplus*. Xylitol yields (0.44 g xylitol/g xylose) were similar to those obtained by García et al. (2011) in the fermentation with *Candida tropicalis* of acid hydrolysates from olive-pruning biomass, when the overall xylitol reported was 53 g xylitol per kg of dry olive debris.

The form in which hemicellulose-derived sugars are found in prehydrolysates from lignocellulose biomass pretreatment is highly dependent on the severity of the pretreatment and the nature of feedstock. Mild pretreatment conditions in terms of pH and temperature may lead to incomplete hydrolysis of hemicellulosic polymers and carbohydrates in form of oligosaccharides, which can be found in prehydrolysates instead of, or together with, monosaccharides. These molecules could be used as prebiotic compounds, of great interest for the food, cosmetic, and pharmaceutical industries. According to Mussatto and Mancilha (2007), oligosaccharides are defined as polymers of monosaccharides with variable degree of polymerization (DP) between 2 and 10. These compounds are considered prebiotics since they are not digestible in the stomach and thus can reach the large intestine without any change and promote the growth and/or activity bacterial associated with health (Gibson et al., 2004).

Oligosaccharides from xylan hydrolysis (xylooligosaccharides) have been found by Reis et al. (2003) in prehydrolysate from partial acid hydrolysis of olive pulp and olive seed hull. Structural characterization of these acidic xylooligosaccharides showed differences between the residues; while those from olive pulp were residues mainly substituted with 4-*O*-methyl glucuronic acid, olive seed hull contained glucuronic acid residues. Recently, Cuevas et al. (2015b) have reported on oligosaccharides production from olive stones pretreated by autohydrolysis at 190°C and 5 min, with production yield of 147 g oligosaccharides/kg olive stones. Lama-Muñoz et al. (2012) further identified some bioactive phenol glucosides and poly- and oligosaccharides from thermally treated olive oil by-products.

Oligosccharides production from OTP has been explored by Cara et al. (2012). They found oligosaccharides in prehydrolysate after autohydrolysis pretreatment at 180°C for 10 min, mainly consisting of gluco-oligosaccharides and xylooligosaccharides, in a total concentration of 37.5 g/L. Further treatment of the hydrolysate by preparative gel filtration chromatography and purification allowed separating a range of oligosaccharides fractions with an average degree of polymerization (DP) from 3 to 25, with gluco-oligosaccharides and xylooligosaccharides as the predominant oligosaccharides. Mateo et al. (2013) also studying oligosaccharides production from OTP biomass by liquid hot water pretreatment, proved a great effect of process conditions in oligosaccharides yield. The most suitable pretreatment conditions for the maximum concentration of oligosaccharides as well as monosaccharides corresponded to experiments carried out at 200°C and 0 min, either with H_2SO_4 at 0.025 M or with ultrapure water.

The production of valuable coproducts, such as xylitol or oligosaccharides coupled with the ethanol production process from olive mill-residues would greatly contribute to improve second generation ethanol economically feasibility, while fulfilling the characteristic features of the biorefinery concept.

3.3.1.2.3 Lignin Conversion

In most current applications of biochemical and/or chemical based technologies to lignocellulosic materials, lignin component is contained in the final residue after biomass fractionation and carbohydrate hydrolysis, as a complex and disperse compound (Vishtal and Kraslawski, 2011). A major part of this lignin-rich residue is incinerated to produce process steam and energy, and only a minor part is derived to the production of other valuable products. However, lignin offers a significant opportunity for enhancing the operation of a lignocellulosic biorefinery by the production of more valuable chemicals, such as resins, composites and polymers, aromatic compounds, or carbon fibers. This is viewed as a medium to a long term opportunity depending on the quality and functionality of the lignin that can be obtained (IEA, 2012). The challenge to use lignin as feedstock material is its complex chemical structure, which is highly influenced by its origin and the method of fractionation/extraction and further processing of the lignocellulose biomass. Thus, it is important to know the particular physicochemical characteristics of the lignin-rich residue to direct process decision making in relation to lignin upgrading. Regarding biomass residues from olive crop and olive-oil industry, residual solid materials from the conversion of OTP biomass to bioethanol at laboratory scale have been characterized for their lignin and carbohydrate content, heating value, ash and inorganic elements in our laboratories. Regardless of whether the pretreated biomass is processed by SSF or SHF, process residues are mainly composed of lignin (68–74%), unrecovered sugars (6.8–9%), unrecovered enzymes, and ash (6%). The high proportion of lignin in the residue is in the basis of the high heating value of 22.4 MJ/kg HHV (dwb) found, which makes the residue a suitable green energy source. However, other options must be explored to revalorize this lignin-rich residue by conversion into valuable products. Following this pathway,

Santos et al. (2015) have characterized the residue from SSF process on OTP by FTIR and 2D-NMR, as a prior step in the development of added-value products. Results show lignin has a strong predominance of S over G units, as well as high content of β-*O*-4′aryl ether linkages, followed by resinols and phenylcoumarans. According to the authors, the high content of native aryl ether linkages would result in a high molecular mass lignin, which could be interesting for dispersants and composites production.

Another way to address lignin revalorization in lignocellulose biomass feedstocks is to extract or isolate lignin before the material is submitted to the conversion process. Several methodologies have been successfully applied to lignocellulosic materials rendering different types of lignin, depending on the separation process (alkali, organosolv, ionic liquids, microwave, etc.) (Vishtal and Kraslawski, 2011; Manara et al., 2014). Particularly for olive derived residues, organosolvent pretreatment using a mixture of ethanol and water has been applied to OTP for lignin production. Toledano et al. (2012) studied not only operational conditions (temperature, time, and ethanol concentration), but also the characteristics of lignin obtained from OTP after organosolv treatment. Optimal organosolv conditions (200°C, 70% ethanol and 90 min.) allowed for the production of higher quality lignin, high acid insoluble lignin content (71.9%) and low contamination (sugars 2.94% and ash 0.39%). Interestingly, lignin seemed to be highly reactive due to its high functionality, which is a promising feature for further applications in the formulation of inks, varnishes and paints (Vishtal and Kraslawski, 2011). Manara et al. (2014) have addressed the study of lignin extraction from olive kernels by different procedures in order to evaluate the effect of extraction conditions and determine lignin chemical structure. A chemical (organosolv-formic/acetic acid) and a physicochemical (water/ethanol with sulphuric acid at 150°C in microwave reactor) pretreatment process were applied and the isolated lignins characterized by FT-IR and TGA. FT-IR spectra profiles were rather similar in both pretreated substrates, indicating a comparable chemical structure of extracted lignins. Results show higher delignification yield after microwave pretreatment, while both procedures rendered high-purity lignins close to 100% purity. On the other hand, the thermal degradation behavior of the derived lignins from TGA analysis gives useful information to explore the potential of conversion to valuable compounds, such as liquid fuels or hydrogen by advanced thermochemical technologies as pyrolysis or gasification, considered of great interest in the context of a biorefinery.

3.3.2 PROCESSES FOR HIGH-MOISTURE RESIDUES

As above mentioned, the process of obtaining olive oil generates highly polluting olive mill wastewater streams (OMWW), comprising vegetation water, soft tissues of the olives and the water used in the different stages of oil production. Due to its low biodegradability and high phytotoxiticy, OMWW is one of the most environmentally concerning food processing effluents in the Mediterranean countries.

Natural evaporation in open-air lagoons, favored by the Mediterranean weather, is the most conventional method for OMWW treatment (Cegarra et al., 1996; Jarboui et al., 2008). Physicochemical treatments (flocculation, advanced oxidative methods, and electrochemical processes, such as the electrocoagulation) are also used to reduce the organic load of OMWW (Azbar et al., 2004). Biological treatments, such as activated sludge, anaerobic treatments, and membrane bioreactors, have also been used for the reduction of its organic load (Ramos-Cormenzana et al., 1996). Other alternative tested is the direct agronomic application, but its high content in phytotoxic phenolic compounds increases soil hydrophobicity and decreases water retention (Kavvadias et al., 2010).

In an integrated biorefinery approach, aimed at producing high added-value products from OMWW, a preprocessing step to extract the remaining high value antioxidants may represent an economically interesting strategy that provides the triple opportunity to obtain high-added value biomolecules, to increase biodegradability and to reduce phytotoxicity of the effluent.

3.3.2.1 Preprocessing

Most of the phenolic compounds in olives and olive oil (hydroxytyrosol, tyrosol, caffeic acid, rutin, luteolin, and flavonoids) are insoluble in oil and, thereby, remain in wastewater. The water-soluble phenolic compound fractions represent 50–72% of the total phenolic compounds in olives (Alu'datt et al., 2013). The selective recovery of these phenolic compounds in a preprocessing step can achieve both the reduction of the intrinsic wastewater environmental toxicity and the production of high added value molecules.

The main separation strategies (Galanakis, 2012) involves the use of liquid–liquid extraction (LLE) (Zafra et al., 2006; Galanakis et al., 2010b), ultrasound-assisted extraction (Japón-Luján et al., 2006; Roselló-Soto et al., 2015b; Zinoviadou et al., 2015), solvent extraction (Obied et al., 2005; Heng et al., 2015), superheated liquid extraction (Japón-Luján et al., 2006), supercritical fluid extraction (Le Floch et al., 1998), and more recently emerging technologies, such as high voltage electrical discharges and pulsed electric field (Galanakis, 2013; Rahmanian et al., 2014; Galanakis and Schieber, 2014; Roselló-Soto et al., 2015a). Although LLE provides efficient results, the necessity of using large volume of organic solvents is increasing the interest to replace LLE by solid-phase extraction (Scoma et al., 2011).

Membrane processes have also been proposed for the selective fractionation and total recovery of polyphenols, water, and organic substances from OMW (Paraskeva et al., 2007a,b; Coskun et al., 2010; De Leonardis et al., 2008; Galanakis et al., 2010a; Galanakis, 2015). Russo (2007), using a preliminary membrane filtration followed by two ultrafiltration steps (6 and 1 kDa membranes, respectively) and a final reverse osmosis treatment, obtained an enriched and purified low molecular weight polyphenols extract to be used for food, pharmaceutical or cosmetic purposes. The remaining liquid can be used as fertilizers or in the production of biogas in anaerobic reactors as detailed later.

3.3.2.2 Agricultural Uses of Extracted OMWW

The reduction of phytotoxicity of OMWW by polyphenols extraction would allow its use as a valuable fertilizer. Its agronomic application to soils is very convenient in Mediterranean countries since water is a scarce and soils usually suffer from low organic matter content. The land utilization of polyphenols-extracted OMMW represents a cheap source of irrigation water and provides soils with nutrients, mainly K and organic matter. Main inconveniences for the application of OMWW to soil are the high salinity and low pH.

Disadvantages of the application of OMWW to soil can be reduced or eliminated by composting or cocomposting with agricultural residues (e.g., olive pomace or olive tree pruning residues). The agronomic use of composting OMWW does not cause negative effects on crop productivity and produces significant increases of organic matter and organic N in soils (Martín-Olmedo et al., 1995). Cabrera et al. (2005) showed that soils treated with olive mill wastewater sludge compost, increased organic matter, and nitrogen in comparison to untreated or mineral fertilized soils. The nitrogen increase in treated soils was even higher than expected by the N added together with compost, and attributed to nonsymbiotic N fixation. Therefore, the organic composting or cocomposting represents one of the most interesting ways of transforming extracted OMWW into fertilizers.

3.3.2.3 Production of Bioenergy and Biofuels

OMWW, especially after increasing its biodegradability by the extraction of phenolic compounds, could be a good substrate for fermentative transformations. Although some studies have been performed for bioethanol (Massadeh and Modallal, 2008) and biohydrogen (Eroglu et al., 2004, 2009) production, studies have been focused on biomethane production by anaerobic digestion (AD). Among bioprocesses to transform OMWW into biofuels, anaerobic digestion is the best choice since all the macromolecules (lipids, proteins, and carbohydrates) are transformed into biogas. Biogas is a versatile energy carrier that can be used for electricity production, heating purposes, vehicle and jet fuel and replacement of natural gas. In addition, biomethane may be considered as starting compounds for biotechnological production of chemicals.

Anaerobic digestion of complex organic substrates, such as OMWW, proceeds through a series of parallel and sequential steps with several groups of microorganism involved. Anaerobic digestion starts with the hydrolysis of high molecular materials and granular organic substrates (lipids, proteins, and carbohydrates) by fermentative bacteria into small molecular materials and soluble organic substrates (fatty acids, amino acids, and glucose), aided by extracellular enzymes (hydrolases), which are excreted by fermentative bacteria. Next, products formed during the hydrolysis are further degraded into volatile fatty acids (e.g., acetate, propionate and butyrate) along with the generation of by-products (e.g., NH_3, CO_2, and H_2S) in a process known as acidogenesis. Finally, the organic substrates produced in the second step are further digested into acetate, H_2, and CO_2 and used by methanogenic archaea for methane production.

The anaerobic degradation of OMWW faces some difficulties due to the high content of hardly degradable cellulosic materials and toxic substance, such as phenols, long-chain fatty acids, ethanol, tannins, etc. (Gunay and Kadarag, 2015). As mentioned earlier, phenols compounds have a strong toxicity on microorganism, hindering their biological degradation by anaerobic bacteria. Combined physical, chemical, and biological methods have been applied prior to anaerobic digestion to eliminate toxic compounds and enhance methane productivity.

Physical pretreatment, such as ultrasound have been used to increase the methane production from OMWW. Recently, Oz and Uzun (2015) have reported 20% more biogas and methane production in anaerobic batch reactor fed with ultrasound pretreated-diluted OMWW compared to control (untreated). Low frequency ultrasound pretreatment resulted in 33% increase of soluble COD and subsequently in biogas production. Other physicochemical alternatives as advanced oxidation technologies also result in a reduction of toxicity and, therefore, in higher biodegradability (Amor et al., 2015).

Another approach to increase biogas production is the aeration of OMWW previously to the anaerobic treatment performance. The consumption of a considerable amount of energy and the removal of a significant part of the organic matter in OMWW are obstacles for its widespread application (Gunay and Kadarag, 2015). Recently, a biological treatment for OMWW has been proposed (González-González and Cuadros, 2015), resulting in 0.39 m^3 methane/kg total chemical organic demand (TCOD) when OMWW was previously aerated for 5 days.

Carbon, nitrogen and phosphorus are the main nutrients for anaerobic bacteria and an appropriate ratio of these compounds is crucial for biogas production. The required optimum C: N: P ratio for enhanced yield of methane has been reported to be 100:5:1 (Steffen et al., 1998). OMMW generally do not have a sufficient C:N:P ratio and codigestion of OMWW with others waste streams as poultry manure or winery residues have been proposed as an interesting technological approach. Alagöz et al. (2015) obtained 17–31% increase in methane production in codigestion of OMWW and wastewater sludge.

3.3.3 OTHER CONVERSION PROCESSES

In addition to the earlier described technologies to convert olive biomass (including solid or liquid wastes) into bioproducts and biofuels, a number of other interesting applications are nowadays being considered at different development stages, such as animal feed, production of activated carbons to be used as biosorbents, or as ingredients in construction materials.

Considering low-moisture residues, such as olive leaves, one of the direct applications is animal feed. For example, feed supplementation in pig diet by olive leaves (up to 25 g/kg) resulted in beneficial effects on the tocopherol content of meat without excessively compromising the growth performance (Paiva-Martins et al., 2014). Olive leaves antioxidants have also been assayed to determine their capacity for preventing oxidation of animal meat (Trebušak et al., 2014).

As far as pomace residues are concerned, the production of biosorbents and the use as construction materials have been identified as potential added-value routes. However, the current valorization method is dominated by the production of olive-pomace oil and the use of the resulting residue as heating biofuel.

Olive stones have been studied as an interesting biosorbent, particularly when used for heavy metal removal. The availability and low cost of olive stones, together with their decontamination potential, are the main advantages for this kind of use (Ronda et al., 2015). Both native and chemically treated olives stones have been studied for the development of new biosorbents (Blázquez et al., 2014).

Producing activated carbon using residual materials has obviously an environmental impact that should also be addressed. The environmental impact of the global process is highly dependent on the different operation steps, such as impregnation, pyrolysis, and drying. Hjaila et al. (2013) recently conducted this kind of study based on the life cycle assessment in Tunisia and determined that impregnation with H_3PO_4 was responsible for the highest environmental impacts for the majority of the indicators (e.g., acidification potential, eutrophication, and fresh water aquatic toxicity and terrestrial toxicity, among others). These authors concluded that those environmental impacts can be reduced by implementing a number of measures in the activated carbon production scheme.

The use of olive oil residues as ingredient in the production of fired clay bricks has been proposed with the double advantage of being a readily available disposal method and, at the same time, reducing the fuel consumption in the clay kiln, because of the effect of the added organic matter present in the olive residue. This application has been studied using olive pomace, OMWW, and the ash obtained from burning olive tree biomass to produce steam, heat, or power. Adding two-phase olive pomace to fired clay brick formulation at different proportions (3–% by weight), resulted in masonry units with several advantages when compared to the conventional products, such as lower densities and higher thermal insulation effectiveness (De la Casa et al., 2012). Partially substitution of process water by two-phase-OMWW produced facing bricks with similar technological properties to conventional bricks. Moreover, heating requirements in the kiln could be reduced in the range 2.4–7.3% depending on the final product, which is an additional advantage to the reduced environmental impact derived from eliminating the OMWW in this application (De la Casa et al., 2009). Three-phase-OMWW was also assessed for partial substitution of fresh water. The resulting bricks showed a significant increase in the volume shrinkage (10%) and the water absorption (12%), while the tensile strength remained constant (Mekki et al., 2008). Olive washing wastewater has also been studied as substitute for fresh water in fired clay ceramic production, resulting in slight improvements in the physical, mechanical, and thermal properties of bricks (Eliche-Quesada et al., 2014). Finally, the ash produced from olive biomass has been assayed as an ingredient for fired clay brick manufacturing and the resulting products fulfill specifications for being used as masonry units (De la Casa and Castro, 2014).

3.4 CONCLUSIONS

The use of different residues derived from olive tree cultivation and olive oil extraction, including waste streams, such as wastewater, olive leaves, or olive stones, in a single installation gives rise to the concept of multifeedstock and multiproduct biorefinery. This modular and intregated biorefinery could transform heterogeneous feedstocks (high- and low-moisture residues) into a number of valuable products (food additives, bioproducts, biomaterials, biofuels, and bioenergy). At the same time, the biorefinery can play an interesting environmental role, since the processes developed at this facility represent an alternative to the classical disposal methods of the residues of the olive oil industry. Depending on the feedstock considered, a wide range of final products have been identified as possible outputs for such an installation. In the case of OTP from the pruning operation, ethanol can be considered the main product to be obtained, although antioxidants, oligosaccharides, xylitol, or electricity can also be produced. High-moisture residues, such as wastewaters produced in several points of the olive oil production scheme, can also be used as raw materials for a number of bioproducts including biogas and phenolics. Despite the strategic relevance of upgrading the residues generated in the olive oil industry, further work is required to address certain issues, such a process optimization, scale up, and process integration before the biorefinery based on olive derived materials can be an industrial reality.

ACKNOWLEDGMENT

The authors wish to acknowledge Ministerio de Economía y Competitividad (Spain) (Projects ref. ENE2014-60090-C2-R-1 and ENE2014-60090-C2-2-R) and the European Commission (ERANET-LAC Project SMIBIO ref. ELAC2014/BEE-0249), for financial support.

REFERENCES

Alagöz, B.A., Yeningün, O., Erdinçler, A., 2015. Enhancement of anaerobic digestion efficiency of wastewater sludge and olive waste: synergistic effect of co-digestion and ultrasonic/microwave sludge pre-treatment. Waste Manag. 46, 182–188.

Alburquerque, J.A., Gonzálvez, J., García, D., Cegarra, J., 2004. Agrochemical characterisation of alperujo, a solid by-product of the two-phase centrifugation method for olive oil extraction. Bioresour. Technol. 91, 195–200.

Alu'datt, M.H., Rababah, T., Ereifej, K., Alli, I., 2013. Distribution, antioxidant and characterisation of phenolic compounds in soybeans, flaxseed and olives. Food Chem. 139, 93–99.

Alvira, P., Tomás-Pejó, E., Negro, M.J., 2010. Pretreatment technologies for an efficient bioethanol production process based on enzymatic hydrolysis: a review. Bioresour. Technol. 101, 4851–4861.

Amor, C., Lucas, M.S., García, J., Dominguez, J.R., De Heredia, J.B., Peres, J.A., 2015. Combined treatment of olive mill wastewater by Fenton's reagent and anaerobic biological process. J. Environ. Sci. Health 50, 161–168.

Aydinoglu, T., Sargin, S., 2013. Production of laccase from *Trametes versicolor* by solid-state fermentation using olive leaves as a phenolic substrate. Bioprocess Biosyst. Eng. 36, 215–222.

Azbar, N., Bayram, A., Filibeli, A., Muezzinoglu, A., Sengul, F., Ozer, A., 2004. A review of waste management options in olive oil production. Crit. Rev. Environ. Sci. Technol. 34, 209–247.

Ballesteros, I., Oliva, J.M., Saéz, F., Ballesteros, M., 2001. Ethanol production from lignocellulosic byproducts of olive oil extraction. Appl. Biochem. Biotechnol. 91–93, 237–252.

Ballesteros, I., Ballesteros, M., Cara, C., Sáez, F., Castro, E., Manzanares, P., Negro, M.J., Oliva, J.M., 2011. Effect of water extraction on sugars recovery from steam exploded olive tree pruning. Bioresour. Technol. 102, 6611–6616.

Benavente, V., Fullana, A., 2015. Torrefaction of olive mill waste. Biomass Bioenerg. 73, 186–194.

Blázquez, G., Calero, M., Ronda, A., Tenorio, G., Martín-Lara, M.A., 2014. Study of kinetics in the biosorption of lead onto native and chemically treated olive stone. J. Ind. Eng. Chem. 20, 2754–2760.

Cabrera, F., Martín Olmedo, P., López Núñez, R., Murillo Carpio, J.M., 2005. Nitrogen mineralization in soils amended with composted olive mill sludge. Nutr. Cycl. Agroecosyst. 71, 249–258.

Campoy, M., Gómez-Barea, A., Ollero, P., Nilsson, S., 2014. Gasification of wastes in a pilot fluidized bed gasifier. Fuel Process. Technol. 121, 63–69.

Cara, C., Ruiz, E., Ballesteros, I., Negro, M.J., Castro, E., 2006. Enhanced enzymatic hydrolysis of olive tree wood by steam explosion and alkaline peroxide delignification. Process Biochem. 41, 423–429.

Cara, C., Romero, I., Oliva, J.M., Sáez, F., Castro, E., 2007. Liquid hot water pretreatment of oilve tree pruning residues. Appl. Biochem. Biotechnol. 137–140, 379–394.

Cara, C., Ruiz, E., Ballesteros, M., Manzanares, P., Negro, M.J., Castro, E., 2008a. Production of fuel ethanol from steam–explosion pretreated olive tree pruning. Fuel 87, 692–700.

Cara, C., Ruiz, E., Oliva, J.M., Sáez, F., Castro, E., 2008b. Conversion of olive tree biomass into fermentable sugars by diluted acid pretreatment and enzymatic saccharification. Bioresour. Technol. 99, 1869–1876.

Cara, C., Ruiz, E., Carvalheiro, F., Moura, P., Ballesteros, I., Castro, E., Girio, F., 2012. Production, purification and characterisation of oligosaccharides from olive tree pruning autohydrolysis. Ind. Crops Prod. 40, 225–231.

Cegarra, J., Paredes, C., Roig, A., Bernal, M.P., García, D., 1996. Use of olive mill wastewater compost for crop production. Int. Biodeter. Biodegr. 38, 193–203.

Christoforou, E., Fokaides, P.A., 2016. A review of olive mill solid wastes to energy utilization techniques. Waste Manag. 49, 346–363.

Coskun, T., Debik, E., Demir, N.M., 2010. Treatment of olive mill wastewaters by nanofiltration and reverse osmosis membranes. Desalination 259, 65–70.

Cuevas, M., Sánchez, S., Bravo, V., Cruz, N., García, J.F., 2009. Fermentation of enzymatic hydrolysates from olive stones by *Pachysolen tannophilus*. J. Chem. Technol. Biotechnol. 84, 61–467.

Cuevas, M., Sánchez, S., García, J.F., Baeza, J., Parra, C., 2015a. Enhanced ethanol production by simultaneous saccharification and fermentation of pretreated olive stones. Renew. Energy 74, 839–847.

Cuevas, M., García, J.F., Hodaifa, G., Sánchez, S., 2015b. Oligosaccharides and sugar production from olive stones by autohydrolysis and enzymatic hydrolysis. Ind. Crops Prod. 70, 100–106.

Daâssi, D., Lozano-Sánchez, J., Borrás, I., Lassaad, B., Woodward, S., Mechichi, T., 2013. Olive oil mill wastewaters: phenolic content characterization during degradation by *Coriolopsis gallica*. Chemosphere 113, 62–70.

De la Casa, J.A., Castro, E., 2014. Recycling of washed olive pomace ash for fired clay brick production. Constr. Build. Mater. 61, 320–326.

De la Casa, J.A., Lorite, M., Jiménez, J., Castro, E., 2009. Valorisation of wastewater from two-phase olive oil extraction in fired clay brick production. J. Hazard. Mater. 169, 271–278.

De la Casa, J.A., Romero, I., Jiménez, J., Castro, E., 2012. Fired clay masonry units production incorporating two-phase olive mill waste (alperujo). Ceram. Int. 38, 5027–5037.

De Leonardis, A., Aretini, A., Alfano, G., Macciola, V., Ranalli, G., 2008. Isolation of a hydroxytyrosol–rich extract from olive leaves (*Olea Europaea* L.) and evaluation of its antioxidant properties and bioactivity. Eur. Food Res. Technol. 226, 653–659.

Dermeche, S., Nadour, M., Larroche, C., Moulti-Mati, F., Michau, P., 2013. Olive mill wastes: biochemical characterizations and valorization strategies. Process Biochem. 48, 1532–1552.

Díaz, M.J., Ruíz, E., Romero, I., Moya, M., Castro, E., 2009. Inhibition of *Pichia stipitis* fermentation of hydrolysates from olive tree cuttings. World J. Microbiol. Biotechnol. 25, 891–899.

Díaz, M.J., Huijgen, W.J.J., Van der Laan, R.R., Reith, J.H., Cara, C., Castro, E., 2011. Organosolv pretreatment of olive tree biomass for fermentable sugars. Holzforschung 65, 177–183.

Díaz-Villanueva, M.J., Cara-Corpas, C., Ruiz-Ramos, E., Romero-Pulido, I., Castro-Galiano, E., 2012. Olive tree pruning as an agricultural residue for ethanol production. Fermentation of hydrolysates from dilute acid pretreatment. Span. J. Agric. Res. 10, 643–648.

Duque, A., Manzanares, P., Ballesteros, I., Negro, M.J., Oliva, J.M., Sáez, F., Ballesteros, M., 2013. Optimization of integrated alkaline-extrusion pretreatment of barley straw for sugar production by enzymatic hydrolysis. Process Biochem. 48, 775–781.

EBTP, 2016. European biofuels technology platform. Available from: http://www.biofuelstp.eu/cellulosic-ethanol.html

El Asli, A., Qatibi, A., 2009. Ethanol production from olive cake biomass substrate. Biotechnol. Bioprocess Eng. 14, 118–122.

Eliche-Quesada, D., Iglesias-Godino, F.J., Pérez-Villarejo, L., Corpas-Iglesias, L.A., 2014. Replacement of the mixing fresh water by wastewater olive oil extraction in the extrusion of ceramic bricks. Construc. Build. Mater. 68, 659–666.

Eroglu, E., Gunduz, U., Yucel, M., Turker, L., Eroglu, I., 2004. Photobiological hydrogen production by using olive mill wastewater as a sole substrate source. Int. J. Hydr. Energy 29, 163–171.

Eroglu, E., Eroglu, I., Gunduz, U., Yucel, M., 2009. Treatment of olive mill wastewater by different physicochemical methods and the utilization of their liquid effluents for biological hydrogen production. Biomass Bioenergy 334, 701–705.

Esteve, C., Marina, M.L., García, M.C., 2015. Novel strategy for the revalorization of olive (*Olea europaea*) residues. Food Chem. 167, 272–280.

FAOSTAT, 2015. http://faostat3.fao.org/

Felizón, B., Fernández-Bolaños, A., Heredia, A., Guillén, R., 2000. Steam-explosion pretreatment of olive cake. J. Am. Oil Chem. Soc. 77, 15–22.

Fernández-Bolaños, J., Rubio-Senent, F., Lama-Muñoz, A., García, A., Rodríguez-Gutiérrez, G., 2014. Production of oligosaccharides with low molecular weights, secoiridoids and phenolic glycosides from thermally treated olive by-products. In: Schweizer, L.S., Krebs, S.J. (Eds.), Oligosaccharides: Food Sources, Biological Roles and Health Implications. Nova Science Publishers, New York, pp. 173–208.

Fernández-Sandoval, M.T., Huerta-Beristaín, G., Trujillo-Martínez, B., Bustos, P., González, V., Bilivar, F., Gosset, G., Martínez, A., 2012. Laboratory metabolic evolution improves acetate tolerance and growth on acetate of ethanologenic *Escherichia coli* under non-aerated conditions in glucose-mineral medium. Appl. Genet. Mol. Biotechnol. 96, 1291–1300.

Galanakis, C.M., 2011. Olive fruit and dietary fibers: components, recovery and applications. Trends Food Sci. Technol. 22, 175–184.

Galanakis, C.M., 2012. Recovery of high added-value components from food wastes: conventional, emerging technologies and commercialized applications. Trends Food Sci. Technol. 26, 68–87.

Galanakis, C.M., 2013. Emerging technologies for the production of nutraceuticals from agricultural by-products: a viewpoint of opportunities and challenges. Food Bioprod. Process. 91, 575–579.

Galanakis, C.M., 2015. Separation of functional macromolecules and micromolecules: from ultrafiltration to the border of nanofiltration. Trends Food Sci.Technol. 42, 44–63.

Galanakis, C.M., Schieber, A., 2014. Editorial of special issue on recovery and utilization of valuable compounds from food processing by-products. Food Res. Int. 65, 299–300.

Galanakis, C.M., Tornberg, E., Gekas, V., 2010a. Clarification of high-added value products from olive mill wastewater. J. Food Eng. 99, 190–197.

Galanakis, C.M., Tornberg, E., Gekas, V., 2010b. Recovery and preservation of phenols from olive waste in ethanolic extracts. J. Chem. Technol. Biotechnol. 85, 1148–1155.

García, J.F., Sánchez, S., Bravo, V., Cuevas, M., Rigal, L., Gaset, A., 2011. Xylitol production from olive–pruning debris by sulphuric acid hydrolysis and fermentation with *Candida tropicalis*. Holzforschung 65, 59–65.

García-Maraver, M., Salvachúa, D., Martínez, M.J., Díaz, L.F., Zamorano, M., 2013. Analysis of the relation between the cellulose, hemicellulose and lignin content and the thermal behavior of residual biomass from olive trees. Waste Manag. 33, 2245–2249.

Gibson, G.R., Probert, H.M., Van Loo, J., Rastall, R.A., Roberfroid, M.B., 2004. Dietary modulation of the human colonic microbiota: updating the concept of prebiotics. Nutr. Res. Rev. 17, 259–275.

González-González, A., Cuadros, F., 2015. Effect of aerobic pretreatment on anaerobic digestion of olive mill wastewater (OMWW): an ecoefficient treatment. Food Bioprod. Process. 95, 339–345.

Gunay, A., Kadarag, D., 2015. Recent developments in the anaerobic digestion of olive mill effluents. Process Biochem. 50, 1893–1903.

Hanandeh, A.E., 2015. Energy recovery alternatives for the sustainable management of olive oil industry waste in Australia: life cycle assessment. J. Clean. Prod. 91, 78–88.

Haro, P., Villanueva Perales, A.L., Arjona, R., Ollero, P., 2014. Thermochemical biorefineries with multiproduction using a platform chemical. Biofuels Bioprod. Biorefin. 8, 155–170.

Heng, W.W., Xiong, L.W., Ramanan, R.N., Hong, T.L., Kong, K.W., Galanakis, C.M., Prasad, K.N., 2015. Two level factorial design for the optimization of phenolics and flavonoids recovery from palm kernel by-product. Ind. Crops Prod. 63, 238–248.

Hjaila, K., Baccar, R., Sarrà, M., Gasol, C.M., Blánquez, C., 2013. Environmental impact associated with activated carbon preparation from olive-waste cake via life cycle assessment. J. Environ. Manag. 130, 242–247.

IEA Bioenergy Report, 2012. Bio-based chemicals. Value added products from biorefineries. Available from: http://www.iea-bioenergy.task42-biorefineries.com

Japón-Luján, R., Luque-Rodríguez, J.M., Luque de Castro, M.D., 2006. Dynamic ultrasound-assisted extraction of oleuropein and related biophenols from olive leaves. J. Chromatogr. 1108, 76–82.

Jarboui, R., Sellami, F., Kharroubi, A., Gharsallah, N., Ammar, E., 2008. Olive mill wastewater stabilization in open–air ponds: impact on clay-sandy soil. Bioresour. Technol. 99, 7699–7708.

Kavvadias, V., Doula, M.K., Komnistsas, K., Liakopoulou, N., 2010. Disposal of olive oil mill wastes in evaporation ponds: effects on soil properties. J. Hazard. Mater. 182, 144–155.

Kazan, A., Celiktas, S.M., Sargin, S., Yesil-Celikta, O., 2015. Bio-based fractions by hydrothermal treatment of olive pomace: process optimization and evaluation. Energy Convers. Manag. 103, 366–373.

Lama-Muñoz, A., Romero-García, J.M., Cara, C., Moya, M., Castro, E., 2014. Low energy-demanding recovery of antioxidants and sugars from olive stones as preliminary steps in the biorefinery context. Ind. Crops Prod. 60, 30–38.

Lama-Muñoz, A., Rodríguez-Gutiérrez, G., Rubio-Senent, F., Gómez-Carretero, A., Fernández-Bolaños, J., 2011. New hydrothermal treatment of alperujo enhances the content of bioactive minor components in crude pomace olive oil. J. Agric. Food Chem. 59, 1115–1123.

Lama-Muñoz, A., Rodríguez-Gutiérrez, G., Rubio-Senent, F., Fernández-Bolaños, J., 2012. Production, characterization and isolation of neutral and pectic oligosaccharides with low molecular weights from olive by-products thermally treated. Food Hydrocoll. 28, 92–104.

Lanfranchi, M., Giannetto, C., De Pascale, A., 2016. Economic analysis and energy valorization of by-products of the olive oil process: "Valdemone DOP" extra virgin olive oil. Renew. Sustain. Energy Rev. 57, 1227–1236.

Le Floch, F., Tena, M.T., Ríos, A., Valcárcel, M., 1998. Supercritical fluid extraction of phenol compounds from olive leaves. Talanta 46, 1123–1130.

López-Linares, J.C., Romero, I., Moya, M., Cara, C., Ruiz, E., Castro, E., 2013. Pretreatment of olive tree biomass with $FeCl_3$ prior enzymatic hydrolysis. Bioresour. Technol. 128, 180–187.

Manara, P., Zabaniotou, A., Vanderghem, C., Richel, A., 2014. Lignin extraction from Mediterranean agro-wastes: impact of pretreatment conditions on lignin chemical structure and thermal degradation behavior. Catal. Today 223, 25–34.

Manzanares, P., Negro, M.J., Oliva, J.M., Sáez, F., Ballesteros, I., Ballesteros, M., Cara, C., Castro, E., Ruiz, E., 2011. Different process configurations for bioethanol production from pretreated olive pruning biomass. J. Chem. Technol. Biotechnol. 86, 881–887.

Martínez-Patino, J.C., Romero-García, J., Ruiz, E., Oliva, J.M., Álvarez, Romero, I., Negro, M.J., Castro, E., 2015. High solids loading pretreatment of olive tree pruning with dilute phosphoric acid for bioethanol production by *Escherichia coli*. Energy Fuels 29, 1735–1742.

Martín-Olmedo, P., Cabrera, F., López, R., Murillo, J.M., 1995. Residual effect of composted olive oil mill sludge on plant growth. Fresen. Environ. Bull. 4, 221–226.

Massadeh, M.I., Modallal, N., 2008. Ethanol production from olive mill wastewater (OMW) pretreated with Pleurotus sajor-caju. Energy Fuels 22, 150–154.

Mateo, S., Puentes, J.G., Sánchez, S., Moya, A.J., 2013. Oligosaccharides and monomeric carbohydrates production from olive tree pruning biomass. Carbohydr. Polym. 93, 416–423.

Mekki, H., Anderson, M., Benzina, M., Ammar, E., 2008. Valorization of olive mill wastewater by its incorporation in building bricks. J. Hazard. Mater. 158, 308–315.

Mussatto, S.I., Mancilha, I.M., 2007. Non-digestible oligosaccharides: a review. Carbohydr. Polym. 68, 587–597.

Negro, M.J., Alvarez, C., Ballesteros, I., Romero, I., Ballesteros, M., Castro, E., Manzanares, P., Moya, M., Oliva, J.M., 2014. Ethanol production from glucose and xylose obtained from steam exploded water-extracted olive tree pruning using phosphoric acid as catalyst. Bioresour. Technol. 153, 101–107.

Negro, M.J., Duque, A., Manzanares, P., Sáez, F., Oliva, J.M., Ballesteros, I., Ballesteros, M., 2015. Alkaline twin-screw extrusion fractionation of olive-tree pruning biomass. Ind. Crops Prod. 74, 336–341.

Obied, H.K., Allen, M.S., Bedgood, Jr., D.R., Prenzler, P.D., Robards, K., Stockmann, R., 2005. Bioactivity and analysis of biophenols recovered from olive mill waste. J. Agric. Food Chem. 53, 823–837.

Oz, N.A., Uzun, A., 2015. Ultrasounds pretreatment for enhance biogas production from olive mill wastewater. Ultrason. Sonochem. 22, 565–572.

Ozveren, U., Ozdogan, Z.S., 2013. Investigation of the slow pyrolysis kinetics of olive oil pomace using thermo-gravimetric analysis coupled with mass spectrometry. Biomass Bioenergy 58, 168–179.

Paiva-Martins, F., Ribeirinha, T., Silva, A., Gonçalves, R., Pinheiro, V., Mourão, J.L., Outor-Monteiro, D., 2014. Effects of the dietary incorporation of olive leaves on growth performance, digestibility, blood parameters and meat quality of growing pigs. J. Sci. Food Agric. 94, 3023–3029.

Paraskeva, C.A., Papadakis, V.G., Kanellopoulou, D.G., Koutsoukos, P.G., Angelopoulos, K.C., 2007a. Membrane filtration of olive mill wastewater (OMW) and OMW fractions exploitation. Water Environ. Res. 79, 421–429.

Paraskeva, C.A., Papadakis, V.G., Tsarouchi, E., Kanellopoulou, D.G., Koutsoukos, P.G., 2007b. Membrane processing for olive mill wastewater fractionation. Desalination 213, 218–229.

Rahmanian, N., Jafari, S.M., Galanakis, C.M., 2014. Recovery and removal of phenolic compounds from olive mill wastewater. J. Am. Oil Chem. Soc. 91, 1–18.

Rahmanian, N., Jafari, S.M., Wani, T.A., 2015. Bioactive profile, dehydration, extraction and application of the bioactive components of olive leaves. Trends Food Sci. Technol. 42, 150–172.

Ramos-Cormenzana, A., Juarez-Jimenez, B., García-Pareja, M.P., 1996. Antimicrobial activity of olive mill wastewaters (alpechín) and biotransformed olive oil mill wastewater. Int. Biodeter. Biodegr. 38, 283–290.

Reis, A., Domíngues, M.R.M., Ferrer-Correia, A.J., Coimbra, M.A., 2003. Structural characterisation by MALDI-MS of olive xylo-oligosaccharides obtained by partial acid hydrolysis. Carbohydr. Polym. 53, 101–107.

Requejo, A., Peleteiro, S., Rodríguez, A., Garrote, G., Parajó, J.C., 2012. Valorization of residual woody biomass (*Olea europaea* trimmings) based on aqueous fractionation. J. Chem. Technol. Biotechnol. 87, 87–94.

Rodríguez, G., Lama, A., Rodríguez, R., Jiménez, A., Guillén, R., Fernández-Bolaños, J., 2008. Olive stones an attractive source of bioactive and valuable compounds. Bioresour. Technol. 99, 5261–5269.

Romero, I., Sánchez, S., Moya, M., Castro, E., Ruiz, E., Bravo, V., 2007. Fermentation of olive tree pruning acid–hydrolysates by *Pachysolen tannophilus*. Biochem. Eng. J. 36, 108–115.

Romero-García, J.M., Niño, L., Martínez-Patiño, C., Álvarez, C., Castro, E., Negro, M.J., 2014. Biorefinery based on olive biomass. State of the art and future trends. Bioresour. Technol. 159, 421–432.

Ronda, A., Martín-Lara, M.A., Calero, M., Blázquez, G., 2015. Complete use of an agricultural waste: application of untreated and chemically treated olive stone asbiosorbent of lead ions and reuse as fuel. Chem. Eng. Res. Des. 104, 740–751.

Roselló-Soto, E., Barba, F.J., Parniakov, O., Galanakis, C.M., Grimi, N., Lebovka, N., Vorobiev, E., 2015a. High voltage electrical discharges, pulsed electric field and ultrasounds assisted extraction of protein and phenolic compounds from olive kernel. Food Bioprocess Technol. 8, 885–894.

Roselló-Soto, E., Galanakis, C.M., Brnĉeiće, M., Orlien, V., Trujillo, F.J., Mawson, R., Knoerzer, K., Tiwari, B.K., Barba, F.J., 2015b. Clean recovery of antioxidant compounds from plant foods, byproducts and algae assisted by ultrasounds processing. Modeling approaches to optimize processing conditions. Trends Food Sci. Technol. 42, 134–149.

Rubio-Senent, F., Rodríguez-Gutiérrez, G., Lama-Muñoz, A., Fernández-Bolaños, J., 2012. New phenolic compounds hydrothermally extracted from the olive oil by-product alperujo and their antioxidative activities. J. Agric. Food Chem. 60, 1175–1186.

Rubio-Senent, F., Rodríguez-Gutiérrez, G., Lama-Muñoz, A., Fernández-Bolaños, J., 2015. Pectin extracted from thermally treated olive oil by-products: characterization, physicochemical properties, in vitro bile acid and glucose binding. Food Hydrocoll. 43, 311–321.

Russo, C., 2007. A new membrane process for the selective fractionation and total recovery of polyphenols, water and organic substances from vegetation waters (VW). J. Membr. Sci. 288, 239–246.

Saleh, M., Cuevas, M., García, J.F., Sánchez, S., 2014. Valorization of olive stones for xylitol and ethanol production from dilute acid pretreatment via enzymatic hydrolysis and fermentation by *Pachysolen tannophilus*. Biochem. Eng. J. 90, 286–293.

Santos, J.I., Martín-Sampedro, R., Fillat, U., Oliva, J.M., Negro, M.J., Ballesteros, M., Eugenio, M.E., Ibarra, D., 2015. Evaluating lignin–rich residues from biochemical ethanol production of wheat straw and olive tree pruning by FTIR and 2D–NMR. Int. J. Polym. Sci. 2015, 11.

Scoma, A., Bertin, L., Zanaroli, G., Fraraccio, S., Fava, F., 2011. A physicochemical-biotechnological approach for an integrated valorization of olive mill wastewater. Bioresour. Technol. 102, 10273–10279.

Steffen, R., Szolar, O., Braun, R., 1998. Feedstocks for anaerobic digestion—AD Nett. Available from: Q:\RODL\PROJEKTE\AD–NETT\FEEDNEW.DOC. 25 pp.

Talebnia, F., Karakashev, D., Angelidaki, I., 2010. Production of bioethanol from wheat straw: an overview on pretreatment, hydrolysis and fermentation. Bioresour. Technol. 101, 4744–4753.

Toledano, A., Serrano, L., Labidi, J., 2011. Enhancement of lignin production from olive tree pruning integrated in a green biorefinery. Ind. Eng. Chem. Res. 50, 6573–6579.

Toledano, A., Serrano, L., Labidi, J., 2012. Process for olive tree pruning lignin revalorization. Chem. Eng. J. 193–194, 396–403.

Toledano, A., Alegría, I., Labidi, J., 2013. Biorefining of olive tree (*Olea europea*) pruning. Biomass Bioenergy 59, 503–511.

Trebušak, T., Levart, A., Salobir, J., Pirman, T., 2014. Effect of *Ganoderma lucidum* (Reishi mushroom) or *Olea europaea* (olive) leaves on oxidative stability of rabbit meat fortified with n-3 fatty acids. Meat Sci. 96, 1275–1280.

Vera, D., Jurado, F., Margaritis, N.K., Grammelis, P., 2014. Experimental and economic study of a gasification plant fueled with olive industry wastes. Energy Sustain. Dev. 23, 247–257.

Vishtal, A., Kraslawski, A., 2011. Challenges in industrial applications of technical lignins. BioResources 6 (3), 3547–3568.

Zafra, A., Juárez, M.J.B., Blanc, A., Navalon, A., Gonzalez, J., Vilchez, J.L., 2006. Determination of polyphenolic compounds in wastewater olive oil by gas chromatography–mass spectrometry. Talanta 70, 213–218.

Zinoviadou, K.G., Barba, F.J., Galanakis, C.M., Brncˇicˊ, M., Trujillo, F., Mawson, R., Knoerzer, K., 2015. Fruit juice sonication: implications on food safety and physicochemical and nutritional properties. Food Res. Int. 77, 743–752.

CHAPTER

ENERGY RECOVERY ALTERNATIVES FOR THE SUSTAINABLE MANAGEMENT OF OLIVE OIL INDUSTRY

4

Paris A. Fokaides

School of Engineering, Frederick University, Nicosia, Cyprus

4.1 INTRODUCTION

The main stream in olive oil extraction is the use of a two- or a three-phase decanter and the extraction of the product through centrifugal and gravitational forces. In this method the olives are crushed to a fine paste, which is then malaxed in order to allow the small olive droplets to agglomerate. Depending on the process (two-phase, three-phase) the main by-products of the olive oil extraction process include wet olive pomace (two-phase) or wastewater and olive kernel (three-phase process). Leaves obtained after the olives washed and cleaned on entering the oil mill also account for 5% of the weight of the olives (Fokaides and Polycarpou, 2013; Christoforou and Fokaides, 2016b). The olive kernel is distinguished in crude and exhausted olive kernel. The crude olive kernel contains the olive kernel shell crushed into fragments, the skin and the crushed pulp, and a remaining quantity of oil. The exhausted olive kernel is the residue obtained after extraction of the oil from the crude olive kernel by a solvent, usually hexane. Exhausted olive kernel differs from crude olive kernel mainly by lower oil content and a smaller water content.

Several applications concerning the extraction of chemicals from crude and exhausted olive kernel have recently been reported including hydroxytyrosol (Federici et al., 2009; Schievano et al., 2015; Ntaikou et al., 2014) 4-vinylphenols (Federici et al., 2009), isocumarines and isochromans (Federici et al., 2009), tyrosol (Federici et al., 2009), polyhydroxyalkanoates (PHAs) (Fernández-Dacosta et al., 2015; Lin et al., 2008), caffeic acid (Ntaikou et al., 2014), rutin (Ntaikou et al., 2014), luteolin (Ntaikou et al., 2014; Tsagaraki et al., 2007), oleanolic acid (Jemai et al., 2008), oleuropein rich extract (Omar, 2010; Ghanbari et al., 2012), tocopherols (MacKellar et al., 1973), elenolic acid derivatives (MacKellar et al., 1973; Alipieva et al., 2014), vanillic aid (MacKellar et al., 1973), diosmetin (MacKellar et al., 1973), verbascoside (Ghanbari et al., 2012; MacKellar et al., 1973; Alipieva et al., 2014), catechol (MacKellar et al., 1973) proteins (Roselló-Soto et al., 2015), and insoluble dietary fiber (Galanakis, 2011). Although these chemicals could potentially be used for the cosmetic, the pharmaceutical, the food processing, and the food products conservation industry, the use of agricultural waste for biofuels production has, in view of the high greenhouse gas emission saving potential, significant environmental benefits (Christoforou and Fokaides, 2015a). This perspective is

Olive Mill Waste. http://dx.doi.org/10.1016/B978-0-12-805314-0.00004-2

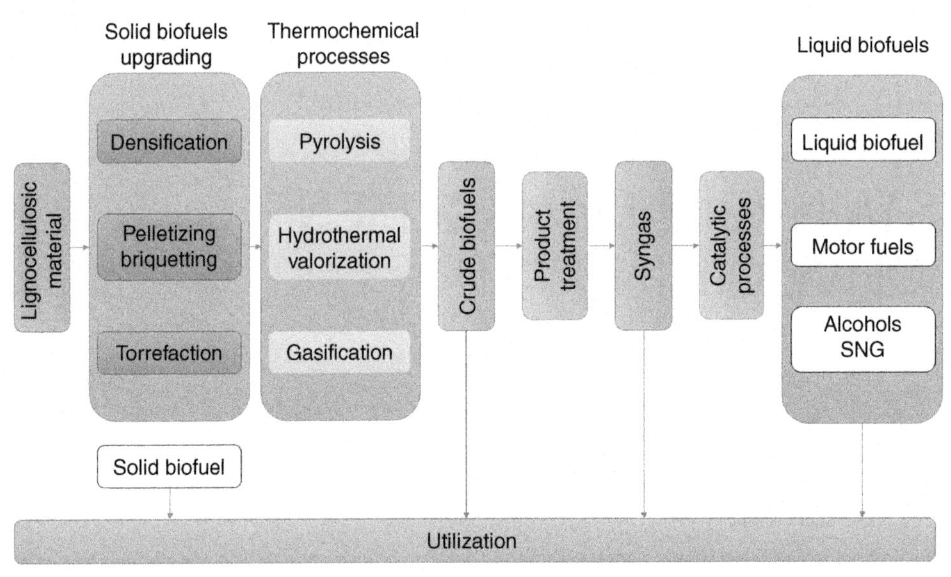

FIGURE 4.1 Lignocellulosic Biomass Conversion Routes

particularly of interest for the olive oil industry, as the native region of olive trees, the Mediterranean Basin, consists an arid environment with scarce water resources, limiting thus the options of indigenous biomass resources (Kylili et al., 2016a; Fokaides et al., 2015). The olive derived biomass releases considerable energy during its utilization and its calorific value largely exceeds that of other wood or agricultural biomass.

The biomass to biofuels conversion routes include the upgrading of solid biomass, the thermochemical and the biochemical conversion routes. Fig. 4.1 depicts the main lignocellulosic biomass conversion routes. The feedstock treatment includes the densification, the pelletizing, and the torrefaction of solid biomass, and the thermochemical conversion include the pyrolysis, the gasification, and the hydrothermal process of biomass. The upgrade of solid biomass results to solid biofuels, and the thermochemical processes to crude biofuels which can be further treated through conditioning, salt separation, and gas cleaning to deliver syngas. Syngas can be directly utilized or it can be converted to liquid biofuels through catalytic processes including the Fischer Tropsch process.

The purpose of this chapter is to present the potential energy recovery alternatives of the solid and liquid residues delivered by the olive oil industry. In this chapter, two different energy recovery alternatives are presented:

- the upgrading of olive oil industry solid waste through pelletizing, briquetting, and torrefaction; and
- the thermochemical conversion of olive oil industry residues, through pyrolysis and gasification (thermochemical routes).

The biochemical conversion of olive oil industry residues through anaerobic digestion (biochemical routes) are discussed in another chapter. Furthermore, the state-of-the-art concerning the energy conversion of biofuels delivered from the olive oil industry residues through combustion is discussed. The advancements in the field of environmental assessment of the alternative energy conversion routes of olive oil industry residues are also presented.

4.2 ELEMENTAL AND PROXIMAL ANALYSIS OF OLIVE MILL SOLID WASTE

4.2.1 STANDARDS FOR SOLID BIOFUELS

In view of the rapidly increasing international trade of solid biofuels, the need for concise and unambiguous criteria for their classification has become imperative. The criteria are both in name and in measure of physicochemical characteristics. The European Committee for Standardization (CEN) has developed a technical committee (TC 335) with the mission to develop standards in the following areas within solid biofuels. The standardization work on solid biofuels is within the scope of products from agriculture and forestry including waste and by-products. The standards concern both the raw and processed materials originating from agriculture, horticulture, and forestry to be used as a source for solid biofuels. The following distinctly different markets are established for solid biofuels:

- wood pellets for supply to large scale production of heat and power;
- packaged pellets for the residential market;
- briquetted wood for residential market and smaller industrial heat producing facilities; and
- agro-pellets, used by domestic business transactions.

Crude and exhausted olive kernel is classified in the 17225 standard series (EN ISO 17225-1:2014) under the by-products and residues from the fruit processing industry class.

As of 2016, the main standards delivered by the TC335 used for the elemental classification of solid biofuels include tests for:

- the moisture content (EN ISO 18134:2015; Sustainable Energy Research Group, 2015c);
- the ash content (550°C) (EN ISO 18122:2015);
- the calorific value (EN 14918:2009; Sustainable Energy Research Group, 2015b);
- the definition of the total content of carbon, hydrogen, and nitrogen (EN ISO 16948:2015; Sustainable Energy Research Group, 2015a)
- the definition of the sulfur and the chlorine content (EN ISO 16994:2015);
- the definition of the major elements—Al, Ca, Fe, Mg, P, K, Si, Na, and Ti (EN ISO 16967:2015);
- the definition of the minor elements—As, Pb, Cd, Cr, Cu, Hg, Ni, and Zn content (EN ISO 16968:2015); and
- the ash melting behavior (CEN/TS 15370:2006).

Furthermore, a series of standards concerning the proximal analysis of the raw and processed biomass were also issued by the TC335 including tests for:

- the mechanical durability (EN ISO 17831:2015);
- the particle size distribution (CEN/TS 15149:2014);
- the length and diameter (EN ISO 17829:2015); and
- the bulk density (EN ISO 17828:2015).

The main tests carried out for the elemental and proximal analysis of solid biofuels, including solid olive kernel residues, are depicted in Fig. 4.2.

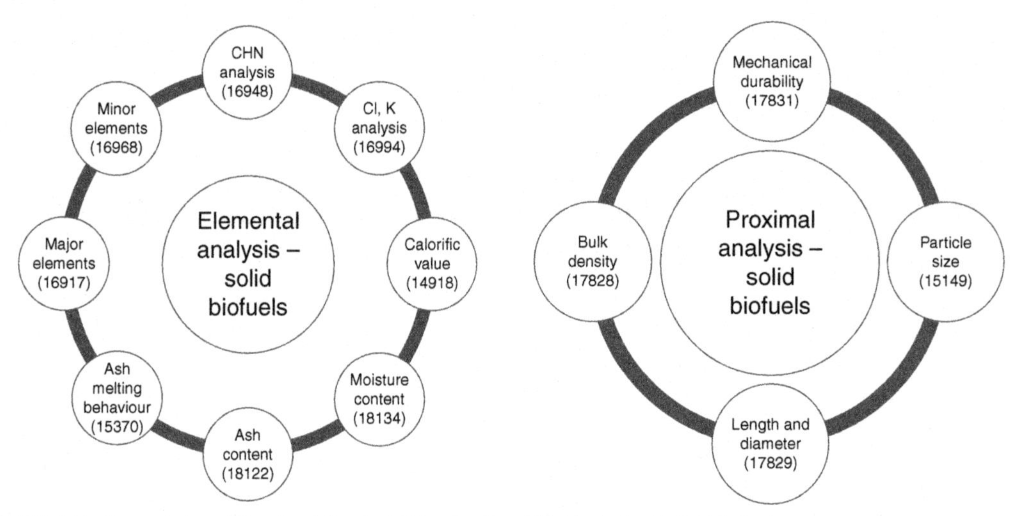

FIGURE 4.2 Elemental and Proximal Analysis Standardized Methods for Solid Biofuels

4.2.2 ELEMENTAL ANALYSIS OF CRUDE OLIVE KERNEL

There are quite few studies, since the early 2000 and the establishment of the European renewable energy penetration directives, which focus on the elemental and proximal properties of crude olive kernel (Demirbas and Demirbas, 2004; Yin, 2011; Christoforou et al., 2014; Niaounakis and Halvadakis, 2006; ECN, 2015). These studies are centered around the potential utilization of the olive oil industry solid wastes as raw material for solid biofuels. Table 4.1 summarizes the results of the elemental analysis of crude olive kernel samples found in the literature. The average net calorific value of crude olive kernel in all studies is documented to be between 21 and 22 MJ/kg. In Table 4.2, the net calorific value of different solid biomass fuels is provided as derived from Fokaides and Polycarpou (2013). Although the net calorific values of crude olive kernel in the standard are 10% lower compared to the literature documented values, its energy content is still considered to be relatively high, compared to the other biomass sources contained within the standard.

Table 4.1 Elemental Analysis of Crude Olive Kernel Samples

Source	C (%)	H (%)	N(%)	Calorific Value (MJ/kg)
Christoforou et al. (2014)	50.67	6.38	0.37	21.645
Niaounakis and Halvadakis (2006)	52.80	6.69	0.45	21.61
Niaounakis and Halvadakis (2006)	48.81	6.23	0.36	21.39
Niaounakis and Halvadakis (2006)	49.59	6.09	0.95	22.09
ECN (2015)	51.38	6.32	0.45	21.60

Table 4.2 Net Calorific Value of Solid Biomass Fuels (Fokaides and Polycarpou, 2013)

Solid Biomass Fuels	Net Calorific Value (MJ/kg)
Virgin wood materials	18.5–19.8
Virgin bark materials	17.5–21.3
Virgin wood materials, logging residues	18.3–20.5
Virgin wood materials, short rotation coppice	17.7–19.0
Virgin straw materials	15.8–19.1
Virgin cereal grain materials	16.5–26.6
Virgin reed canary grass	16.5–16.6
Virgin grass in general (hay) and miscanthus	17.1–17.6
Crude olive kernel	18.1–20.7
Exhausted olive kernel	13.9–19.2
Olive kernels	17.3–19.3
Crude grape kernel	16.7
Exhausted grape kernel	19.0
Apricot, peach, cherry fruit stone	19.5–22.9
Almond, hazelnut, pinenut shells	17.5–19.0
Oil palm shell, nut, fiber	18.0–24.8
Rice husk	14.5–16.2
Cotton stalks	15.8–18.3
Cotton gin trash	16.4–17.5
Sunflower husk	17–22
Pensylvanian malva	17.7

A number of analytical equations were derived for the calculation of the net calorific value of crude and exhausted olive kernel, based on the elemental analysis results. The common ground of the existing models is based on the assumption of a linear relation between the carbon and/or hydrogen content and the net calorific value (Vargas-Moreno et al., 2012). Considering though the complex chemical kinetics of combustion and the fact that the overall reaction is a consequence of a large number of elementary reactions (Warnatz et al., 2001) methods, such as linear regression, widely used in the literature (Demirbas and Demirbas, 2004; Yin, 2011; Tillman, 1978; Ebeling and Jenkins, 1985; Friedl et al., 2005; Sheng and Azevedo, 2005) produce models that fail to provide well-grounded results.

Christoforou et al. (2014) introduced a novel approach, based on Monte Carlo parametric modeling, for calculating the calorific value of crude olive kernel using elemental analysis measurements. The model, described in resulting to an exponential model.

$$L = 987.1628C^{0.7587} + 683.0607H^{0.3645} + 105.5334N^{3.2688} + 862.0001$$

Table 4.3 Monte Carlo Parameter Identification Algorithm for the Calculation of the Net Calorific Value Based on Elemental Analysis Measurements (Christoforou et al., 2014)

- Solution initialization: $\alpha_{l,k} \sim U\left[\alpha_{l,min}, \alpha_{l,max}\right], l = 1, \ldots, 7, k = 1, \ldots, K.$
- Calorific value calculation: $L_{\underline{\alpha_k}} = \alpha_{1,k}C^{\alpha_{2,k}} + \alpha_{3,k}H^{\alpha_{4,k}} + \alpha_{5,k}N^{\alpha_{6,k}} + \alpha_{7,k}, k = 1, \ldots, K$
- Error calculation: $E_{\alpha,k} = \sqrt{\sum_{m=1}^{M}\left(L_{\underline{\alpha_k}} - L_m\right)^2}, k = 1, \ldots, K$
- Set $E_{min} = \min E_{\alpha,k}, \underline{\alpha_{best}} = \mathbf{argmin}\, E_{\alpha,k}$

Repeat until error remains unchanged:

- For $l = 1, \ldots, 7$
- Global solution sampling: $\alpha_{l,k} \sim U\left[\alpha_{l,min}, \alpha_{l,max}\right], k = 1, \ldots, K/2$
- Local solution sampling: $\alpha_{l,k} \sim N\left(\alpha_{l,best}, \sigma_{\alpha,l}^2\right), \; k = K/2+1, \ldots, K$
- If $\alpha_{l,k} < \alpha_{l,min}$ OR $\alpha_{l,max} < \alpha_{l,k}$, let $\alpha_{l,k} = (\alpha_{l,max} - \alpha_{l,min})/2$
- Calorific value calculation: $L_{\underline{\alpha_k}} = \alpha_{1,k}C^{\alpha_{2,k}} + \alpha_{3,k}H^{\alpha_{4,k}} + \alpha_{5,k}N^{\alpha_{6,k}} + \alpha_{7,k}, \; k = 1, \ldots, K$
- Error calculation: $E_{\alpha,k} = \sqrt{\sum_{m=1}^{M}\left(L_{\underline{\alpha_k}} - L_m\right)^2}, k = 1, \ldots, K$
- If $\min_k E_{\alpha,k} < E_{min}$ then let $E_{min} = \min_k E_{\alpha,k}, \underline{\alpha_{best}} = \operatorname{argmin}_k E_{\alpha,k}$

The model, presented in Table 4.3, was validated by calculating the root mean square values of the calculated to the actual calorific value and comparing the results with the root mean square values derived by six other predictive models (Demirbas and Demirbas, 2004; Yin, 2011; Tillman, 1978; Ebeling and Jenkins, 1985; Friedl et al., 2005; Sheng and Azevedo, 2005), reducing the deviation of the calculated to the actual calorific value by 40%.

4.3 UPGRADED SOLID BIOFUELS

Upgraded solid biofuels are solid fuels, which are produced through a technical process, using biomass as feedstock. For the production of upgraded solid biofuels, two pathways are mainly used: agglomeration, including briquetting and pelletizing, and torrefaction (Klemm et al., 2013).

4.3.1 PELLETIZING

Pelletizing is the process of compressing or moulding crude or exhausted olive kernel into the shape of a pellet. Olive kernel pellets is densified biofuel made from grinded or milled olive kernel with or without additives and unitized as cylinders, usually diameter <25 mm, random length and typically 3, 15–40 mm with broken ends, obtain by mechanical compression. Nonwoody pellets including olive kernel pellets have high ash, chlorine, nitrogen, and sulfur content and major element contents. To this

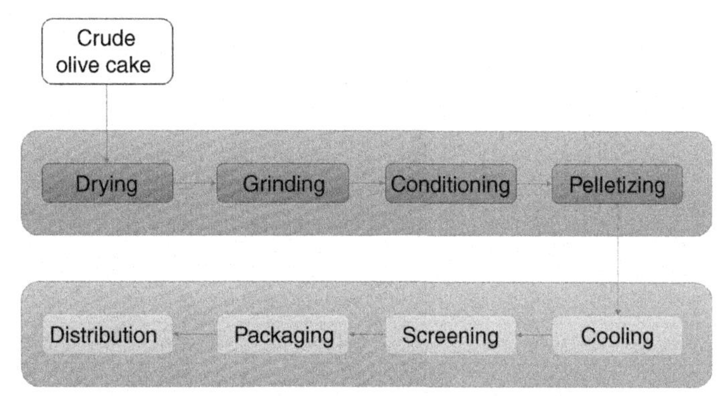

FIGURE 4.3 Crude Olive Cake (Kernel) Pelletizing Process (Kylili et al., 2016b)

end olive kernel pellets are recommended to be used in appliances, which are specially designed or adjusted for this kind of pellet. (Christoforou and Fokaides, 2015a). The main steps of the pelleting process are presented in Fig. 4.3.

Kylili et al. (2016b) presented a novel pelletizing center designed for crude olive kernel. The pelletizing process is based on a typical series of stages, which may vary depending on the condition of the raw material: reception of raw material: drying; grinding; conditioning; pelletizing; cooling; screening; packaging. Crude olive kernel is collected at the bottom end of a dryer and is automatically fed to a grid-connected milling machine. The main parts of the milling machine are the machine frame and engine assembly, and the cutting and crushing rotor assembly. Eighteen cutting knives are used for the milling of dried olive mills solid waste giving a capacity of 100–500 kg/h depending on the sieves used (4–25 mm). In this plant, a 4 mm sieve was used and a capacity of 150 kg/h was achieved. The crushed, dried olive mills solid waste is transferred by compressed air to a collection tank with a total capacity of 300 L, which is connected to a 2.5 kW cyclone for the separation of the finest particles. The grinded material is manually transferred to a rotating roller (i.e., three rollers) pelletizing machine which is powered by a 22 kW three-phase motor. The diameter of the produced pellets is 6 mm (Peksa-Blanchard et al., 2007) and a maximum capacity of 500 kg/h can be achieved based on the processed feedstock and the feeding rate which can be adjusted. The produced pellets are finally transferred to a silo using a perforated belt conveyor, which allows the simultaneous cooling and dust removal from the produced pellets. The collection of silo is connected to a weighing and packaging assembly which delivers the final, tradable product. Table 4.4 provides information regarding the installed and operating pellet production equipment of the olive mills solid waste pellet plant. The operation of the crude olive kernel pelletizing plant is presented in Sustainable Energy Research Group (2015d). Fig. 4.4 presents pelletized crude olive kernel.

Christoforou et al. (2016) provided evidence that the exploitation of pelletized crude olive kernel can have a significant input on the contribution of renewable energy sources in a country' energy balance, at an affordable competitive price. By considering the elemental composition and the available crude olive kernel quantities in Cyprus, it was revealed that the exploitation of pelletized crude olive kernel for heat generation purposes across the households, industrial, and agricultural sectors in a 40:30:30% ratio could contribute significantly by 1114 toe, 934 toe, and 835 toe, respectively.

Table 4.4 Technical Data of Olive Mills Solid Waste Pellet Plant Equipment (Kylili et al., 2016b)

Equipment	Unit
Dryer	
Manufacturer	Litsakis Pantelis & Antonios O.E. (GR)
Model	Phaethon 1
Power	3 kW
Capacity	Maximum 250 kg/h
Milling machine	
Manufacturer	KOVO NOVÁK (CZ)
Model	RS 650
Power	7.5 kW
Number of blades	18
Capacity	100–500 kg/h
Sieves size	4 mm
Cyclone power	2.5 kW
Pellet Mill	
Manufacturer	Laizhou Chengda Machinery Co., Ltd. (CN)
Model	AMP360c
Type	Rotating roller (3 rollers)
Power	22 kW
Capacity	220–500 kg/h
Pellet diameter	6 mm

FIGURE 4.4 Crude Olive Kernel Pellets

Furthermore, the findings of the economic feasibility analysis based on the assumption of a baseline case study revealed that the costs for collecting the raw biomass, manufacturing and distributing crude olive kernel pellets stands at 0.372 €/kg. A sensitivity analysis though showed that a significant drop of the selling price of the pellets down to 142 €/ton could be achieved by increasing the pellet plant's annual operating hours and capacity, being competitive with the retail price of conventional wood pellets, reported in Europe for the year 2013 at 280 €/ton (European Biomass Association, 2013).

4.3.2 TORREFACTION

Torrefaction is a mild pyrolysis process performed at temperatures between 200 and 300°C within an inert atmosphere. Torrefied biomass contains typically 60–70% of the initial mass and 90% of the initial lower heating value. Benefits of torrefied biomass compared with green biomass are increased energy density and homogeneity, improved grindability, pronounced hydrophobic character, and enhanced reactivity (Klemm et al., 2013).

Benavente and Fullana (2015) published the physicochemical properties of torrefied crude olive kernel produced by a two-phase olive mill for temperatures ranging from 150 to 300°C for 2 h (Table 4.5). The weight fraction of C, defined in percentage as wt.%, improved from 56 to 68 wt.% and the high heating value rose from 26.4 to 30.0 MJ/kg as torrefaction temperature increased, reaching typical values of subbituminous coal. The best results at 200°C in terms of maximizing the heating value and minimizing the energy losses. Volpe et al. (2015) introduced a downdraft gasifier designed for olive tree trimmings and olive pulp. The gasifier was also applied for torrefaction experiments under nitrogen to temperatures between 200 and 325°C. Elemental analyses of the chars showed a consistent linear increase of carbon to values around 75%, a linear decrease of oxygen to values near 10%, and constant values for H-content up to about 300°C.

Table 4.5 Ultimate Analysis (% Dry Basis), Ash Content (wt.%), Olive Oil Content (wt.%) and High Heating Values (HHV, MJ/kg) of Torrefied Crude Olive Kernel Samples (Benavente and Fullana, 2015)

Values (% Dry Basis)		150°C	200°C	250°C	300°C
Ultimate analysis					
C	56.1	59.9	66.3	67.2	67.7
H	7.4	8	8.9	8	4.1
N	0.8	1.1	1.7	1.6	1.6
O	29.9	25	16.2	14	10.1
S	<0.1	<0.1	<0.1	<0.1	<0.1
Ash content (wt.% d.b.)	5.5	5.6	6.5	8.7	15.7
Olive oil content (wt.% d.b.)	26	25	24.1	19.9	0.1
HHV (MJ/kg)	26.4	26.5	28.7	29.4	30
HHV rise (%)	—	0.4	8.6	11	13.4

During torrefaction, in addition to the water removal, the major decomposition reactions affect the hemicellulose. Cellulose and lignin degrade slightly depending on the torrefaction temperature (Alonso et al., 2016). Torrefied biomass retains 60–80% of the initial mass, loses its hygroscopic properties and preserves 70–90% of its energy contents (Van der Stelt et al., 2011). The torrefied biomass has a lower O/C (oxygen-to-carbon) ratio, high energy density, and hydrophobic character (Bridgeman et al., 2008). Thermogravimetric analysis is widely used in the field of crude oil pulp as a method of thermal analysis in which changes in physical and chemical properties of materials are measured as a function of increasing temperature (with constant heating rate), or as a function of time (with constant temperature or constant mass loss). Guizani et al. (2016) investigated the combustion characteristics and kinetics of torrefied olive kernel. The results showed a decrease in the torrefied olive kernel reactivity with the increase of the torrefaction temperature, which was attributed to the degraded proportion of hemicellulose, cellulose, and lignin.

4.4 THERMOCHEMICAL CONVERSION

4.4.1 PYROLYSIS

In pyrolysis, crude olive kernel is heated up within a nearly oxygen-free environment. At temperatures up to about 200°C, the biomass is first dried and water is evaporated. With increasing temperatures up to roughly 500°C, the organic matter is almost decomposed. The share and composition of the different products or product groups depends, for example, on the temperature range the pyrolytic decomposition is realized at, the presence of catalysts, the heating up rate of the biomass etc. (Kaltschmitt, 2012).

Zabaniotou et al. (2015) presented the performance of a pyrolysis–biochar fed with crude olive kernel from a two-phase olive mill. For a temperature range of 335–600°C, using olive kernels as feedstock material, weight losses increased with temperature and volatiles conversion of ≈70 wt.% was obtained at 580–600°C. Most of devolatilization occurred between 400 and 550°C, and liquids yield went through a maximum of ≈50 wt.% at about 550°C. The gaseous product mainly constituted of CO (99%) and a minor fraction of CO_2 and CH_4 resulting in a low enthalpy gases mixture with low energy value. In Hmid et al. (2014), the yield and properties of biochar derived from pyrolysis of crude olive was investigated for three pyrolysis temperatures (430°C, 480°C, and 530°C) and three heating rates (25°C/min, 35°C/min, and 45°C/min). The highest biochar yield was obtained at low pyrolysis temperature (430°C) and low heating rate (25°C/min), characterized by a high heating value (31 MJ/kg) that makes it a possible fuel candidate. Manyà et al. (2013) investigated the pyrolysis of two-phase olive mill waste, reaching the general conclusion that biochar yield from pyrolysis of two-phase olive mill waste decreases when both peak temperature and pressure increases. Also an increase of both peak temperature and pressure resulted in a higher fixed-carbon yield; and a significant increase of the overall devolatilization rate was observed for experiments conducted at intermediate pressure values. The catalytic fast pyrolysis of biomass has been recently paid a tremendous interest. In this process, biomass is pyrolyzed in a fluidized bed in the presence of an acid catalyst. Increasing the pyrolysis temperature of dried or exhausted olive kernel from 500 to 800°C leads to a decrease in the solid and liquid yields and to an increase in the gas yield in both catalytic and noncatalytic pyrolysis. The presence and amount of catalyst may cause a significant decrease in the liquid phase yield and a high increase in the gas phase yield giving rise to a vast rise in hydrogen production (Encinar et al., 2009).

4.4.2 GASIFICATION

Gasification of biomass could be a process of choice for producing heat and power with a higher efficiency of power production compared to combustion. The gasification process can be broken down into three phases (IFRF Online Combustion Handbook, 2016).

- The first phase is an endothermic process of pyrolysis during which the biomass is converted by heat into char and volatile matter, such as steam, methanol, acetic acids, and tars.
- The second phase is an exothermic reaction in which part of the carbon is oxidized to carbon dioxide.
- In the third phase, part of the carbon dioxide, the volatile compounds, and the steam are reduced to carbon monoxide, hydrogen, and methane. This mixture of gases diluted with nitrogen from the air and unreduced carbon dioxide is known as syngas.

Concerning the energy value of the produced syngas, the olive pit gasification provides the highest calorific value (5.4 MJ/kg) followed by tree pruning (4.8 MJ/kg) (Vera et al., 2014).

Cogasification of bagasse wastes mixed with coal has proven to be a suitable process to manage olive kernel. Both laboratory and pilot-scale experiments showed that it is quite possible to incorporate olive dried and in a coal gasification installation and change bagasse content according to seasonal variations, without significant changes in the installation. However, bagasse content incorporation should not exceed 40% (w/w) to guarantee gasification stabilization and to prevent the formation of high amounts of tars and heavier hydrocarbons. On the other hand, bagasse high contents of silica, calcium, and potassium may demand purging the bed more frequently to prevent bed sintering (André et al., 2005).

4.5 COMBUSTION

4.5.1 COMBUSTION BOILERS MINIMUM LEGISLATIVE REQUIREMENTS

The simplest way to exploit olive solid residues for energy production is by direct combustion. A typical example of dried or exhausted kernel combustion boiler is shown in Fig. 4.5. Combustion type of boilers gives off their heat to radiators in exactly the same way as, for example, an oil-fired one. These boilers are equipped with a silo containing olive dried kernel or exhausted kernel. A screw feeder feeds the fuel simultaneously with the output demand of the dwelling. A solid biomass boilers laboratory is presented Sustainable Energy Research Group (2015e).

Based on the requirements of directive 2009/125/EC (European Commission, 2009) to set ecodesign requirements for energy-related products, the European Commission has carried out a preparatory study to analyze the technical, environmental, and economic aspects of the solid fuel boilers used in households and for commercial purposes. The environmental aspects of solid fuel boilers that have been defined for the purposes of this Regulation include the energy consumption in the use phase and the emissions of particulate matter (dust), organic gaseous compounds, carbon monoxide, and nitrogen oxides in the use phase. Nonwoody biomass boilers were though exempted from this regulation, due to the insufficient European-wide information to determine appropriate levels for the ecodesign requirements for them and they may have further significant environmental impacts, such as furan and dioxin emissions, whereas "nonwoody biomass" includes both olive stones and olive kernels. The decision of

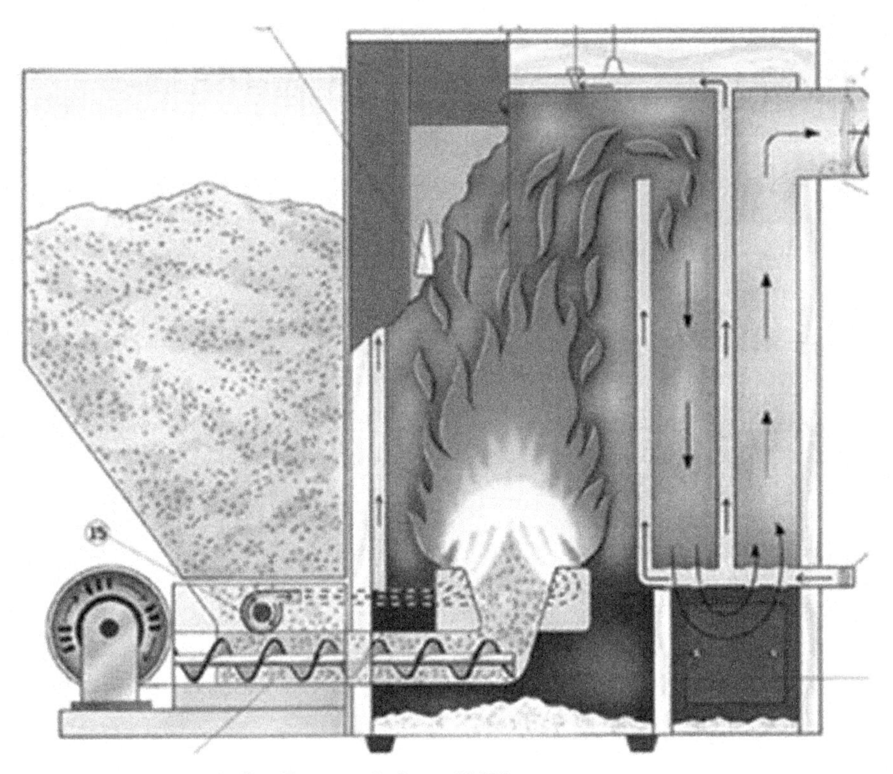

FIGURE 4.5 Olive Kernel Pellets Boiler (Samaras Boilers, 2015)

the Commission is to review this Regulation in the light of technological progress and present the result of that review to the Consultation Forum by 2022, in order to assess whether to include nonwoody biomass boilers, with ecodesign requirements for their specific types of pollutant emissions (European Commission, Regulation 2015/1189).

4.5.2 COMBUSTION EMISSIONS

The European Standard EN 303-5 (EN, 2012) concerns the testing and marking of heating boilers up to a nominal heat output of 500 kW which are designed for the burning of solid fuels only. The standard contains requirements and test methods for combustion quality, operating characteristics, marking, and maintenance of heating boilers. Table 4.6 presents the emission limits for biogenic fuels for carbon monoxide, organic gaseous carbon, and dust.

Lajili et al. (2015) studied the combustion characteristics of a domestic boiler (12 kW) fired with biofuels prepared by blending pine sawdust and crude olive kernel. It was found that the main exhaust gases (CO, CO_2 C_xH_y, and NO_x) were emitted in acceptable concentrations comparing to other biofuels in the literature. The analysis of bottom ash compositions showed the presence of three oxides, namely CaO, K_2O, and SiO_2 as well as a significant amount of unburned carbon, revealing

Table 4.6 Emission Limits for Biogenic Fuels (European Commission, Regulation 2015/1189)

		Emission Limits								
		CO			OGC			Dust		
		mg/m³ at 10% O_2								
		Class	Class	Class	Class	Class	Class	Class	Class	Class
Stoking	**Nominal Heat Output (kW)**	3	4	5	3	4	5	3	4	5
Manual	*≤50*	5000	1200	700	150	50	30	150	75	60
	>50 ≤ 150	2500			100					
	>150 ≤ 500	1200			100					
Automatic	*≤50*	3000	1000	500	100	30	20	150	60	40
	>50 ≤ 150	2500			80					
	>150 ≤ 500	1200			80					

that the boiler control through the primary and secondary air injections should be adapted to the different pellets properties.

Miranda et al. (2007) investigated the combustion performance of olive mill waste water with the support of different fractions of solid waste. A fuel mixture of 30–70% (% semiliquid fraction-% solid fraction) allowed a combustion process with no significant alterations, increasing the thermal power of the furnace. This finding was confirmed in (Kraiem et al., 2014). In this study, sawdust was used for the olive mill waste water impregnation. The blend was densified into pellets and combusted in a domestic combustor. Combustion efficiencies, gaseous, and particulate matter emissions as well as ash contents were evaluated, leading to an increase of energy content through the heating values increase. An increase of the impregnated samples reactivity was observed and assigned to the potassium catalytic effect. Combustion performances showed that the olive mill waste water addition did not had a negative effect on the firing quality.

4.5.3 COMBUSTION KINETICS

The chemistry that drives combustion is an extremely complicated network of reactions. To describe the combustion of crude olive kernel compounds in full chemical kinetics, detailed knowledge of hundreds of chemical species that participate in thousands of individual chemical reactions is required. Each of these reactions, in turn, has a detailed description in terms of fundamental physical principles (Warnatz et al., 2001).

Garcia-Maraver et al. (2015) delivered thermogravimetric curves in air, measured for the different types of agricultural residues from olive trees (leaves, pruning, and wood) at different heating rates (5, 10, 20, 40, 100 K/min), confirming that the combustion process is divided into three stages: water removal, roasting phase, and char decomposition. The three combustion stages were also confirmed for the case of cocombustion of different olive oil production chain residues blends in Buratti et al. (2016).

In this study, the minimum apparent activation energy for the examined blends was found 77.8 kJ/mol. With the increase of olive tree pruning content in the blends, it was proved that the reactivity was improved, revealing nonadditivity behavior.

Zhong et al. (2015) simulated the combustion of olive kernel in a circulating fluidized bed reactor by employing a gas–solid flow, chemical reaction, heat transfer, and mass transfer three-dimensional numerical model. Simulations were carried out in a circulating fluidized bed riser of 0.125 m diameter and 1.8 m height at atmosphere. The whole bed was found to be in a turbulent fluidized state, and a downward movement of solid particles was detected in the near-wall region. The gas and solid temperatures increased gradually along the height of the riser, revealing that the burning of volatiles and char primarily occurs in the upper region of the riser. The increase in the superficial velocity of the circulating fluidized bed and the excess air ratio induced more NO formation from HCN and NH_3 oxidation and enhanced char combustion.

4.6 ENVIRONMENTAL ASSESSMENT OF TECHNOLOGY ALTERNATIVES

With biofuels currently gaining momentum, the entire biomass to biofuel process chain has become popular subject of investigation in the past recent years to assess the environmental performance of energy recovery alternatives for the sustainable management of olive oil industry. Life cycle assessment (LCA) methodology has proven to be one of the most effective tools for carrying out environmental impact analysis of any process or system. LCA is a technique to assess environmental impacts associated with all the stages of a product's life from cradle to grave. The application of LCA provides a useful insight regarding the environmental pollution caused by such processes, as well as enables the comparison of different scenarios and the identification of "hot spots" for potential improvement of the process chain.

El Hanandeh (2015) analyzed five alternatives for managing solid waste and wastewater streams generated by the olive oil industry. Using the waste to manufacture pellets for domestic water heating is the alternative that is likely to deliver the highest environmental benefits. Briquette manufacturing for use in domestic stoves for space heating ranked second. The best practices of the Australian industry which promotes composting of the waste ranked fourth in the global warming potential, eutrophication potential and fifth in the ozone depletion potential. In Kylili et al. (2016b), LCA methodology to assess the environmental impacts of the pelletizing process using crude olive kernel for the production of pellets for domestic heat purposes. The results of the analysis indicated that the transportation aspect and the exploitation of RES for energy generation and consumption are both crucial points influencing the environmental impact of pelletizing process chains. It has been concluded that decentralized scenarios are less harmful to the environment than their respective centralized scenarios, establishing the decentralized production line as the most environmental friendly option. The decrease noted across all impact categories, 30% in global warming potential and acidification potential, 50% in ozone depletion potential, 32% in resource depletion, mineral, fossil, and renewable elements (abiotic resources depletion potential), and 28% in photochemical ozone formation. The incorporation of renewable energy sources for energy generation decreased by approximately 85% of the impact of the pelletizing process on climate change, acidification, and photochemical ozone creation, compared to the basic centralized scenario where conventional fossil fuelled technologies were employed.

Although the benefits of torrefied biomass compared with green biomass include increased energy density and homogeneity, improved grindability, pronounced hydrophobic character, and enhanced reactivity (Kaltschmitt, 2012), the torrefaction process per se requires a notable amount of thermal energy. Christoforou and Fokaides (2016a) examined alternative scenarios regarding the required thermal energy for crude olive kernel torrefaction using LCA. The use of alternative renewable energy sources at the drying stage contributed to the mitigation of the negative environmental impact, leading to a significant improvement of the environmental performance of torrefaction for olive mill solid wastes processing. In Christoforou and Fokaides (2015b), the environmental performance of crude olive kernel torrefaction and hydrothermal carbonization was compared. The ratio of total energy consumption for the production of 1 kg of dry biochar over the calorific value of the final product was calculated for both systems. A ratio of 0.218 and 0.263 was obtained for the hydrothermal carbonization and torrefaction system respectively, which indicated that the first is energetically more feasible than the second under the specific examined conditions.

4.7 CONCLUSIONS

In this chapter, the main stream in the energy recovery practices of olive oil industry residues were presented. As seen in the literature, there are currently numerous studies and applications revealing that the olive oil industry can serve as a major supplier of the solid biofuels industry. Emphasis was given on the production and exploitation of solid biofuels, which is currently the key application in the field. The standardized tests which are employed to characterize the quality of the produced solid biofuel, as well as the minimum requirements for accepting olive kernel derived biofuels were discussed. An adequate analysis of the required elements of a pelletizing plant for olive kernel were presented, based on a pilot in operation, designed and tailored for olive kernel (Kylili et al., 2016b). The main kinetics of torrefaction, a process of improving the properties of olive kernel and delivering biochar, were also introduced and recent literature findings were provided. The thermochemical routes for the processing of olive kernel, were also presented, based on the scientific state-of-the-art. Pyrolysis and gasification applications of olive kernel, a field which is still under development, were also given. Best practices for the combustion of pelletized and briquetted olive kernel were discussed, based on the European Regulations on the minimum requirements of the applied boilers. The chapter is concluded with the environmental assessment of selected practices and routes for the energy conversion of olive kernel.

REFERENCES

Alipieva, K., Korkina, L., Orhan, I.E., Georgiev, M.I., 2014. Verbascoside—a review of its occurrence, (bio) synthesis and pharmacological significance. Biotechnol. Adv. 32 (6), 1065–1076.

Alonso, E.R., Dupont, C., Heux, L., Perez, D.D.S., Commandre, J.M., Gourdon, C., 2016. Study of solid chemical evolution in torrefaction of different biomasses through solid-state 13 C cross-polarization/magic angle spinning NMR (nuclear magnetic resonance) and TGA (thermogravimetric analysis). Energy 97, 381–390.

André, R.N., Pinto, F., Franco, C., Dias, M., Gulyurtlu, I., Matos, M.A.A., Cabrita, I., 2005. Fluidised bed co-gasification of coal and olive oil industry wastes. Fuel 84 (12), 1635–1644.

Benavente, V., Fullana, A., 2015. Torrefaction of olive mill waste. Biomass Bioenerg. 73, 186–194.

Bridgeman, T.G., Jones, J.M., Shield, I., Williams, P.T., 2008. Torrefaction of reed canary grass, wheat straw and willow to enhance solid fuel qualities and combustion properties. Fuel 87 (6), 844–856.

Buratti, C., Mousavi, S., Barbanera, M., Lascaro, E., Cotana, F., Bufacchi, M., 2016. Thermal behaviour and kinetic study of the olive oil production chain residues and their mixtures during co-combustion. Bioresour. Technol. 214, 266–275.

CEN/TR 15149, 2014. Solid biofuels—determination of particle size distribution.

CEN/TS 15370, 2006. Solid biofuels—method for the determination of ash melting behaviour. Chem. Technol. Biotechnol. 84(6), 895–900.

Christoforou, E.A., Fokaides, P.A., 2015a. A review of quantification practices for plant-derived biomass potential. Int. J. Green Energ. 12 (4), 368–378.

Christoforou, E.A., Fokaides, P.A., 2015b. Life cycle assessment of torrefaction and hydrothermal carbonization of olive husk. In: Proceedings of Twenty-Third European Biomass Conference, Vienna, Austria.

Christoforou, E.A., Fokaides, P.A., 2016a. Life cycle assessment (LCA) of olive husk torrefaction. Renew. Energ. 90, 257–266.

Christoforou, E., Fokaides, P.A., 2016b. A review of olive mill solid wastes to energy utilization techniques. Waste Manag. 49, 346–363.

Christoforou, E.A., Fokaides, P.A., Kyriakides, I., 2014. Monte Carlo parametric modeling for predicting biomass calorific value. J. Therm. Anal. Calorim. 118 (3), 1789–1796.

Christoforou, E., Kylili, A., Fokaides, P.A., 2016. Technical and economical evaluation of olive mills solid waste pellets. Renew. Energ. 96, 33–41.

Demirbas, A., Demirbas, H., 2004. Estimating the calorific values of lignocellulosic fuels. Energ. Explor. Exploit. 22 (2), 135–143.

Ebeling, J.M., Jenkins, B.M., 1985. Physical and chemical properties of biomass fuels. Trans. ASAE 28 (3), 898–902.

ECN, 2015. Phyllis2, database for biomass and waste.

El Hanandeh, A., 2015. Energy recovery alternatives for the sustainable management of olive oil industry waste in Australia: life cycle assessment. J. Clean. Prod. 91, 78–88.

EN 14918, 2009. Solid biofuels—determination of calorific value. Brussels. CEN.

EN 303-5, 2012. E Heating boilers—Part 5 Heating boilers for solid fuels, manually and automatically stoked, nominal heat output of up to 500 kW—Terminology, requirements, testing and marking CEN.

EN ISO 17225-1, 2014. Solid biofuels—Fuel specifications and classes—Part 1 General requirements. CEN.

EN ISO 16948, 2015. Solid biofuels—Determination of total content of carbon, hydrogen and nitrogen. CEN.

EN ISO 16967, 2015. Solid biofuels—Determination of major elements—Al, Ca, Fe, Mg, P, K, Si, Na and Ti. CEN.

EN ISO 16968, 2015. Solid biofuels—Determination of minor elements. CEN.

EN ISO 16994, 2015. Solid biofuels—Determination of total content of sulfur and chlorine. CEN.

EN ISO 17828, 2015. Solid biofuels—Determination of bulk density.

EN ISO 17829, 2015. Solid biofuels—Determination of length and diameter of pellets.

EN ISO 17831, 2015. Solid biofuels—Determination of mechanical durability of pellets and briquettes.

EN ISO 18122, 2015. Solid biofuels—determination of ash content. Brussels. CEN.

EN ISO 18134, 2015. Solid biofuels—determination of moisture content. Brussels. CEN.

Encinar, J.M., Gonzalez, J.F., Martinez, G., Roman, S., 2009. Catalytic pyrolysis of exhausted olive oil waste. J. Anal. Appl. Pyrol. 85 (1), 197–203.

European Biomass Association, European Bioenergy Outlook, 2013. Belgium.

European Commission, 2009. European Commission, Directive, 2009/125/EC of 21 October 2009 establishing a framework for the setting of eco-design requirements for energy-related products.

European Commission, 2015. European Commission, Commission Regulation 2015/1189 of 28 April 2015 implementing Directive 2009/125/EC of the European Parliament and of the Council with regard to ecodesign requirements for solid fuel boilers.

Federici, F., Fava, F., Kalogerakis, N., Mantzavinos, D., 2009. Valorisation of agro-industrial by-products, effluents and waste: concept, opportunities and the case of olive mill wastewaters. J. Chem. Technol. Biotechnol. 84 (6), 895–900.

Fernández-Dacosta, C., Posada, J.A., Kleerebezem, R., Cuellar, M.C., Ramirez, A., 2015. Microbial community-based polyhydroxyalkanoates (PHAs) production from wastewater: techno-economic analysis and ex-ante environmental assessment. Bioresour. Technol. 185, 368–377.

Fokaides, P.A., Polycarpou, P., 2013. Exploitation of Olive Solid Waste for Energy Purposes. Renewable Energy, Economies, Emerging Technologies and Global Practices. Nova Science Publishers, Inc, New York, 163-178.

Fokaides, P.A., Tofas, L., Polycarpou, P., Kylili, A., 2015. Sustainability aspects of energy crops in arid isolated island states: the case of Cyprus. Land Use Policy 49, 264–272.

Friedl, A., Padouvas, E., Rotter, H., Varmuza, K., 2005. Prediction of heating values of biomass fuel from elemental composition. Analyt. Chim. Acta 544 (1), 191–198.

Galanakis, C.M., 2011. Olive fruit dietary fiber: components, recovery and applications. Trends Food Sci. Technol. 22 (4), 175–184.

Garcia-Maraver, A., Perez-Jimenez, J.A., Serrano-Bernardo, F., Zamorano, M., 2015. Determination and comparison of combustion kinetics parameters of agricultural biomass from olive trees. Renew. Energ. 83, 897–904.

Ghanbari, R., Anwar, F., Alkharfy, K.M., Gilani, A.H., Saari, N., 2012. Valuable nutrients and functional bioactives in different parts of olive (*Olea europaea* L.)—a review. Int. J. Mol. Sci. 13 (3), 3291–3340.

Guizani, C., Haddad, K., Jeguirim, M., Colin, B., Limousy, L., 2016. Combustion characteristics and kinetics of torrefied olive pomace. Energy 107, 453–463.

Hmid, A., Mondelli, D., Fiore, S., Fanizzi, F.P., Al Chami, Z., Dumontet, S., 2014. Production and characterization of biochar from three-phase olive mill waste through slow pyrolysis. Biomass Bioenerg. 71, 330–339.

IFRF Online Combustion Handbook, 2016. Available from: http://www.handbook.ifrf.net/handbook/

Jemai, H., Bouaziz, M., Fki, I., El Feki, A., Sayadi, S., 2008. Hypolipidimic and antioxidant activities of oleuropein and its hydrolysis derivative-rich extracts from Chemlali olive leaves. Chem. Biol. Interact. 176 (2), 88–98.

Kaltschmitt, M., 2012. Biomass as Renewable Source of Energy, Possible Conversion Routes. Encyclopedia of Sustainability Science and Technology. Springer, New York.

Klemm, M., Schmersahl, R., Kirsten, C., Weller, N., 2013. Biofuels biofuel: upgraded New Solids biofuel upgraded new solids. Renewable Energy SystemsSpringer, New York, pp. 138–160.

Kraiem, N., Jeguirim, M., Limousy, L., Lajili, M., Dorge, S., Michelin, L., Said, R., 2014. Impregnation of olive mill wastewater on dry biomasses: impact on chemical properties and combustion performances. Energy 78, 479–489.

Kylili, A., Christoforou, E., Fokaides, P.A., Polycarpou, P., 2016a. Multicriteria analysis for the selection of the most appropriate energy crops: the case of Cyprus. Int. J. Sustain. Energ. 35 (1), 47–58.

Kylili, A., Christoforou, E., Fokaides, P.A., 2016b. Environmental evaluation of biomass pelleting using life cycle assessment. Biomass Bioenerg. 84, 107–117.

Lajili, M., Jeguirim, M., Kraiem, N., Limousy, L., 2015. Performance of a household boiler fed with agropellets blended from olive mill solid waste and pine sawdust. Fuel 153, 431–436.

Lin, Y., Shi, R., Wang, X., Shen, H.M., 2008. Luteolin, a flavonoid with potentials for cancer prevention and therapy. Curr. Cancer Drug Targets 8 (7), 634.

MacKellar, F.A., Kelly, R.C., Van Tamelen, E.E., Dorschel, C., 1973. Structure and stereochemistry of elenolic acid. J. Am. Chem. Soc. 95 (21), 7155–7156.

Manyà, J.J., Roca, F.X., Perales, J.F., 2013. TGA study examining the effect of pressure and peak temperature on biochar yield during pyrolysis of two-phase olive mill waste. J. Anal. Appl. Pyrol. 103, 86–95.

Miranda, M.T., Cabanillas, A., Rojas, S., Montero, I., Ruiz, A., 2007. Combined combustion of various phases of olive wastes in a conventional combustor. Fuel 86 (3), 367–372.

Niaounakis, M., Halvadakis, C.P., 2006, second ed. Olive Processing Waste Management: Literature Review and Patent Survey, vol. 5. Elsevier, Oxford.

Ntaikou, I., Peroni, C.V., Kourmentza, C., Ilieva, V.I., Morelli, A., Chiellini, E., Lyberatos, G., 2014. Microbial bio-based plastics from olive-mill wastewater: generation and properties of polyhydroxyalkanoates from mixed cultures in a two-stage pilot scale system. J. Biotechnol. 188, 138–147.

Omar, S.H., 2010. Oleuropein in olive and its pharmacological effects. Sci. Pharm. 78 (2), 133–154.

Peksa-Blanchard, M., Dolzan, P., Grassi, A., Heinimö, J., Junginger, M., Ranta, T., Walter, A., 2007. Global wood pellets markets and industry: policy drivers, market status and raw material potential. IEA. Bioenergy Task 40, 2007.

Roselló-Soto, E., Barba, F.J., Parniakov, O., Galanakis, C.M., Lebovka, N., Grimi, N., Vorobiev, E., 2015. High voltage electrical discharges, pulsed electric field, and ultrasound assisted extraction of protein and phenolic compounds from olive kernel. Food Bioprocess Technol. 8 (4), 885–894.

Samaras Boilers, N., 2015. Available from: http://www.nsamaras.gr/

Schievano, A., Adani, F., Buessing, L., Botto, A., Casoliba, E.N., Rossoni, M., Goldfarb, J.L., 2015. An integrated biorefinery concept for olive mill waste management: supercritical CO_2 extraction and energy recovery. Green Chem. 17 (5), 2874–2887.

Sheng, C., Azevedo, J.L.T., 2005. Estimating the higher heating value of biomass fuels from basic analysis data. Biomass Bioenerg. 28 (5), 499–507.

Sustainable Energy Research Group, 2015a. Frederick University and Frederick Research Center . Elemental Analysis of Solid Biofuels. Available from: https://www.youtube.com/watch?v=hHPseal7puQ

Sustainable Energy Research Group, 2015b. Sustainable Energy Research Group, Frederick University and Frederick Research Center. Calorific Value Measurement of Solid Biofuels using Bomb Calorimetry. Available from: https://www.youtube.com/watch?v=RzAPQPWOlNI

Sustainable Energy Research Group, 2015c. Frederick University and Frederick Research Center. Moisture Content Measurement of Solid Biofuels. Available from: https://www.youtube.com/watch?v=6UR_5zF83Uo

Sustainable Energy Research Group, 2015d. Frederick University and Frederick Research Center. Olive Mill Solid Waste Management Center for the Production of Solid Biofuels. Available from: https://www.youtube.com/watch?v=2QId0Qs1LEY

Sustainable Energy Research Group, 2015e. Frederick University and Frederick Research Center. Frederick University Boilers Laboratory. Available from: https://www.youtube.com/watch?v=BHqak2ZpqfU

Tillman, D.A., 1978. Wood as an Energy Resource. Elsevier, New York.

Tsagaraki, E., Lazarides, H.N., Petrotos, K.B., 2007. Olive mill wastewater treatment. Utilization of By-Products and Treatment of Waste in the Food IndustrySpringer, US, pp. 133–157.

Van der Stelt, M.J.C., Gerhauser, H., Kiel, J.H.A., Ptasinski, K.J., 2011. Biomass upgrading by torrefaction for the production of biofuels: a review. Biomass Bioenerg. 35 (9), 3748–3762.

Vargas-Moreno, J.M., Callejón-Ferre, A.J., Pérez-Alonso, J., Velázquez-Martí, B., 2012. A review of the mathematical models for predicting the heating value of biomass materials. Renew. Sustain. Energ. Rev. 16 (5), 3065–3083.

Vera, D., Jurado, F., Margaritis, N.K., Grammelis, P., 2014. Experimental and economic study of a gasification plant fuelled with olive industry wastes. Energ. Sustain. Dev. 23, 247–257.

Volpe, R., Messineo, A., Millan, M., Volpe, M., Kandiyoti, R., 2015. Assessment of olive wastes as energy source: pyrolysis, torrefaction and the key role of H loss in thermal breakdown. Energy 82, 119–127.

Warnatz, J., Maas, U., Dibble, R.W., Warnatz, J., 2001. Combustion, vol. 3. Springer, Berlin.

Yin, C.Y., 2011. Prediction of higher heating values of biomass from proximate and ultimate analyses. Fuel 90 (3), 1128–1132.

Zabaniotou, A., Rovas, D., Libutti, A., Monteleone, M., 2015. Boosting circular economy and closing the loop in agriculture: case study of a small-scale pyrolysis–biochar based system integrated in an olive farm in symbiosis with an olive mill. Environ. Dev. 14, 22–36.

Zhong, W., Xie, J., Shao, Y., Liu, X., Jin, B., 2015. Three-dimensional modeling of olive cake combustion in CFB. Appl. Therm. Eng. 88, 322–333.

CHAPTER

REUSE OF OLIVE MILL WASTE AS SOIL AMENDMENT

5

Luca Regni*, Giovanni Gigliotti, Luigi Nasini*, Evita Agrafioti[†], Charis M. Galanakis[†], Primo Proietti***

**Department of Agricultural, Food and Environmental Sciences, University of Perugia, Perugia, Italy;*
***Department of Civil and Environmental Engineering, University of Perugia, Perugia, Italy;*
[†]Department of Research & Innovation, Galanakis Laboratories, Chania, Greece

5.1 INTRODUCTION

The environmental impact of olive oil production is important as oil extraction can require a considerable amount of water and generates huge quantities of olive mill wastes (OMW). The latter are generated in a limited time period of 3–4 months of olive harvesting (Ntougias et al., 2013; Mechri et al., 2011; Ouzounidou et al., 2010; Morillo et al., 2009; Alburquerque et al., 2004, 2007; Paraskeva and Diamadopoulos, 2006; Roig et al., 2006).

By applying pressure (traditional method) or three-phase centrifugation oil extraction, system effluents are olive pomace and olive mill wastewater (OMWW). On the other hand, by applying a two-phase centrifugation oil extraction system, the effluent is wet olive pomace (Barbera et al., 2013; Suzzi and Tofalo, 2009; Boubaker and Ridha, 2007; Kapellakis et al., 2006; El Hadrami et al., 2004). In many olive-producer countries (e.g., Italy) OMW could be legally spread directly in the field under specific requirements (Proietti et al., 2015; Nasini et al., 2013; Altieri and Esposito, 2008). Such application has raised interest not only because of its low cost, but also due to its potential to enhance soil properties. Intensive farming leads to a gradual soil organic matter depletion deteriorating its physicochemical and biological properties and leading to degenerative phenomena, such as erosion and loss of fertility (Toscano and Montemurro 2012). Toward that direction, OMW high organic matter content could contribute to the restoration of such soils and thus to the mitigation of the aforementioned adverse phenomena. In addition, as demonstrated by numerous studies, OMWW and pomace soil application could positively affect soil productivity, since not only they could increase soil organic matter but also to serve as a valuable source of nutrients (N, P, and K) (Ouzounidou et al., 2010; Karpouzas et al., 2009; Brunetti et al., 2007; Piotrowska et al., 2006; Amirante, 2003; Rinaldi et al., 2003; Lessage-Meessen et al., 2001; Mulinacci et al., 2001; Sierra et al., 2001; Zenjari and Nejmeddine, 2001). This is of particular interest for soils found in most of the Mediterranean agricultural regions, which are typically poor in organic matter and nutrients (Toscano and Montemurro 2012; López-Piñeiro et al., 2007; Rinaldi et al., 2003). In fact, the organic matter of OMW is able to improve soil aggregates and, thus its porosity and water retention capability. However, an incorrect OMW implementation may cause temporary immobilization of mineral nitrogen, causing consequently crop yield reduction (Toscano and Montemurro, 2012).

Apart from the aforementioned agronomic effects, OMWW and pomace application to soils could also have significant environmental benefits. More specifically, considering their high organic content,

Olive Mill Waste. http://dx.doi.org/10.1016/B978-0-12-805314-0.00005-4

their application into soil could sequester significant amounts of carbon (C), thus reducing the atmospheric carbon dioxide (CO_2) concentration (Lozano-Garcia et al., 2011; Mondini and Sequi, 2008; Sánchez-Monedero et al., 2008).

Indeed, although the soil and vegetation potential of subtracting C is not sufficient to compensate CO_2 emissions, the storage capacity of C in the biosphere could be an essential measure to stabilize emissions in the coming years. Thus a technique of agricultural land management that maximizes the accumulation of C in the soil is concentrating a lot of interest, as stated by the provisions of art. 3.4 of the Kyoto Protocol* (countries have pledged to reduce levels of CO_2 emissions) (Proietti et al., 2014, 2016). On the other hand, different interactions (i.e., "priming effect") due to OMW spreading may cause the extra release of soil-derived C and/or N, leading to a less (compared to what is expected) or null increase in soil organic matter.

Spreading of OMW should be implemented with rationality since the not-well stabilized organic materials may inhibit or reduce the development of crops. This can happen due to:

1. the presence of tannins, fatty acids, and phenols;
2. the high C/N ratio leading to a nitrogen competition among the microorganisms of the soil and the roots; and
3. the anoxia of the roots caused by microorganisms' oxygen consumption (Barbera et al., 2013; Suzzi and Tofalo 2009; Amaral et al., 2008; Saadi et al., 2007; Roig et al., 2006; Ayed et al., 2005; El Hadrami et al., 2004).

   In particular, the high concentration of polymeric phenols is known to have a bacteriostatic effect on microorganisms and phytotoxic effects on crops. The antimicrobial and phytotoxic activity is mainly due to the ability of phenolic compounds to combine with other organic components (e.g., proteins), thus altering cell membrane permeability and intercellular transfer mechanisms (Peri and Proietti, 2014). In addition, a careful management is needed due to the high content of OMW salts and its acidic nature. On the other hand, high amounts of phosphorous, potassium, and organic matter could contribute to the fertilization of the cultivated soil (Amirante, 2003).

Indeed, many studies have shown the positive effects of the rational use of OMW for soil amendment and have excluded significant risks for crops and environment, but there are other studies that show negative effects on soil and adjacent environmental compartments, and in particular surface and ground-water. These disagreements may be attributed to the different experimental conditions of the assayed studies, for example, to different doses used, methods of spreading, type of soil, phenological stage of the crop, depth, and nature of groundwater, cultural practices, and climatic conditions. Since all these factors can influence the effects of OMW spreading, to avoid problems, a proper amendment method and a rational use of OMW (also respecting specific regulations) should be applied (Belaqziz et al., 2016).

5.2 AGRONOMIC USE OF OMWW

In the Mediterranean countries, more than 30×10^6 m^3 of OMWW are produced during the harvest and processing season (Barbera et al., 2013). OMWW consists of water, that derives from the olives, (~40–50% of the processed fruit weight), dilution of the paste (40–60% of the processed olives weight) used in some continuous oil extraction systems, and the washing of olives and plants (10–15% of the processed olive weight) (Suzzi and Tofalo 2009). Therefore, the total amount of OMWW produced every 100 kg of processed olives is 50–60, 10–20, 90–130, and 30–45 kg, using the pressure, the two-phase, the three-phase, and the three-phase recycling water oil extraction system, respectively (Peri and Proietti, 2014; Amirante, 2003). Obviously, the addition of different water volumes in the different extraction systems determine a different compound concentration which can vary from 5% to 12% of the OMWW (Amirante, 2003).

The chemical-physical characteristics of OMWW depend on the climatic conditions of the cultivation area, the cultivar, the cultural practices, and the ripening stage of the olives. These parameters influence not only the composition of olives, but also the type of processing adopted (Ilarioni and Proietti, 2014; Nasini and Proietti, 2014; Aviani et al., 2012; Suzzi and Tofalo, 2009; Jarboui et al., 2008; Borja et al., 2006; Ammar et al., 2005). OMWW constitutes a suspension containing many organic compounds such as sugars, amino acids, tannins, organic fatty acids, lipids, alcohols, pectins, carotenoids, and phenols (Jarboui et al., 2008; Borja et al., 2006; Rinaldi et al., 2003; Lessage-Meessen et al., 2001; Mulinacci et al., 2001; Sierra et al., 2001; Zenjari and Nejmeddine, 2001). OMWW contains also significant amounts of inorganic salts (mainly potassium but also phosphates, sulfates and chlorides). Most of these salts are soluble, whereas a fraction of 20% is insoluble (e.g., carbonates, phosphates, and silicates) (Mekki et al. 2013; Justino et al., 2012; Ouzounidou et al., 2008; Rinaldi et al., 2003; Lessage-Meessen et al., 2001; Paredes et al., 1999). The dark brown color of OMWW has been related to the presence of polymeric phenolic compounds (~10% of the organic matter) that display a lignin-like structure (Justino et al., 2012; Zbakh and El Abbassi, 2012; Tsioulpas et al., 2002). However, it does not contain synthetic chemicals or additives since it is exclusively derived by a physical-mechanical extraction process. Several analysis of OMWW samples have shown the absence of toxic substances or pathogenic microorganisms and viruses (typically contained in urban wastewaters) (Bedbabis et al., 2015; Zbakh and El Abbassi, 2012; Vivaldi et al., 2013). Nevertheless, the high content of phenols (caffeic, protocatechuic, *p*-coumaric and *p*-hydroxybenzoic acids, hydroxytyrosol, and tyrosol) induces a strong antimicrobial activity (Barbera et al., 2013; Mekki et al., 2013; Karpouzas et al., 2010; Quaratino et al., 2007). Despite its latter activity, OMWW contains many (up to 105 cells/mL) bacteria, aerobic and facultative anaerobic, yeasts, and molds (Suzzi and Tofalo, 2009). Indeed, their presence in the OMWW storage tanks promotes detoxification processes. For instance, some microorganisms, such as those belonging to the *Pseudomonas* genus are able to grow in the presence of phenolic compounds (Rabenhorst, 1996). This fact results in a decrease of BOD (biological oxygen demand) and total suspended solids values of OMWW (Hanafi et al., 2011; Suzzi and Tofalo, 2009; McNamara et al., 2008; Roig et al., 2006; Mantzavinos and Kalogerakis, 2005). Besides, fungi species (e.g., *Pleurotus*) can detoxify OMWW removing oxidizing aromatic compounds with laccases (Ammar et al., 2005; Aggelis et al., 2003; Tsioulpas et al., 2002).

OMWW has a high pollution index due to its acidic nature as well as the high concentrations of salts, organic matter, and phenolic compounds (catechol, hydroxytyrosol, tyrosol, and oleuropein). For instance, BOD values of OMWW ranges between 12,000 and 63,000 mg/L (but can reach more than 100,000), while COD (chemical oxygen demand) values range between 80,000 and 200,000 mg/L (Tardioli et al., 1997). These values are 200–400 fold higher than municipal water. For agricultural use, an effluent of good quality should be characterized by much lower BOD and COD values, for example, 10–20 and 30–60 mg/L, respectively (Amirante, 2003; Suzzi and Tofalo, 2009). Due to this pollution index, the discharge of OMWW into water streams is prohibited because it may result in a significant damage to the aquatic flora and fauna. In addition, the mucilage present in suspension (10,000–15,000 mg/L) increases the viscosity and the adhesiveness of OMWW. Last but not least, OMWW has a low biodegradability, since the BOD/COD ratio ranges from 0.25 to 0.30, due to the presence of organic substances that are not biologically degradable in short times (Amirante, 2003). The oxygen demand induced by the distribution of OMWW in the soil can be satisfied by the dissolved oxygen that comes into the ground or through the air that is naturally diffused in the soil surface layers. However, most of the available oxygen remains involved in the transformation of ammonia into nitrate, while only a small part is used to break down the BOD value of the OMWW (Amirante, 2003). In any case, the decomposition of OMWW organic matter is attributed to soil bacteria rather than to enzymes and bacteria present in the OMWW. This means that the disposal of such water through fertigation is, in general, the most effective way to reduce its BOD value (Moraetis et al., 2011; Mechri et al., 2008; Mekki et al., 2006, 2007; Sierra et al., 2001; Zenjari and Nejmeddine, 2001; Morisot, 1979). For these reasons, national laws concerning the protection of surface and ground-waters against pollution (e.g., in Italy the law n. 319/76 and subsequent modifications) ban the discharge of the OMWW or urban sewage in case are not proper treated. The seasonality of OMWW production and the characteristics of biological toxicity make difficult their management treatment in the urban wastewater plants (Toscano and Montemurro, 2012). To reduce the polluting potential of OMWW, various wastewater treatment technologies have been proposed (incineration, ultrafiltration, concentration, etc.) which, however, are not able to reduce in an economically sustainable way the pollutant rate to the levels set by law (Toscano and Montemurro, 2012; Yay et al., 2012; Hanafi et al., 2011; Moraetis et al., 2011; Paraskeva and Diamadopoulos, 2006). Furthermore, these treatment systems produce sludges difficult to dispose and, in addition, their energy consumption is very high (Yay et al., 2012; Paraskeva and Diamadopoulos, 2006). Finally, although the lagooning treatment does not require large investments, it is not a functional practice, as it has a very slow activity and produces bad smells that diffuse on large surfaces (McNamara et al., 2008; Roig et al., 2006).

5.2.1 THE AGRONOMIC PROPERTIES OF THE OMWW

OMWW spreading is not only considered as a simple disposal system, but also as an agronomic technique aimed at improving the physical and chemical properties of soil and plant nutrition (Chartzoulakis et al., 2010; Belaqziz et al., 2008; Mekki et al., 2006; Amirante, 2003). From the agronomic point of view, the most important properties of OMWW are the low pH value, the presence of nitrogen (almost exclusively in organic form), the high concentrations of potassium and phenols, and also the presence of yeasts, fungi, and especially, cellulolytic bacteria. The beneficial effect of phenols is due to their antioxidant and bacteriostatic effect, which can affect the oxidative cycles of soil organic and mineral nutrients, such as the oxidation of ammonium to nitrite and nitrate (Isidori et al., 2005). Besides, the dissolved organic matter (DOM) of OMWW is contained in both stabilized and not stabilized forms. This is an important characteristic, as organic fertilizers stability and DOC concentration are strongly correlated (Said-Pullicino and Gigliotti, 2007). DOC may affect the turnover of soil organic matter, as well as the microbial activity of amended soils (Said-Pullicino et al., 2007). Pezzolla et al. (2015) demonstrated that the addition of a not-well stabilized organic fertilizer to soil caused a significant increase of DOC and CO_2 emissions, thus suggesting that the labile C causes a rapid increase of microbial respiration after amendment.

As noted earlier, OMWW spreading could also be accompanied with negative effects due to its composition. However, the soil acts as a filter that retains suspended solids and its clay, humus and newly formed organic colloids fix the mineral salts (carbonates, sulfates, calcium humate, hydroxides of iron, and aluminum, etc. are formed) (Di Bene et al., 2013; Tamburino and Barbagallo, 1989).

In some cases, soil microorganisms favor the rapid decomposition of organic constituents, such as lipids and phenols (Riolfatti, 1983). Indeed, the degradation of phenols is enhanced by their exposure to light and air and, after 1–4 months, their levels in the soil could be reduced to normal values (Barbera et al., 2013; Chartzoulakis et al., 2010; Saadi et al., 2007; Proietti et al., 1995; Potenz et al., 1980). In any case, phenols are typically retained by the soil colloids and even under heavy rainfall, there are no leaching; only in fast-draining soil and in the case of a very rainy season, phenolic substances could move in the deeper soil layers (Piotrowska et al., 2011; Chartzoulakis et al., 2010; Saadi et al., 2007; Gamba et al., 2005; Amirante, 2003; Sierra et al., 2001). Phenolic compounds constitute a pollution factor only for surface and ground water bodies. In the case of soil, they are precursors of humic substances. The latter are the most active fractions of soil organic matter; they have an important role in soil fertility and protect plant growth from contaminants like pesticides and heavy metals (Toscano and Montemurro, 2012). Finally, the odor release following OMWW spreading is very low and less offensive than other waste materials (Proietti et al., 1995).

5.2.1.1 Fertigation

Clay-loam soils, with a high percentage of clay and high cation-exchange capacity (CEC), seem to be the best choice for fertigation with OMWW. In addition, the carbonate content of the soil is important in order to exert efficient buffer power and avoid drastic changes in the pH value. The latest could cause the immobilization of different macro- and micronutrients and accelerate changes in the microbiological activity of the soil (IOOC, 2004). In order to evaluate the effects of OMWW fertigation, it is necessary to study not only crop production, but also its effects on the soil chemical, physical, and biological properties (Barbera et al., 2013; Amirante, 2003; Cicolani et al., 1992; Morisot, 1979). For example, after few weeks of OMWW spreading, some physical soil characteristics, such as porosity, stability of aggregates, as well as water retention and movement, may change.

In particular, OMWW spreading may initially decrease the water infiltration due to the water repellency of the fatty particles adsorbed in the soil surface layers. However, this action attenuates after the decomposition of these substances in a short time. Subsequently, the porosity is improved, while air and water dynamics in the soil are increased and surface soil crust formation is reduced. These facts lead to reduce erosion in the sloping land and water logging in the plain (Mahmoud et al., 2010; Papini et al., 2000).

The ability of the soil to retain water increases few weeks after OMWW spreading. This is attributed to the enrichment of the organic matter and the increase of soil porosity (Barbera et al., 2013; Tarchitzky et al., 2007). Soil aeration and drainage improve due to the increase of stability of the aggregates after few weeks from the application. The binding substances (e.g., carbohydrates contained in the DOM) derive from the decomposition of the OMWW organic fraction (Pagliai et al., 2001; Papini et al., 2000).

OMWW has an acidic nature (pH value ~5) and thus causes slight but temporary modifications of the soil pH value after spreading. However, this parameter returns to normal or slightly higher values after some months (Piotrowska et al., 2011; Proietti et al., 1995). This change is attributed to the production of ammonia following bacterial breakdown of organic matter present in the OMWW (Potenz et al., 1980; Della Monica et al., 1978).

The richness of OMWW in potassium, nitrogen, phosphorus, magnesium, and organic matter can improve soil fertility (Montemurro et al., 2011; Chartzoulakis et al., 2010; Mechri et al., 2008; Sierra et al., 2001, 2007; Proietti et al., 1995). When OMWW is spread on soil, the organic matter shows initially strong modifications. After that, due to the microbial activity, the degradation of organic matter leads to improved physicochemical properties of the soil. At the same time, organic nitrogen evolves into mineral forms (Della Monica et al., 1978), whereas the available phosphorus and the exchangeable potassium increase (Proietti et al., 1988, 1995). In conclusion, fertigation with OMWW can substitute some of the nutrients provided with the chemical fertilization. For instance, Di Giovacchino and Seghetti (1990) estimated that the spreading of 80 m^3 OMWW s^{-1} provides to soil approximately 3000–6000 kg of dry organic matter, 25–50 kg of nitrogen, 15–30 kg of phosphorus, and 80–160 kg of potassium. Consequently, the implementation of OMWW allows to reduce phosphorus, potassium, and nitrogen fertilization.

Although a number of studies have demonstrated the effects of OMWW on soil nitrogen content, relatively poor is the literature investigating the chemical nitrogen forms of OMWW and their transformation once in the soil. OMWW typically contain organic forms of nitrogen and only traces are present in its mineral forms (both nitrate and ammonia). This organic N is mainly water-soluble and then is rapidly available for soil microorganisms. After a short time, this organic N is mineralized becoming available for plant nutrition. The relatively rapid transformation of organic N to nitrate in soil may cause some environmental problems, related to the nitrate transport in surface water or its leaching to groundwater. When a plant nutrient, like N, reaches surface water, eutrophication may occur, as a consequence of the aquatic vegetation increase. Whereas, the leaching of nitrates to groundwater leads to the contamination of drinking water. EU is particularly sensitive to this issue, for example, the so called "Nitrate Directive" was emitted (1991) to prevent drinking water pollution. When OMWW is applied to the soil, leaching of N is avoided, as is present mainly in the organic form or as ammonium which, positively charged, is adsorbed by the soil colloids. Thereafter, the ammonium is oxidized to nitrate that, negatively charged, is subjected to leach. This phenomenon occurs when nitrate concentration in the soil exceeds the needs of the plant.

5.2.1.2 Biological Activity

OMWW spreading influences also the biological activity of the soil, however, information about changes in soil microbial communities are generally lacking (Karpouzas et al., 2009). There are only few studies dealing with the identification of microbial communities in OMWW and thus its biotechnological potential has not been fully exploited (Ntougias et al., 2013). Initially, OMWW amendment causes only a slight decrease of the total microflora, which indicates a substantial absence of biological toxicity phenomena. Subsequently, the microflora increases, mainly due to the contribution of organic matter together with an enrichment of nitrogen-fixing bacteria (Kotsou et al., 2004; Mekki et al., 2006; Piotrowska et al., 2006; Gamba et al., 2005; Di Serio et al., 2008). The changes, however, are transitory. Some studies have used cultivation-enumeration methods or enzymatic activity to study the effect of OMWW on soil microorganism. A substantial increase in coryneform bacteria and a decrease in *Bacillus spp.* has been reported by Paredes et al. (1987). Mekki et al. (2006) reported a significant increase in soil actinomycetes, spore-forming bacteria and soil fungi and a decrease in bacterial nitrifiers. Proietti et al. (1995) by spreading 400 m^3 OMWW ha^{-1}, in the upper soil layers (0–20 cm depth) observed moderate changes in microbiological activity with respect to the control plots. In particular, the autochthon microorganisms increased after 35 days of treatment, but after 65 days the difference between treated and untreated soils decreased and the relative ratio (treated/untreated) between the beginning and the end of experimental period was very low. Paredes et al. (1987) observed a significant increase of denitrifying and nitrifying bacteria after OMW soil application, whereas García-Barrionuevo et al. (1992) reported a stimulatory effect of OMW on nitrogen-fixing bacteria. Mekki et al. (2006) found a significant reduction in the number of soil nitrifying bacteria at the highest OMWW dose (400 m^3/ha) applied. Using a biochemical approach, Mechri et al. (2007) observed that the addition of more than 30 m^3 OMWW ha^{-1} resulted in an increase of soil fungi, Gram-negative bacteria and actinomycetes, while Gram positive bacteria decreased significantly after 1 year. Kotsou et al. (2004) reported that OMWW application resulted in a shift of the bacterial community toward copiotrophic bacteria (r-selected), which comprise the first colonizers of newly added organic matter. Interestingly, OMWW amendment caused minor long-term effects on soil microflora, whereas there was no evidence of any inhibitory effect on the growth of soil microorganisms.

Using high doses of OMWW in orchards and crops, a positive effect on weed control has been observed due to its herbicide action (Proietti et al., 1988; Boz et al., 2003, 2009; Albay and Boz, 2003). In particular, the gallotannins, *p*-cumaric, etc. have an inhibiting effect on germination, growth, and development of different weeds (Lodhi, 1976). OMWW suppressed purslane, redroot pigweed, and jungle rice in okra, and little seed canary grass, annual bluegrass, wild chamomile, and shepherd's purse in faba bean and onion. In general, a rate of 10–20 t/ha OMWW are adequate for weed control and crop safety. In an experiment carried out in the semi-arid environment of Southern Italy, the application of OMWW without preliminary treatments, positively affected the yield of ryegrass and proteic pea. The clover crop showed a species-specific sensitiveness, but the OMWW applications increased the protein content compared to the untreated plots (Toscano and Montemurro, 2012).

5.2.2 TECHNIQUES OF STORAGE AND SPREADING OF OMWW

In order to avoid unwanted side effects of OMWW spreading, it is necessary to follow the correct spreading techniques (IOOC, 2004) according to the corresponding national legislation. For instance, in Italy the producer who intends to fertigate with OMWW (or to amend soil with pomace) must submit annually a technical report to the authority at least 30 days before starting OMWW spreading. The technical report must contain information on geomorphological, hydrologic, and other characteristics of the receiver, spreading time planned, as well as on systems used to ensure a suitable distribution (Barbera et al., 2013; IOOC, 2004).

5.2.2.1 Spreading Load

In some countries, the national or local laws establish the maximum spreading load allowed. In Italy, the agronomic use of OMWW derived from extraction system by pressure (traditional method) is allowed up to 50 m^3/ha annually, whereas the use of those derived from the extraction system by centrifugation of the paste (modern method) are allowed up to 80 m^3/ha. To determine the spreading load, it is necessary to take account of water infiltration of the soil. If it is low (<5 mm/h), low spreading load should be applied to prevent surface runoff or phenomena of water logging in the plains (Chiesura et al., 2005). Spreading load have to be reduced even for soils with high water infiltration (> 150 mm/h), since excessive percolation of OMWW organic fraction can cause pollution of groundwater. In general, a good suggestion is to establish a spreading load value depending on the polluting charge of the OMWW, which may be considered as proportional to the total solids content. In order to

Table 5.1 The Rule of 12 for OMWW Spreading Loads (Peri and Proietti, 2014)

Total solids (%)	3	4	5	6	7	8	9	10
Tens of cubic meters of OMWW ha^{-1}	9	8	7	6	5	4	3	2

quantify the spreading load as a function of total solids content in OMWW, the "rule of 12" (Table 5.1) can be applied as a rough indication (Peri and Proietti, 2014).

The first line in the Table 5.1 indicates the total solids concentration in the OMWW, while the second line indicates the suggested quantity of OMWW that can be spread on the soil. Spreading load can be increased by 50% on tree crops, namely olive groves. Spreading load can be substantially increased when a physical-chemical pretreatment of OMWW occurs. The latest can partially remove suspended solids, dispersed macromolecules, and emulsified oil. In general, the elimination of suspended solids also results in a significant reduction of polymeric phenolic compounds leading to attenuation of their antimicrobial and phytotoxic effects (Peri and Proietti, 2014).

5.2.2.2 Physical and Physical-Chemical Treatments of OMWW

In some cases, it may be appropriate to pretreat the OMWW by sedimentation processes in order to facilitate the operations of spreading and reducing antimicrobial and phytotoxic effect (Peri and Proietti, 2014). Sedimentation is the separation of solid particles from the liquid in which they are suspended, under the action of gravity, due to the difference in density (Khoufi et al., 2007). Particles of very small diameter (in the range of 10 μm or less) form stable dispersed systems and can be eliminated only through the application of flocculation, flotation, or filtration (Achak et al., 2009). Flocculation is a process of adhesion where the suspended solid particles form large-sized clusters. This phenomenon takes place spontaneously in OMWW, but at a very slow rate. Nevertheless, it can be greatly accelerated by the addition of flocculating agents, which are synthetic polymers with charged molecular chains (polyelectrolytes) that are able to link particles and single macromolecules into clusters. Flotation is a process in which air is finely bubbled through the OMWW. Tiny air bubbles adhere to the suspended materials due to surface adhesion, so that the bubble-solid complexes have a low "apparent" density and can readily surface as a floating air-solid layer (Peri and Proietti, 2014). This layer is separated by a skimming and overflow mechanism. Both flocculation and flotation tend to drag oil droplets along in their movement with further reduction of the polluting potential. Coarse filtration through sand layers or fine particulate media can also be used to remove suspended materials. In addition, when charcoal or activated carbon are used in the filter media, polyphenols can be absorbed and removed very effectively, with substantial reduction of antimicrobial and phytotoxic effects (Peri and Proietti, 2014). The solid residues from wastewater pretreatments may be mixed with pomace during composting or spread on the soil directly. The water can be used for irrigation. Another possible treatment, as reported before, is spontaneous evaporation in ponds in the open (lagoons). It is one of the oldest methods of OMWW treatment and is still used in some countries with a warm climate, such as in southern Europe and Northern Africa. Although lagooning is very simple and inexpensive, it is a process very slow and there are some serious drawbacks:

1. contamination of groundwater can take place if the lagoon is not properly isolated;
2. decomposition of organic materials produces significant odor pollution; and
3. lagoons attract insects and contribute to insect (and rodent) infestation.

Consequently, lagooning is rapidly disappearing (Roig et al. 2006; Peri and Proietti, 2014).

5.2.2.3 Risks of Groundwater Pollution

The application of high amounts of OMWW in a short time and the presence of channels in the soil could induce leaching of ammonia, ureic, and other organic nitrogen forms, which are generally retained by the adsorption complex of the soil and are round only in traces in drainage water (Tamburino and Barbagallo, 1989). Catalano e De Felice (1989) found a significant metabolizing activity in soil treated with OMWW, as noted by the degradation of organic load (75–80%) 60 days after spreading. In another study (Morisot, 1979), the high organic load of OMWW degraded relatively quickly in agricultural soil, when the annual distribution rate was less than 100 m^3/ha.

5.2.2.4 Time of Spreading

Time of OMWW spreading should comply with regional laws (if any). Based on the results of numerous studies, spreading should be performed 2–3 months before sowing for the spring-summer crops (especially maize and sunflower). For the autumn-winter cereals (wheat and barley), spreading could be performed with crops in the tillering phase, preferably without exceeding the dose of 40 m^3/ha per year (Barbera et al., 2013). For tree crops (especially olive), the major limit is represented by the slope of treated soils that in rainy periods can cause runoff or accessibility problems for spreading machineries. To prevent leaching, OMWW must not be applied on days when there is rain or the soil is frozen (IOOC, 2004).

5.2.2.5 Storage and Transport

OMWW can be collected from the decanters in mobile tanks, spread directly on the field or stored in basins (IOOC, 2004). The basin is the simplest and economic system of storage, but risks of soil and subsoil water contamination exist (Toscano and Montemurro, 2012). For this reason, storage of OMWW requires prior notice to the competent authorities, and should be implemented in tanks or basins suitably waterproofed according to specific rules (size, shape, materials, coverage, etc.). It allows avoiding spreading until the rains persist and the soil is saturated with water. Storage tanks or basins should be large enough to contain OMWW during periods when spreading could be prevented by agronomic or climatic causes or regulations (Roig et al., 2006; McNamara et al., 2008). In general, legislation prohibits mixing OMWW (or pomace) with other effluents and waste. During storage, the concentration of pollutants in the OMWW decreases by ~15% due to the fermentation of organic compounds (sugars) by microorganisms. Indeed, microorganisms belonging to the genus *Pseudomonas* are able to grow in the presence of phenols (Amirante, 2003; Rabenhorst, 1996; Proietti et al., 1995). Consequently, the BOD, the amount of suspended solids and the nitrogen concentration tend to decrease, whereas pH value increases (Amirante, 2003). Finally, the transport of OMWW (as well as wet pomace) should be conducted with appropriate means of transport to prevent spills and hygienic-sanitary incidents.

5.3 AGRONOMIC USE OF POMACE

5.3.1 CHARACTERISTICS OF POMACE

The pomace is the solid by-product, originating from the processing of olives, consisting of the fibrous part of the fruit, pit, ~5% of residual oil, and an amount of water ranging from 25–30% using pressure oil extraction system (traditional method), to 50–55%, 55–70%, and 50% using "three-phase," "two-phase," and "water saving" centrifugation systems, respectively (Toscano and Montemurro, 2012; Suzzi and Tofalo, 2009; Amirante, 2003). The two-phase system reduces the consumption of water and does not produce OMWW, but just a wet pomace with a slightly acidic pH, a high concentration of organic matter (especially fibers), potassium, oil, and phenols. This pomace has a wet soft consistency that restricts its transportation (Suzzi and Tofalo, 2009). The total amount of pomaces produced per 100 kg of olives is around 87 and 62 kg in "two-phase" (called wet olive pomace) and "three-phase" (called olive pomace) centrifugation system, respectively (Peri and Proietti, 2014). With the exception of the water content, on a dry bases, the pomace has similar characteristics in the pressure and centrifugation oil extraction systems. Pomaces have an acid pH (around 5.2), a high content of phenolic substances (1–3% of dry matter), and of fats (8–14% of dry matter) (Mechri et al., 2011; Alburquerque et al., 2004).

5.3.2 THE AGRONOMIC PROPERTIES OF THE POMACE

Pomace has a similar composition, compared to a soil organic amendment, and thus can be used for agronomic purposes (Toscano and Montemurro, 2012; Amirante, 2003). As a soil amendment, olive pomace can be directly incorporated in the soil. This is obviously a very simple and direct utilization and hence the very frequent one (Federici et al., 2009; Alburquerque et al., 2007; Roig et al., 2006). The use of pomace as organic amendment has shown some problems due to their high organic loads and mineral salt contents, low pH and the presence of phytotoxic compounds (Proietti et al., 2015; Gigliotti et al., 2012; Del Buono et al., 2011; Canet et al., 2008; Nastri et al., 2006). Pomace may also be composted (mixing them with other by-products from food chains such as stalks, orchard pruning residues, straw, etc.) prior their agronomic use. These matrices increase the porosity of the mass and making it suitable for composting. To a greater degree than pomace, compost improves the physicochemical characteristics of the soil, adding more stabilized organic matter and plant nutrients (especially potassium, nitrogen, and phosphorous) in a consistent proportion (Proietti et al., 2015; Gigliotti et al., 2012; Del Buono et al., 2011; Sellami et al., 2008). Besides, the agronomic use of the pomaces allows to increase soil organic C content, and so C sequestration.

5.3.3 SOIL AMENDMENT

The use of pomace on grass-covered soils or incorporated at a depth of 10–25 cm in tilled ones is a practice widespread in olive groves since, in addition being economical and easily practicable, it has positive effects on the characteristics of the soil and on crops (Lozano-Garcìa et al., 2011; Niaounakis and Halvadakis, 2006; Amirante, 2003). In Italy, spreading of pomace on agricultural soils can be made in accordance with the law no. 574 of 1996.

5.3.3.1 Physical Characteristics

The spreading of organic amendments improves soil physical properties, particularly if the fertilizer is well stabilized and shows a high content of humic-like substances. In this case, an increase in soil porosity, field and water-holding capacity, as well as a decrease in bulk density are expected (Giusquiani et al., 1995). Kavdir and Killi (2007) demonstrated that the distribution of pomace on soil also improves its structure and stability and increases the water-holding capacity. Moreover, the increase in aggregate stability makes the amended soils less susceptible to the erosion. The increased aggregate stability may be attributed to an increase in the organic matter, particularly in carbohydrate deriving by the application of pomace to the soils (López-Piñeiro et al., 2007).

5.3.3.2 Chemical Characteristics

Soil amendment with pomace favors over time an increase of soil chemical fertility, as it increases organic matter and nutrient content (total nitrogen, available phosphorus and exchangeable potassium) without changing significantly its pH value and salinity (Ferarra et al., 2012; Chartzoulakis et al., 2010; Mechri et al., 2008; Uygur and Karabatak, 2009; Cucci et al., 2008; López-Piñeiro et al., 2008; Montemurro et al., 2004). For example, in an experiment conducted in central Italy, spreading of large amounts of pomace (50 t/ha) in an olive grove for 4 consecutive years caused a slight decrease in pH, an increase in organic matter content, total nitrogen, exchangeable potassium, and magnesium as well as in available phosphorus (Nasini et al., 2013). In a short and medium term experiment, López-Piñeiro et al. (2008) observed a significant increase of organic C in a soil treated for 3 years with a two-phase OMW. This result was in consistent with those obtained by other authors (Madejón et al., 2003; Montemurro et al., 2004).

Another study was carried out for 3 years in an olive grove (Central Italy) to investigate the effects of OMW-amendment and its derived-compost on some soil chemical properties (Proietti et al., 2015). In both soils amended with fresh or composted pomace, no increase in organic C in the soil was found, indicating a probable priming effect. Nasini et al. (2013) observed the same results in a trial where a different typology of pomaces was tested. In both experiments an increase in the microbial activity was found, demonstrating the rapid utilization of the organic matter by soil microorganisms. These results suggest that the use of pomace may improve soil organic C content, but it is necessary to keep in consideration the quantity and quality of organic matter added with the amendments, as well as the techniques of application of pomace. In fact, organic matter mineralization is enhanced when pomace is distributed on soil surface without soil incorporation (Proietti et al., 2015).

Brunetti et al. (2005) founded that the application of pomaces (fresh and exhausts) for 2 years led to a significant linear increase of soil total N. In the first year, the greatest increase was attributable to organic N, but in the second year also inorganic N concentration increased with increasing rates of wastes, compared to the control. These results were also confirmed in a 2-year greenhouse experiment. Besides, when an amendment is applied to the soil, an increase in available P content is expected, as the main result in the chemical fertility improvement. The available P increase is the consequence of phospho-humic complexes formation, which avoids P insolubilization in both acidic and calcareous soils (Giusquiani et al., 1995). Significant increases in soil available P were also found when pomaces are used as soil amendments. These results were obtained by some authors by adding different kind of pomaces, composted or not (López-Piñeiro et al., 2008; Paredes et al., 2005; Madejón et al., 2003). Proietti et al. (2015) also founded that the P increase occurs just in the first 15 cm of the treated soil, whereas no differences were accounted in the 15–30 cm layer between the soil control and treated soils. However, other studies (Nasini et al. 2013; Brunetti et al., 2005) indicated that the application of different types of olive pomace did not affect the available P content.

Olive pomace shows always a high quantity of K salts. Therefore, an increase in exchangeable K content is expected in pomace-amended soils. For instance, Proietti et al. (2015), demonstrated an increase in exchangeable K compared to the control after 3 years of treatment, 4.9-fold greater in soil amended with fresh pomace and just 1.3-fold in soil amended with the composted pomace. The increase in available P and exchangeable K obtained in pomace-treated soils confirms that olive pomace could be used as an alternative source of P and K in soils. This is important from an agronomic and economic point of view. In fact, P applied via olive pomaces is "protected" by insolubilization phenomena and results more effective in soil P enrichment than mineral fertilizers. On the other hand, an adequate

K fertilization allows better tolerance to drought, which is very frequent under Mediterranean conditions. The proper use of pomaces may be considered a valuable method to increase soil chemical fertility, as interesting rates of N, P, and K are added to the soil with a cost lower than chemical fertilizers.

5.3.3.3 Biological Activity

It is generally accepted that most potential changes in soil chemistry and microbiology are likely to occur during the first weeks of organic amendments application (Blagodatskaya and Kuzyakov, 2008). Nasini et al. (2013), after some months of amendment, found that the addition of pomace did not dramatically increase the number of total microorganisms, thus suggesting that this organic amendment could be only a short-term substrate for the soil microbiota. Indeed, spreading of olive mill by-products may represent a temporary enrichment with a readily available C source for microbial growth (Di Serio et al., 2008; Mekki et al., 2006; Perucci et al., 2006; Saadi et al., 2007). OMW are known to have antimicrobial properties, mainly ascribed to their high phenolic content (Roig et al., 2006; Linares et al., 2003). However, Sampedro et al. (2009) showed that the reduction in cultivable bacteria following addition of pomace to soil was limited to the first 48 h of incubation. Environmental concerns related to the use of pomace as a soil amendment can be excluded considering that at the 15–30 cm depth, the chemical and microbiological differences between treated and control soil were minimal or completely absent (Nasini et al., 2013). Thus, it can be hypothesized that pomace spread on the surface of the soil almost exclusively affects the upper layer (Ferri et al., 2002; Cayuela et al., 2008).

5.3.3.4 Weed Control

Cayuela et al. (2008) studied the allelopathic effects of the extracts of pomace derived from two-phase centrifugation oil extraction system, both fresh or composted, on seed germination of *Amaranthus retroflexus* L., *Solanum nigrum* L., *Chenopodium album* L., and *Sorghum halepense* (L.) Pers., which are highly invasive and globally distributed weeds. The extracts of two-phase pomace and immature compost substantially inhibited germination of *A. retroflexus* and *S. nigrum*, whereas extracts of mature composts only partially reduced the germination of *S. nigrum*. Camposeo and Vivaldi (2011) in a 2-year-long experiment, used deoiled olive pomace for soil management in a young super high-density olive orchard. In particular, they compared row mulching with deoiled olive pomace, plastic mulching materials (polypropylene tissue, polyethylene film), chemical and mechanical weeding. During the test, deoiled olive pomace mulching remained as solid layer and increased significantly the available K soil content that doubled with respect to other soil management methods. In the second year after application, where mulching material remained solid, soil mulching significantly improved stomata conductance, net assimilation and water use efficiency with respect to both chemical and mechanical weeding. Nonetheless, olive tree growth, fruit per plant, and olive oil content in the mulched plots were equivalent to those in weeded plots.

5.3.3.5 Effect of Pomace-Amendment on Crops

There are few studies concerning the effect of fresh pomace as an organic amendment on soil (Toscano and Montemurro, 2012; Rodrìguez-Lucena et al., 2010; Cabrera et al., 2009; López-Piñeiro et al., 2008). After amendment with different amount of pomace, crop yield improvement has been recorded in most of them. For example, López-Piñeiro et al. (2006) in winter wheat crop grown in greenhouse, observed an increase in crop yield up to 198%. Similarly, Brunetti et al. (2005) found an increase in *Triticum turgidum* L. production related to kernel weight, kernel number per square meter, and soil organic matter content after pomace amendment (10 and 20 t/ha). Tajada and Gonzalez (2004), applying 0, 10, 20, 30, and 40 t/ha of wet olive pomace on a maize crop for 2 years, found that yield parameters in the second experimental season were better than those of the first experimental season, due to the residual effect of the organic matter applied in the first season. An increase in grain gross protein, grain soluble carbohydrate content, numbers of grain per corncob and yield was observed. Nasini et al. (2013) found an increase in the shoot growth, canopy volume and pruning weight in pomace amended (50 t/ha) olive trees after 4 years of application. Alburquerque et al. (2007) reported a similar increase in plant growth due to the fertilizing effects of a compost made with pomace and cotton gin waste. Nasini et al. (2013) found also an increase in fruit growth and fruit yield per tree, without negative effects on the oil content. A positive effect on olive yield as a consequence of the increase in organic matter content, total N, and available P and K, was also found by López-Piñeiro et al. (2008), with both oiled and deoiled wet olive pomace used as soil amendment. Proietti et al. (2015) treated for 3 consecutive years, an olive grove with wet olive pomace (composted or not) and observed an increase on the vegetative activity and productivity of the olive trees. The overall absence of effects on the oil chemical characteristics observed in this case indicated that wet olive pomace application (composed or not) to the soil did not lead to a negative impact on oil quality.

5.3.4 MODALITIES OF STORAGE AND SPREADING OF POMACE

The Italian law (no. 574 of 1996 and subsequent modification and integration) states that the rate of pomace allowed to be spread is 50 m^3 when it is produced by pressure (traditional method) olive mills and 80 m^3 when produced by centrifugation (modern method) olive mills. Nasini et al. (2013) and Proietti et al. (2015) did not observed negative effects on soil and tree in olive groves after spreading 50 t/ha of pomace. Lower dosage were used for crop, for example, Brunetti et al. (2005) applied 20 t/ha on *T. turgidum* and Tajada and Gonzalez (2004) applied 40 t/ha on maize. Both pomaces and OMWW contain N, mainly as ammonium salts. Therefore, when applied to soils, ammonium (positively charged) is adsorbed by soil colloids and leaching is avoided. After that, the ammonium is oxidized to nitrate (negatively charged and then not adsorbed by soil colloids), being subjected to leaching. Nitrate leaching occurs when its quantity in soil solution exceeds the plant demand. The best period for the pomace spreading is autumn-winter (López-Piñeiro et al., 2007) and in any case, before vegetative growth resumes. Spreading should be avoided in rainy days, as well as on water saturated or frozen soils. Due to the presence of sugar, pomace is susceptible to ferment, although with difficulty due to the presence of polyphenols. As in the case of OMWW, the storage and transport of the olive pomace must be done in compliance with current legislation, as well as in compliance with safety requirements. For instance, the containers for the storage of wet pomace must be properly waterproofed and covered with tarpaulins to avoid percolation and infiltration.

5.4 CONCLUSIONS

Most of the experiments carried out on the fertigation with OMWW showed no real risks of irreversible soil degradation or toxic effects on crops, except in the case of excessive doses or irrational distributions. Consequently, the controlled spreading of OMWW on agricultural soil can be considered as an alternative technique to the use of chemical fertilizers, since it provides soil with fertilizing substances.

Table 5.2 The Main Strengths, Weaknesses, Threats, and Opportunities Related to the Reuse of OMW as Soil Amendment

Strengths	Weakness
• Spreading is regulated and permitted by specific laws in different countries. • Timely disposal of the waste. • Simple equipment and technology are required. • Low costs. • Agronomic and environmental advantages documented, including the storage of C (with potential possibility of acquisition of "C credits").	• Environmental risks if conducted in unsuitable conditions and using irrational techniques. • Need of suitable agricultural soils. • Poor fertilizer value of OMWW and not composted pomace and agronomic short-term effect for OMWW. • Possible phytotoxic actions by irrational spreading of OMWW. • Spreading logistics are complicated due to the low accessibility of fields in some periods of the year.
Threats	Opportunities
• Environmental pollution with irrational spreading.	• Increase in agronomic benefits with composting. • Valorization of compost derived from pomace in nurseries after a partial or total replacement of the peat. • Accounting of "C credits".

Similar to OMWW, olive pomaces (both pomace and wet pomace) can be spread on agricultural soils, even though it is preferable to use after composting (Amirante, 2003; Proietti et al., 2015). In addition, the reuse of OMW leads to a decrease in pollutant wastes, improves the chemical, biological, and physical soil properties and can enhance crop yields. The application of OMWW and olive pomace compost is an answer to limit loss of soil fertility when soil organic matter content falls below 1% (IOOC, 2004). This is very important where the environmental conditions (hot and arid) make particularly hard to maintain soil fertility with the dangerous consequences of desertification (IOOC, 2004) (Table 5.2).

REFERENCES

Achak, M., Mandi, L., Ouazzani, N., 2009. Removal of organic pollutants and nutrients from olive mill wastewater by a sand filter. J. Environ. Manag. 90, 2771–2779.

Aggelis, G., Iconomou, D., Christou, M., Bokas, D., Kotzailias, S., Christou, G., Tsagou, V., Papanikolaou, S., 2003. Phenolic removal in a model olive oil mill wastewater using *Pleurotus ostreatus* in bioreactor cultures and biological evaluation of the process. Water Res. 37 (16), 3897–3904.

Albay, F., Boz, Ö., 2003. An Investigation into the Efficacy of Olive Processing Waste and Corn Gluten Meal for Controlling Weeds in Strawberries. 7th Mediterranean Symposium, May 6-9. European Weed Research Society, Adana/Turkey, pp. 77–78.

Alburquerque, J.A., Gonzalvez, J., Garcìa, D., Cegarra, J., 2004. Agrochemical characterization of "alperujo", a solid by-product of the two phase centrifugation. Bioresour. Technol. 91 (2), 195–200.

Alburquerque, J.A., Gonzalvez, J., Garcìa, D., Cegarra, J., 2007. Effects of a compost made from the solid by-product (alperujo) of the two-phase centrifugation system for olive oil extraction and cotton gin waste on growth and nutrient content of ryegrass (*Lolium perenne* L.). Bioresour. Technol. 98 (4), 940–945.

Altieri, R., Esposito, A., 2008. Olive orchard amended with two experimental olive mill wastes mixtures: effects on soil organic carbon, plant growth and yield. Bioresour. Technol. 99 (17), 8390–8393.

Amaral, C., Lucas, M.S., Coutinho, J., Crespí, A.L., Anjos, M.R., Pais, C., 2008. Microbiological and physicochemical characterization of olive mill wastewaters from a continuous olive mill in Northeastern Portugal. Bioresour. Technol. 99 (15), 7215–7223.

Amirante, P., 2003. I sottoprodotti della filiera olivicola-olearia. In: Fiorino, P. (Ed.), Olea, Trattato di Olivicoltura. Edagricole, Bologna, pp. 291–303.

Ammar, E., Nasri, M., Medhioub, K., 2005. Isolation of Enterobacteria able to degrade simple aromatic compounds from the wastewater of olive oil extraction. World J. Microbiol. Biotechnol. 21, 253–259.

Aviani, I., Raviv, M., Hadar, Y., Saadi, I., Dag, A., Ben-Gal, A., Yermiyahu, U., Zipori, I., Laor, Y., 2012. Effects of harvest date, irrigation level, cultivar type and fruit water content on olive mill wastewater generated by a laboratory scale 'Abencor' milling system. Bioresour. Technol. 107, 87–96.

Ayed, L., Assas, N., Sayadi, S., Hamdi, M., 2005. Involvement of lignin peroxidase in the decolourization of black olive mill wastewaters by *Geotrichum candidum*. Lett. Appl. Microbiol. 40, 7–11.

Barbera, A.C., Maucieri, C., Cavallaro, V., Ioppolo, A., Spagna, G., 2013. Effects of spreading olive mill wastewater on soil properties and crops, a review. Agric. Water Manag. 119, 43–53.

Bedbabis, S., Trigui, D., Ben Ahmed, C., Clodoveo, M.L., Camposeo, S., Vivaldi, G.A., Ben Ruina, B., 2015. Long-terms effects of irrigation with treated municipal wastewater on soil, yield and olive oil quality. Agric. Water Manag. 160, 14–21.

Belaqziz, M., Lakhal, E.K., Mbouobda, H.D., Hadrami, I.E., 2008. Land spreading of olive mill wastewater: effect on maize (*Zea mays*) crop. J. Agron. 7, 297–305.

Belaqziz, M., El-Abbassi, A., Lakhal, E.K., Agrafioti, E., Galanakis, C.M., 2016. Agronomic application of olive mill wastewaters: effects on maize production and soil properties. J. Environ. Manag. 171, 158–165.

Blagodatskaya, E., Kuzyakov, Y., 2008. Mechanism of real and apparent priming effect and their dependence on soil microbial biomass and community structure: critical review. Biol. Fert. Soil 45, 115–131.

Borja, R., Sánchez, E., Raposo, F., Rincón, B., Jiménez, A.M., Martín, A., 2006. Study of the natural biodegradation of two-phase olive mill solid waste during its storage in an evaporation pond. Waste Manag. 26, 477–486.

Boubaker, F., Ridha, B.C., 2007. Anaerobic co-digestion of olive mill wastewater with olive mill solid waste in a tubular digester at mesophilic temperature. Bioresour. Technol. 98 (4), 769–774.

Boz, Ö., Doğan, M.N., Albay, F., 2003. Olive processing waste as a method of weed control. Weed Res. 43, 439–443.

Boz, Ö., Ogüt, D., Kır, K., Doğan, M.N., 2009. Olive processing waste as a method of weed control for okra, faba bean, and onion. Weed Technol. 23 (4), 569–573.

Brunetti, G., Plaza, C., Senesi, N., 2005. Olive pomace amendment in Mediterranean conditions: effect on soil and humic acid properties and wheat (*Triticum turgidum* L.). J. Agric. Food Chem. 53, 6730–6737.

Brunetti, G., Senesi, N., Plaza, C., 2007. Effects of amendment with treated and untreated olive oil mill wastewaters on soil properties, soil humic substances and wheat yield. Geoderma 138, 144–152.

Cabrera, A., Cox, L., Fernández-Hernández, A., García-Ortiz Civantos, C., Cornejo, J., 2009. Field appraisement of olive mills solid waste application in olive crops: effect on herbicide retention. Agric. Ecosyst. Environ. 132, 260–266.

Camposeo, S., Vivaldi, G.A., 2011. Short-term effects of de-oiled olive pomace mulching application on a young super high-density olive orchard. Sci. Hort. 129 (4), 613–621.

Canet, R., Pomares, F., Cabot, B., Chaves, C., Ferrer, E., Ribó, M., Albiach, M.R., 2008. Composting olive mill pomace and other residues from rural southeastern Spain. Waste Manag. 28 (12), 2585–2592.

Catalano, M., De Felice, M., 1989. L'utilizzazione delle acque reflue come fertilizzante. Atti del seminario internazionale tecnologie e impianti per il trattamento dei reflui dei frantoi oleari, Lecce.

Cayuela, M.L., Millner, P.D., Meyer, S.L.F., Roig, A., 2008. Potential of olive mill waste and compost as biobased pesticides against weeds, fungi, and nematodes. Sci. Total Environ. 399, 11–18.

Chartzoulakis, K., Psarras, G., Moutsopoulou, M., Stefanoudaki, E., 2010. Application of olive mill wastewater to a Cretan olive orchard: effects on soil properties, plant performance and the environment. Agric. Ecosyst. Environ. 138 (3–4), 293–298.

Chiesura, A., Marano, V., De Francesco, P., Maraglino, A., 2005. Verso la sostenibilità della filiera olivicola: trattamento, recupero e valorizzazione dei sottoprodotti oleari. UNASCO, Roma.

Cicolani, B., Seghetti, L., D'Alfonso, S., Di Giovacchino, L., 1992. Spargimento delle acque di vegetazione dei frantoi oleari su terreno coltivato a grano: effetti sulla pedofauna. In: Conte, (Ed.), Trattamento e riutilizzazione dei reflui agricoli e dei fanghi, Lecce, pp. 187–198.

Cucci, G., Lacolla, G., Caranfa, L., 2008. Improvement of soil properties by application of olive oil waste. Agron. Sustain. Dev. 28, 521–526.

Del Buono, D., Said-Pullicino, D., Proietti, P., Nasini, L., Gigliotti, G., 2011. Utilization of olive husks as plant growing substrates: phytotoxicity and plant biochemical responses. Compost Sci. Util. 19, 52–60.

Della Monica, M., Potenz, D., Rigetti, E., Volpicella, M., 1978. Effetto inquinante delle acque reflue della lavorazione delle olive su terreno agrario. Inquinamento 10, 81–87.

Di Bene, C., Pellegrino, E., Debolini, M., Silvestri, N., Bonari, E., 2013. Short- and long-term effects of olive mill wastewater land spreading on soil chemical and biological properties. Soil Biol. Biochem. 56, 21–30.

Di Giovacchino, L., Seghetti, L., 1990. Lo smaltimento delle acque di vegetazione delle olive su terreno agrario destinato alla coltivazione di grano e mais. L'informatore agrario 45, 58–62.

Di Serio, M.G., Lanza, B., Mucciarella, M.R., Russi, F., Iannucci, E., Marfisi, P., Madeo, A., 2008. Effects of olive mill wastewater spreading on the physico-chemical and microbiological characteristics of soil. Int. Biodeterior. Biodegradation 62 (4), 403–407.

El Hadrami, M., Belaqziz, M., El Hassni, S., Hanifi, A., Abbad, R., Capasso, L., Gianfreda, I., 2004. Physico-chemical characterization and effects of olive oil mill wastewaters fertirrigation on the growth of some Mediterranean crops. J. Agron. 3, 247–254.

Federici, F., Fava, F., Kalogerakis, N., Mantzavinos, D., 2009. Valorisation of agroindustrial by-products, effluents and waste: concept, opportunities and the case of olive mill wastewaters. J. Chem. Technol. Biotechnol. 84, 895–900.

Ferarra, G., Fracchiolla, M., Al Chami, Z., Camposeo, S., Lasorella, C., Pacifico, A., Aly, A., Montemurro, P., 2012. Effects of mulching materials on soil performance of cv. nero di troia grapevines in the Puglia region, southeastern Italy. Am. J. Enol. Vitic. 63 (2), 269–276.

Ferri, D., Convertini, G., Montemurro, F., Rinaldi, M., Rana, G., 2002. Olive wastes spreading in Southern Italy: effects on crops and soil. In: Proceedings of Twelth ISCO Conference, May 26–31, Beijing, China, 593–600.

Gamba, C., Piovanelli, C., Papini, R., Pezzarossa, B., Ceccarini, L., Bonari, E., 2005. Soil microbial characteristics and mineral nitrogen availability as affected by olive oil waste water applied to cultivated soil. Comm. Soil Sci. Plant Anal. 36, 937–950.

García-Barrionuevo, A., Moreno, E., Quevedo-Sarmiento, J., González-López, J., Ramos-Cormenzana, A., 1992. Effect of wastewaters from olive oil mills (alpechin) on Azotobacter nitrogen fixation in soil. Soil Biol. Biochem. 24, 281–283.

Gigliotti, G., Proietti, P., Said-Pullicino, D., Nasini, L., Pezzolla, D., Rosati, L., Porceddu, P.R., 2012. Co-composting of olive husks with high moisture contents: organic matter dynamics and compost quality. Int. Biodeterior. Biodegradation 67, 8–14.

Giusquiani, P.L., Pagliai, M., Gigliotti, G., Businelli, D., Benetti, A., 1995. Urban waste compost: effects on physical, chemical and biochemical soil properties. J. Environ. Qual. 24, 175–182.

Hanafi, F., Belaoufi, A., Mountadar, M., Assobhei, O., 2011. Augmentation of biodegradability of olive mill wastewater by electrochemical pre-treatment: effect on phytotoxicity and operating cost. J. Hazard. Mater. 190 (1–3), 94–99.

Ilarioni, L., Proietti, P., 2014. Olive tree cultivars. In: Peri, C. (Ed.), The Extra-Virgin Olive Oil Handbook. John Wiley & Sons, Ltd, Oxford, pp. 59–67.

IOOC, 2004. Recycling of vegetable water & olive pomace on agicultural land (CFC/IOOC/04). Achievements of the project CFC/IOOC/04 "Recycling of vegetable water & olive pomace on agricultural land" Good practice in vegetable water and compost spreading on agricultural land: case of olive growing. Project executing agency CFC/IOOC/04 Agro-pôle Olivier ENA Meknès.

Isidori, M., Lavorgna, M., Nardelli, A., Parrella, A., 2005. Model study on the effect of 15 phenolic olive mill wastewater constituents on seed germination and Vibrio fischeri metabolism. J. Agric. Food Chem. 53, 8414–8417.

Jarboui, R., Sellami, F., Kharroubi, A., Gharsallah, N., Ammar, E., 2008. Olive mill wastewater stabilization in open-air ponds: impact on clay–sandy soil. Bioresour. Technol. 99 (16), 7699–7708.

Justino, C.I.L., Pereira, R., Freitas, A.C., Rocha-Santos, T.A.P., Panteleitchouk, T.S.L., Duarte, A.C., 2012. Olive oil mill wastewaters before and after treatment: a critical review from the ecotoxicological point of view. Ecotoxicology 21 (2), 615–629.

Kapellakis, I.E., Tsagarakis, K.P., Avramaki, Ch., Angelakis, A.N., 2006. Olive mill wastewater management in river basins: a case study in Greece. Agric. Water Manag. 82 (3), 354–370.

Karpouzas, D.G., Ntougias, S., Iskidou, E., Rousidou, C., Papadopoulou, K.K., Zervakis, G.I., Ehaliotis, C., 2010. Olive mill wastewater affects the structure of soil bacterial communities. Appl. Soil Ecol. 45 (2), 101–111.

Karpouzas, D.G., Rousidou, C., Papadopoulou, K.K., Bekris, F., Zervakis, G., Singh, B., Ehaliotis, C., 2009. Effect of continuous olive mill wastewater applications, in the presence and absence of N fertilization, on the structure of the soil fungal community. FEMS Microbiol. Ecol. 70, 56–69.

Kavdir, Y., Killi, D., 2007. Influence of olive oil solid waste applications on soil pH, electrical conductivity, soil nitrogen transformations, carbon content and aggregate stability. Bioresour. Technol. 99, 2326–2332.

Khoufi, S., Feki, F., Sayadi, S., 2007. Detoxification of olive mill wastewater by electrocoagulation and sedimentation processes. J. Hazard. Mater. 142, 58–67.

Kotsou, M., Mari, I., Lasaridi, K., Chatzipavlidis, I., Balis, C., Kyriacou, A., 2004. The effect of olive oil mill wastewater (OMW) on soil microbial communities and suppressiveness against Rhizoctonia solani. Appl. Soil Ecol. 26, 113–121.

Lessage-Meessen, L., Navarro, D., Maunier, S., Sigoillot, J.C., Lorquin, J., Delattre, M., Simon, J.L., Asther, M., Labat, M., 2001. Simple phenolic content in olive oil residues as a function of extraction systems. Food Chem. 75 (4), 501–507.

Linares, A., Caba, J.M., Ligero, F., De la Rubia, T., Martinez, J., 2003. Detoxification of semisolid olive-mill wastes and pine-chip mixtures using *Phanerochaete flavido-alba*. Chemosphere 51, 887–891.

Lodhi, A.K., 1976. Role of allelopathy as expressed by dominating trees in a lowland foresi in controlling the productivity and patterns of herbaceous growth. Am. J. Bot. 63, 1–8.

López-Piñeiro, A., Albarràn, A., Rato Nunes, J.M., Barreto, C., 2008. Short and medium term effects of two-phase olive mill waste application on olive grove production and soil properties under semiarid Mediterranean conditions. Bioresour. Technol. 99, 7982–7987.

López-Piñeiro, A., Fernández, J., Nunes, R., Manuel, J., García Navarro, A., 2006. Response of soil and wheat crop to the application of two-phase olive mill waste to Mediterranean agricultural soils. Soil Sci. 171, 728–736.

López-Piñeiro, A., Murillo, S., Barreto, C., Muñoz, A., Rato, J.M., Albarrán, A., García, A., 2007. Changes in organic matter and residual effect of amendment with two-phase olive-mill waste on degraded agricultural soils. Sci. Total Environ. 378 (1–2), 84–89.

Lozano-García, B., Parras-Alcántara, L., del Toro Carrillo de Albornoz, M., 2011. Effects of oil mill wastes on surface soil properties, runoff and soil losses in traditional olive groves in southern Spain. CATENA 85 (3), 187–193.

Madejón, E., Burgos, P., López, R., Cabrera, F., 2003. Agricultural use of three organic residues: effect on orange production on properties of a soil of the Comarca Costa de Huelva (SW Spain). Nutr. Cycl. Agroecosyst. 65, 281–288.

Mahmoud, M., Janssen, M., Haboub, N., Nassour, A., Lennartz, B., 2010. The impact of olive mill wastewater application on flow and transport properties in soils. Soil Tillage Res. 107 (1), 36–41.

Mantzavinos, D., Kalogerakis, N., 2005. Treatment of olive mill effluents: Part I. Organic matter degradation by chemical and biological processes—an overview. Environ. Int. 31 (2), 289–295.

McNamara, C.J., Anastasiou, C.C., O'Flaherty, V., Mitchell, R., 2008. Bioremediation of olive mill wastewater. Int. Biodeterior. Biodegradation 61 (2), 127–134.

Mechri, B., Ben Mariem, F., Baham, M., Ben Elhadj, S., Hammami, M., 2008. Change in soil properties and the soil microbiological community following land spreading of olive mill wastewater affects olive trees key physiological parameters and the abundance of arbuscular mycorrhizal fungi. Soil Biol. Biochem. 40, 152–161.

Mechri, B., Cheheb, H., Boussadia, O., Attia, F., Ben Mariem, F., Braham, M., Hammami, M., 2011. Effects of agronomic application of olive mill wastewater in a field of olive trees on carbohydrate profiles, chlorophyll a fluorescence and mineral nutrient content. Environ. Exp. Bot. 71 (2), 184–191.

Mechri, B., Echbili, A., Issaoui, M., Braham, M., Elhadj, S.B., Hammami, M., 2007. Short-term effects in soil microbial community following agronomic application of olive mill wastewaters in a field of olive trees. Appl. Soil Ecol. 36 (2–3), 216–223.

Mekki, A., Dhouib, A., Sayadi, S., 2006. Changes in microbial and soil properties following amendment with treated and untreated olive mill wastewater. Microbiol. Res. 161 (2), 93–101.

Mekki, A., Dhouib, A., Sayadi, S., 2007. Polyphenols dynamics and phytotoxicity in a soil amended by olive mill wastewaters. J. Environ. Manag. 84 (2), 134–140.

Mekki, A., Dhouib, A., Sayadi, S., 2013. Review: effects of olive mill wastewater application on soil properties and plants growth. Int. J. Recycl. Org. Waste Agric. 2, 15–21.

Mondini, C., Sequi, P., 2008. Implication of soil C sequestration on sustainable agriculture and environment. Waste Manag. 28 (4), 678–684.

Montemurro, F., Convertini, G., Ferri, D., 2004. Mill wastewater and olive pomace compost as amendments for rye-grass. Agronomie 24, 481–486.

Montemurro, F., Diacono, M., Vitti, C., Ferri, D., 2011. Potential use of olive mill wastewater as amendment: crops yield and soil properties assessment. Comm. Soil Sci. Plant Anal. 42, 2594–2603.

Moraetis, D., Stamati, F.E., Nikolaidis, N.P., Kalogerakis, N., 2011. Olive mill wastewater irrigation of maize: impacts on soil and groundwater. Agric. Water Manag. 98 (7), 1125–1132.

Morillo, J.A., Antizar-Ladislao, B., Monteoliva-Sánchez, M., Ramos-Cormenzana, A., Russell, N.J., 2009. Bioremediation and biovalorisation of olive-mill wastes. Appl. Microbiol. Biotechnol. 82, 25–39.

Morisot, A., 1979. Utilization des margines par épan-dagc. L'olivier 19, 8–13.

Mulinacci, N., Romani, A., Galardi, C., Pinelli, P., Giaccherini, C., Vincieri, F.F., 2001. Polyphenolic content in olive oil waste waters and related olive samples. J. Agric. Food Chem. 49 (8), 3509–3514.

Nasini, L., Gigliotti, G., Balduccini, M.A., Federici, E., Cenci, G., Proietti, P., 2013. Effect of solid olive-mill waste amendment on soil fertility and olive (*Olea europaea* L.) tree activity. Agric. Ecosyst. Environ. 164, 292–297.

Nasini, L., Proietti, P., 2014. Olive harvesting. In: Peri, C. (Ed.), The Extra-Virgin Olive Oil Handbook. John Wiley & Sons, Ltd, Oxford, pp. 89–105.

Nastri, A., Ramieri, N.A., Abdayem, R., Piccaglia, R., Marzadori, C., Ciavatta, C., 2006. Olive pulp and its effluents suitability for soil amendment. J. Hazard. Mater. 138 (2), 211–217.

Niaounakis, M., Halvadakis, C.P., 2006. Olive Processing Waste Management Literature Review and Patent Survey, second ed. Elsevier Ltd., Kidlington, Oxford.

Ntougias, S., Gaitis, F., Katsaris, P., Skoulika, S., Iliopoulos, N., Zervakis, G.I., 2013. The effects of olives harvest period and production year on olive mill wastewater properties—evaluation of Pleurotus strains as bioindicators of the effluent's toxicity. Chemosphere 92 (4), 399–405.

Ouzounidou, G., Asfi, M., Sotirakis, N., Papadopoulou, P., Gaitis, F., 2008. Olive mill wastewater triggered changes in physiology and nutritional quality of tomato (*Lycopersicon esculentum* Mill.) depending on growth substrate. J. Hazard. Mater. 158 (2–3), 523–530.

Ouzounidou, G., Zervakis, G.I., Gaitis, F., 2010. Raw and microbiologically detoxified olive mill waste and their impact on plant growth. Terr. Aquat. Environ. Toxicol. 4, 21–38.

Pagliai, M., Pellegrini, S., Vignozzi, N., Papini, R., Mirabella, A., Piovanelli, C., Gamba, C., Miclaus, N., Castaldini, M., De Simone, C., Pini, R., Pezzarossa, B., Sparvoli, E., 2001. Influenza dei reflui oleari sulla qualità del suolo. L'informatore Agrario 50, 13–18.

Papini, R., Pellegrini, S., Vignozzi, N., Pezzarossa, B., Pini, R., Ceccarini, L., Pagliai, M., Bonari, E., 2000. Impatto dello spandimento di reflui oleari su alcune caratteristiche chimiche e fisiche del suolo. In: Atti XVIII Convegno Nazionale della Società Italiana di Chimica Agraria, 20–22 September, Catania, Italy, pp. 226–234.

Paraskeva, P., Diamadopoulos, E., 2006. Technologies for olive mill wastewater (OMW) treatment: a review. J. Chem. Technol. Biotechnol. 81, 1475–1485.

Paredes, C., Cegarra, J., Bernal, M.P., Roig, A., 2005. Influence of olive mill wastewater in composting and impact of the compost on a Swiss chard crop and soil properties. Environ. Int. 31 (2), 305–312.

Paredes, C., Cegarra, J., Roig, A., Sánchez-Monedero, M.A., Bernal, M.P., 1999. Characterization of olive mill wastewater (alpechin) and its sludge for agricultural purposes. Bioresour. Technol. 67 (2), 111–115.

Paredes, M.J., Moreno, E., Ramos-Cormenzana, A., Martínez, J., 1987. Characteristics of soil after pollution with waste waters from olive oil extraction plants. Chemosphere 16, 1557–1564.

Peri, C., Proietti, P., 2014. Olive mill waste and by-products. In: Peri, C. (Ed.), The Extra-Virgin Olive Oil Handbook. John Wiley & Sons, Ltd, Oxford, pp. 283–302.

Perucci, P., Dumontet, S., Casucci, C., Schnitzer, M., Dinel, H., Monaci, E., Vischetti, C., 2006. Moist olive husks addition to a silty clay soil: Influence on microbial and biochemical parameters. J. Environ. Sci. Health B 41, 1019–1036.

Pezzolla, D., Marconi, G., Turchetti, B., Zadra, C., Agnelli, A., Veronesi, F., Onofri, A., Benucci, G.M.N., Buzzini, P., Albertini, E., Gigliotti, G., 2015. Influence of exogenous organic matter on prokaryotic and eukaryotic microbiota in an agricultural soil. A multidisciplinary approach. Soil Biol. Biochem. 82, 9–20.

Piotrowska, A., Iamarino, G., Rao, M.A., Gianfreda, L., 2006. Short-term effects of olive mill waste water (OMW) on chemical and biochemical properties of a semiarid Mediterranean soil. Soil Biol. Biochem. 38 (3), 600–610.

Piotrowska, A., Rao, M.A., Scotti, R., Gianfreda, L., 2011. Changes in soil chemical and biochemical properties following amendment with crude and dephenolized olive mill waste water (OMW). Geoderma 161 (1–2), 8–17.

Potenz, D., Rigetti, V., Valpolicella, M., 1980. Effetto inquinante delle acque reflue della lavorazione delle olive su terreno agrario. Inquinamento 3 (2), 65–68.

Proietti, P., Cartechini, A., Tombesi, A., 1988. Influenza delle acque reflue di frantoi oleari su olivi in vaso ed in campo. L'informatore Agrario 45, 87–91.

Proietti, P., Federici, E., Fidati, L., Scargetta, S., Massaccesi, L., Nasini, L., Regni, L., Ricci, A., Cenci, G., Gigliotti, G., 2015. Effects of amendment with oil mill waste and its derived-compost on soil chemical and microbiological characteristics and olive (*Olea europaea* L.) productivity. Agric. Ecosyst. Environ. 207, 51–60.

Proietti, P., Paliotti, A., Tombesi, A., Cenci, G., 1995. Chemical and microbiological modifications of two different cultivated soils induced by olive oil waste water administration. Int. J. Agric. Sci. 125, 160–171.

Proietti, P., Sdringola, P., Brunori, A., Ilarioni, L., Nasini, L., Regni, L., Pelleri, F., Desideri, U., Stefania Proietti, S., 2016. Assessment of carbon balance in intensive and extensive tree cultivation systems for oak, olive, poplar and walnut plantation. J. Clean. Prod. 112 (4), 2613–2624.

Proietti, S., Sdringola, P., Desideri, U., Zepparelli, F., Brunori, A., Ilarioni, L., Nasini, L., Regni, L., Proietti, P., 2014. Carbon footprint of an olive tree grove. Appl. Energ. 127, 115–124.

Quaratino, D., D'Annibale, A., Federici, F., Cereti, C.F., Rossini, F., Fenice, M., 2007. Enzyme and fungal treatments and a combination thereof reduce olive mill wastewater phytotoxicity on *Zea mays* L. seeds. Chemosphere 66, 1627–1633.

Rabenhorst, J., 1996. Production of methoxyphenol-type natural aroma chemicals by biotransformation of eugenol with a new *Pseudomonas* sp. Appl. Microbiol. Biotechnol. 46, 470–474.

Rinaldi, M., Rana, G., Introna, M., 2003. Olive-mill wastewater spreading in southern: effects on a durum wheat crop. Field Crops Res. 84 (3), 319–326.

Riolfatti, M., 1983. Criteri di valutazione igienico-sanitaria per corsi di acqua per l'irrigazione, con riferimento ai fiumi Guà, Fratta-Gorzone e Frassine, della regione Veneto. L'igiene moderna 80, 163–174.

Rodrìguez-Lucena, P., Hernandez, D., Hernandez-Apaolaza, L., Lucena, J.J., 2010. Revalorization of a two-phase olive mill waste extract into a micronutrient fertilizer. J. Agric. Food Chem. 58, 1085–1092.

Roig, A., Cayuela, M.L., Sánchez-Monedero, M.A., 2006. An overview on olive mill wastes and their valorisation methods. Waste Manag. 26 (9), 960–969.

Saadi, I., Laor, Y., Raviv, M., Medina, S., 2007. Land spreading of olive mill wastewater: effects on soil microbial activity and potential phytotoxicity. Chemosphere 66 (1), 75–83.

Said-Pullicino, D., Erriquens, F.G., Gigliotti, G., 2007. Changes in the chemical characteristics of water-extractable organic matter during composting and their influence on compost stability and maturity. Bioresour. Technol. 98, 1822–1831.

Said-Pullicino, D., Gigliotti, G., 2007. Oxidative biodegradation of dissolved organic matter during composting. Chemosphere 68, 1030–1040.

Sampedro, I., Giubilei, I., Cajthaml, T., Federici, E., Federici, F., Petruccioli, M., D'annibale, A., 2009. Short-term impact of dry olive mill residue addition to soil on the resident microbiota. Bioresour. Technol. 100, 6098–6106.

Sánchez-Monedero, M.A., Aguilar, M.I., Fenoll, R., Roig, A., 2008. Effect of the aeration system on the levels of airborne microorganisms generated at wastewater treatment plants. Water Res. 42 (14), 3739–3744.

Sellami, F., Jarboui, R., Hachicha, S., Medhioub, K., Ammar, E., 2008. Co-composting of oil exhausted olive-cake, poultry manure and industrial residues of agro-food activity for soil amendment. Bioresour. Technol. 99, 1177–1188.

Sierra, J., Martí, E., Garau, M.A., Cruañas, R., 2007. Effects of the agronomic use of olive oil mill wastewater field experiment. Sci. Total Environ. 378, 90–94.

Sierra, J., Martí, E., Montserrat, G., Cruañas, R., Garau, M.A., 2001. Characterisation and evolution of a soil affected by olive oil mill wastewater disposal. Sci. Total Environ. 279 (1–3), 207–214.

Suzzi, G., Tofalo, R., 2009. Trattamento dei reflui. In: Pisante, M., Inglese, P., Lercker, G. (Eds.), Collana Coltura & Cultura - L'ulivo e l'olio. Bayer CropScience, Bologna, pp. 690–695.

Tamburino, V., Barbagallo, S., 1989. Effetti dell'irrigazione con acque reflue su terreno, colture e acque sotterranee. Irrigazione e drenaggio 4, 44–50.

Tarchitzky, J., Lerner, O., Shani, U., Arye, G., Lowengart-Aycicegi, A., Brener, A., Chen, Y., 2007. Water distribution pattern in treated wastewater irrigated soils: hydrophobicity effect. Eur. J. Soil Biol. 58, 573–588.

Tardioli, S., Bánn, E.T.G., Santori, F., 1997. Species-specific selection on soil fungal population after olive mill waste-water treatment. Chemosphere 34 (11), 2329–2336.

Tajada, M., Gonzalez, J.L., 2004. Effects of application of a by-product of the two-step olive oil mill process on maize yield. Agron. J. 96, 692–699.

Toscano, P., Montemurro, F., 2012. Olive Mill By-Products Management. In: Muzzalupo, I. (Ed.), Olive germplasm—The Olive Cultivation, Table Olive and Olive Oil Industry in Italy,. InTech Open Access Publisher, pp. 173–200.

Tsioulpas, A., Dimou, D., Iconomou, D., Aggelis, G., 2002. Phenolic removal in olive oil mill wastewater by strains of Pleurotus spp. in respect to their phenol oxidase (laccase) activity. Bioresour. Technol. 84, 251–257.

Uygur, V., Karabatak, I., 2009. The effect of organic amendments on mineral phosphate fractions in calcareous soils. J. Plant Nutr. Soil Sci. 172, 336–345.

Vivaldi, G.A., Camposeo, S., Rubino, P., Lonigro, A., 2013. Microbial impact of different types of municipal wastewaters used to irrigate nectarines in Southern Italy. Agric. Ecosyst. Environ. 181, 50–57.

Yay, A.S.E., Oral, H.V., Onay, T.T., Yenigün, O., 2012. A study on olive oil mill wastewater management in Turkey: a questionnaire and experimental approach. Resour. Conserv. Recycl. 60, 64–71.

Zbakh, H., El Abbassi, A., 2012. Potential use of olive mill wastewater in the preparation of functional beverages: a review. J. Funct. Foods 4 (1), 53–65.

Zenjari, B., Nejmeddine, A., 2001. Impact of spreading olive mill wastewater on soil characteristics: laboratory experiments. Agronomie 21 (8), 749–755.

CHAPTER

6 INDUSTRIAL CASE STUDIES ON THE DETOXIFICATON OF OMWW USING FENTON OXIDATION PROCESS FOLLOWED BY BIOLOGICAL PROCESSES FOR ENERGY AND COMPOST PRODUCTION

Apostolos G. Vlyssides, George K. Lamprou, Anestis Vlysidis

Laboratory of Organic Chemical Technology, School of Chemical Engineering, National Technical University of Athens, Zografou, Greece

6.1 INTRODUCTION

Depending on the applying process, olive oil is extracted by pressing or centrifugation. Pressing is the traditional method of extracting olive oil, which has been gradually replaced by the centrifugation process over the last decades (Roig et al., 2006). For the centrifugation process, there are two different systems depending, whether or not, on the use of water during the process: the two-phase and the three-phase olive mills. In three-phase mills, apart from oil, olive pomace, and wastewater are also produced as water is used in large quantities during the crushing of olives, the malaxation process, and the centrifugation process where the separation of the solid olive pomace takes place (Vlyssides et al., 1998). The liquid stream containing oil and olive mill wastewaters (OMWW) is then processed in a disc centrifugal where the separation of the two phases occurs (Niaounakis and Halvadakis, 2004). Pomace is handled as a by-product and is used as raw material in pomace oil industries where the residual oil is extracted using hexane as solvent.

Traditional mills that are still using the pressing process produce products and by-products similar to three-phase olive mills. This method is more expensive, as it requires more labor, however it gives a little higher quality with less but more concentrated OMWW (100–120 g COD/L). The obtained olive pomace, as in three-phase mills, is utilized in pomace processing industries for the extraction of olive stone oil and the production of solid biofuels (Vlyssides et al., 2002). After the extraction, the residual solids (olive stones and dried olive pulp) are used as biofuel (e.g., pellets) (Vlyssides et al., 2002) or according to (Christodoulou et al., 2007, 2008), it can be also used as animal feed.

In two-phase olive mills, water is only used during the washing of the olives prior to crushing and not in the olive extraction process. Hence, there are only two streams produced after the centrifugation process, the olive oil and the olive pomace. However, the olive pomace from two-phase olive mills has higher moisture content, compared to the one generated from three-phase or pressing mills, and the cost

Olive Mill Waste. http://dx.doi.org/10.1016/B978-0-12-805314-0.00006-6

for transportation and processing is higher. Although, the production cost is lower in two-phase olive mills as the processed volumes are smaller due to the nonaddition of water, there is a 7% less yield in the production of olive oil compared to the three-phase mills. There is also a small amount of OMWW produced due to the use of water during the oil washing unit. Together with the olive pomace, there are also approximately 5 tons of olive leaves generated together with the olive pomace which account for all types of olive mills (Vlyssides et al., 1998, 2015).

The difficulties in treating olive mill effluents are mainly related to (1) its high organic loading, (2) the seasonal operation of olive mills (typically from November to February), (3) its high territorial scattering, (4) the high cost, especially for small- and middle-scale producers for installing a waste treatment facility, and (5) the presence of organic compounds which are hard to biodegrade, such as long-chain fatty acids and phenolic compounds. Paraskeva and Diamadopoulos (2006) claimed that 1 m^3 of OMWW has the same environmental impact as 100–200 m^3 of domestic liquid wastes (Paraskeva and Diamadopoulos, 2006), while their disposal to rivers, lakes or sea contributes to the eutrophication potential and have an excessive impact on aquatic ecosystems (DellaGreca et al., 2001).

Many different processes have been proposed to treat OMWW, such as physical methods (open evaporating ponds, concentration by evaporation, filtration, and ultrafiltration/reverse osmosis), chemical methods (Fenton oxidation, wet oxidation, electrolytic oxidation, combustion, and neutralization with lime), physicochemical methods (flocculation, coagulation, electrocoagulation, and ion exchange), and biological methods (activated sludge, anaerobic digestion, biofiltration, lagooning, or direct watering on fields, and composting) (Rozzi and Malpei, 1996). The well-known classical physical and physicochemical treatment methods have been proved unsustainable while the biological treatment methods although they could be effective and viable, they have been proved inapplicable due to the very slow degradation rates of OMWW. Very few from the earlier proposed methods have been materialized from laboratory investigations to applied procedures in real olive mills plants and none of them was proved to be technical feasible and simultaneously cost effective (Dogruel et al., 2009). However, the application of these technologies has indicated that a combination of methods needs to be implemented so as to have a satisfactory wastewater treatment outcome. The implementation of biological treatment methodologies (anaerobic digestion, aerobic treatment, and composting) could be viable solutions for treating OMWW only if the high toxicity of this waste is drastically reduced. The latter can be acquired with advanced oxidation processes (Mantzavinos and Psillakis, 2004).

6.2 AEROBIC AND ANAEROBIC TREATMENT OF OMWW

Traditional treatment methods applied in OMWW are biological processes, such as anaerobic and aerobic processes. Both of them use microorganisms to dissimilate the complex organic material of OMWW. In aerobic processes, there is provision of oxygen either from agitating the system or from supplying oxygen to the aerobic reactor. The organic load in aerobic reactors should be much lower than the COD of OMWW (i.e., 1 g/L instead of 60–120 g/L) in order to effectively treat the wastes. Therefore, the waste needs to be diluted several times in order to be fed to the aerobic reactor and this results to the generation of a huge amount of secondary sludge, which needs to undergo further treatment before its disposal. As a consequence, aerobic treatment becomes too costly, especially for small- and middle-scale olive mills. Most importantly, aerobic treatment cannot detoxify the inhibitory compounds of OMWW, such as lipids and polyphenols (Mantzavinos and Kalogerakis, 2005). Another weakness for applying aerobic treatment methods is the seasonal operation of olive mills, which operate only few months per year from November to January. This small period does not allow the biological ecosystem to adjust to the high requirements of this facility. Furthermore, considering the need of at least 1 kWh for every kilogram of COD fed to the aerobic bioreactor, the energy power needed to treat the OMWW of a middle olive oil mill due to air supply requirements would be around to 100 kW resulting in a noneconomical viable solution.

In anaerobic treatment, the main target is to convert the organic compounds of the wastes via a number of different microbiological reactions to carbon dioxide and methane. The processes are driven only by bacteria and have the following major steps: hydrolysis, acidogenesis, and methanogenesis. The produced biogas can be then used for the energy requirements of the olive oil mill. Anaerobic digestion generates less sludge than aerobic treatment while it allows higher organic loads and it is more easily restarted after a stop period. Moreover, anaerobic treatment has the advantage of low energy requirements. Therefore, anaerobic digestion is generally preferred from aerobic methods for treating OMWW (Mantzavinos and Kalogerakis, 2005). However, even if anaerobic digestion shows some advantages over aerobic treatment, it is a high sensitive process to the toxic inhibitors of OMWW, such as phenolic compounds and lipids which inhibit the biological reactions (Niaounakis and Halvadakis, 2004). There is a number of studies that have used two-stage anaerobic digesters (Fezzani and Cheikh, 2010; Rincon et al., 2010) demonstrating that this design gives more promising results than the one-stage anaerobic treatment. In the first reactor, acidogenic bacteria bioconvert organic macromolecules, such as protein and lipids into volatile fatty acids, which are fed under controlled concentrations into the second digester where they are converted into methane and carbon dioxide by methanogenic bacteria. Two stage anaerobic digesters have shown double CH_4 production (32 L/L OMWW) compared to one stage anaerobic treatment (Fezzani and Cheikh, 2010). OMWW has been also codigested with other agricultural residues like cow or poultry manure giving attractive results (Dareioti et al., 2010; Gelegenis et al., 2007).

An anaerobic digestion unit of OMWW was first studied in a pilot-scale installation of 260 L active volume bioreactor in Peza, Heraklion Crete (Georgacakis et al., 1987). The digester was constructed inside the facilities of the local cooperative olive oil mill. The system consisted of a continuous stirred bioreactor with sludge recycling followed by a sequential anaerobic-aerobic lagooning system. After one year of operation, the optimum obtained methane production was 4.86 m^3-CH_4/m^3 of 1:5 diluted OMWW/day when the COD loading was around 15 kg-COD/m^3 of reactor/day. The final effluent continued to have a significant toxicity as the phenolic compounds present in the OMWW were only reduced about 40%. These results were in agreement with research findings by (Filidei et al., 2003).

6.3 COMPOSTING AND COCOMPOSTING OF OMWW

Another biological method that is highly used for the treatment of OMWW is its cocomposting with agricultural by-products, one of them being Olive Kernel (Vlyssides, 2000). Composting is a controlled biochemical aerobic degradation process where organic solid waste materials are stabilized producing carbon dioxide and microbial mass. Composting has three consecutive stages: (1) the initial activation stage, (2) the thermophilic stage which is characterized by a sudden temperature increase up to 70°C because of biooxidation of carbon content substances, and (3) the mesophilic stage (35–45°C), where the organic material cools down to ambient temperature because of the reduction of degradable carbon

(Haug, 1980). Heat is generated due to exothermic microbial metabolic activities which transform the organic matter into biomass and carbon dioxide. At the end of the composting process, a stable humus-rich complex mixture is produced (Cooperband, 2002).

The composting process is controlled by some physicochemical parameters, such as pH, electrical conductivity, moisture, bulk density, porosity, and particle size. However, the most important parameters are the temperature of the compost, the moisture content, essential nutrients, such as nitrogen and phosphorous and oxygen concentration. The moisture content of the composting material must be between 30% and 50%, while the organic carbon to organic nitrogen ratio (C/N) must be between 5 and 30 and organic carbon to phosphorous ratio (C/P) must be less than 500. Also, in order to fulfill the microbial respiration needs, the oxygen concentration into the composting gas phase must be more than 5%. The necessary amount of oxygen is usually supplied by providing air by a compressing device or by agitation. Under these conditions and if the biodegradable carbon content of the composting biomass is more than 25% in dry basis, the temperature increases rapidly up to 60°C due to high microbial activity. This temperature is the upper limit for an effective thermophilic composting. Above this temperature, the microbial activity decreases considerably and above 70°C it becomes almost zero (Finstein et al., 1986; Haug, 1980).

Factors that influence composting can be classified into two categories. The first category includes the conditions depending on the preparation of the compost, such as pH, particle size, porosity, total volume pile, initial moisture, and nutritional balance. The second category involves the operational conditions, such as moisture content, temperature, and aeration (Muktadirul Bari Chowdhury et al., 2013). According to (de Bertoldi et al., 1983), the temperature norm in compost piles is a critical factor to the microbial activity and has a major impact on the composting process. An ideal temperature range of a composting process is 40–65°C while temperatures above 55°C are necessary to kill pathogens (de Bertoldi et al., 1983). Control of the heat output is a key factor as the main purpose during the thermophilic period is to retain the temperature in a range where the thermophilic microbes survive and continue the biological processes (de Bertoldi et al., 1983). Commonly, the temperature is controlled by supplying air. The latter is also needed for respiration reasons (MacGregor et al., 1981). Generally, the supplied air for cooling needs is around 9 times more than the air needed for respiration (Finstein et al., 1986). Recorded temperatures on composting experiments conducted with olive mill wastes (OMWs) were ranged from 43 to 70°C (Muktadirul Bari Chowdhury et al., 2013). Thermophilic phase lasts from 16 to 70 days and the total composting duration including maturation and humification periods from 60 to 365 days. Variations between experiments are attributed to different types of OMW, bulking agents, aeration systems used, and volume of the compost pile. As it is recorded, the addition of bulking agents and the mechanical or forced aeration promote the increase on the temperature at the thermophilic zone. The presence of a bulking agent also increases compost porosity, and hence, oxygen availability facilitating the microbial activities (Vasiliadou et al., 2015).

During composting, a significant quantity of water, contained in the biomass, is evaporated due to temperature increase and air supply. This quantity is lost and must be replaced with fresh water in order to keep the moisture content at desirable levels (30–50%). Instead of fresh water, OMWW can be supplied to the compost piles. The advantage here is that supplying OMWW not only helps in retaining the water balance constant but also provides organic carbon that is lost as carbon dioxide due to microbial activity. Therefore, Olive Kernel can be effectively cocomposted with OMWW where the organic matter and water is stabilized continuously. At the end of the composting process, the excess of water is evaporated leaving a solid residue that after maturation and humification can be released to the market as an organohumic product.

According to Haug (1980), a composting process is characterized from first order kinetics:

$$\frac{d(BM)}{dt} = k(BM) \Rightarrow \frac{BM_e}{BM_o} = e^{-kt}$$

where *BM* represents the biodegradable portion of the volatile material in percentage of VM; *k* represents the degradation rate of the *BM* per unit of *BM* contained in the composting system in kg/kg/d; *o* is the initial product at time zero; and *e* is end product after time *t*.

According to the above equation and assuming that a convenient compost period can last about two weeks where 70% of the compostable material has been degraded then the expected mean value of *k* must be about 0.08 d^{-1}. The target of cocomposting OMWW with Olive Kernel is that the latter needs to be able to absorb all the generated liquid wastes during the olive mill operation period, which lasts 2–3 months. A successful cocomposting must approach at least a ratio of OMWW/Olive Kernel equal to 2.5 (a typical three-phase olive mill OMWW/Olive Kernel ratio) which corresponds to a minimum biodegradation rate of $k = 0.16$ d^{-1}. The rate of the generated thermal energy at this biodegradation rate will be able to evaporate the amount of water, which is fed and replaced by the feeding rate of OMWW (Vlyssides et al., 2009).

After the thermophilic composting phase, a mesophilic maturation period is followed. During this phase, the products of composting transformed into humic substances. The humification index (HI) is the key parameter for evaluating and controlling the results of this stage (De Nobili and Petrussi, 1988). Higher HI values indicate higher quality of produced organic fertilizers. A HI value more than 4 designates good quality compost. The duration of humification period depends on the performance of stabilization period. So, the maturation-humification period could be ranged from 1 month to some years depending on the residual toxicity of the compost.

6.4 CASE STUDIES OF OMWW COCOMPOSTING WITH AGRICULTURAL SOLID RESIDUES

Tomati et al. (1995) used wheat straws to cocompost OMWW in an aerated static pile. Urea was added so as to keep the C/N ratio at 35. The addition of OMWW to solid wheat straws was reached 1:1 when the thermophilic phase started. At the end of the thermophilic stage, lignin was degraded approximately 70% while the final product was not phytotoxic having a humification index of 0.28 (Tomati et al., 1995). Papadimitriou et al. (1997) used OMWW in composting olive pomace in a pilot-scale container facility so as to replace the water loss due to evaporation. The addition of OMWW reached 2.1 L/kg initial dry olive pomace. The authors indicated that the increase of salinity was the restricting factor of using OMWW instead of process water. However, a total reduction in volatile suspended solids equal to 45% was obtained (Papadimitriou et al., 1997).

A pilot-scale unit of cocomposting OMWW with olive mill solid wastes (Olive Kernel) was also studied by Vlyssides et al. (1989) in 1 m^3 batch reactor without agitation. Olive Kernel was composted inside the vessel while OMWW was continuously added to the reactor for cocomposting according to temperature levels. When the mean temperature of the compost was between 45 and 65°C, the OMWW was supplied to the reactor at a rate of 2 kg/h. The experiment lasted for 40 days and during this period the total consumption of OMWW was around 1.34 kg-OMWW/kg of OMSW. The unit was operated

FIGURE 6.1 Cocomposting of OMWW with Olive Kernel at the Agricultural Cooperative Olive Mill in Sitia, Crete

(A) Composting pile and (B) force draft aeration system.

for 3 months and it was installed inside the Agricultural University of Athens (Greece). The average biodegradation rate (k) was very little and equal to 0.003 d^{-1} due to insufficient aeration and lack of an agitation system (Vlyssides et al., 1989).

The above pilot scale trial was repeated in a "real" middle-scale olive mill using mechanical agitation and satisfactory supply of air. The cocomposting unit was installed in the olive mill of the agricultural cooperation of Sitia, in Crete. The cocomposting of OMWW and Olive Kernel was carried out using the method of windrows with a parallel provision of air from the bottom of the piles (Fig. 6.1). The unit was designed according to the experimental results from the pilot scale installation in the Agricultural University of Athens and it operated for one olive oil processing period (i.e., 3 months). At the end of the process, 140 tons of Olive Kernel and 280 tons of OMWW (OMWW/Olive Kernel = 2) were cocomposted leading to an average biodegradation ratio (k) of 0.008 d^{-1}. The final product was matured in 4 months reaching a HI of 0.17. The final results of composting according to the criteria discussed earlier were not satisfactory mainly due to the fact that the moisture content of the piles could not be controlled properly as the provision of OMWW was implemented manually (Vlyssides et al., 1989).

In another study of Vlyssides et al. (1995), an industrial scale cocomposting unit was constructed in a three-phase olive mill, in Crete island. In this case study, the OMWW were continuously supplied in a composting reactor containing solid residues from the olive mill process. A 237.6 m^3 open vessel (18 m long, 6 m wide, and 2.2 m height) was constructed with an active volume of 195 m^3. The schematic diagram and two photos of the process are illustrated in Figs. 6.2 and 6.3, respectively. The open vessel was agitated through a helical type agitator from the top of the reactor mounted in a travelling bridge. The biomass was aerated from the bottom of the reactor using three air movers and a solution of urea was also dosing in the system according to OMWW feeding. The agitator was moving continuously in two dimensions agitating the whole material in a time cycle of 2 h. The OMWW feeding was controlled by a temperature recorder which followed the agitation moving. At the end of the thermophilic stage, which lasted 23 days, the system was fed with 263 m^3 OMWW corresponding to 2.9 kg-OMWW per kg of Olive Kernel. Applying this method to the entire olive mill period, it was estimated that the whole amount of OMWW generated from this small olive mill can be consumed in five batches. The final product was maturated in approximately 3 months. Although the OMWW/Olive Kernel ratio was proved to be satisfactory, the biodegradation rate k, even if it was an order of magnitude higher than

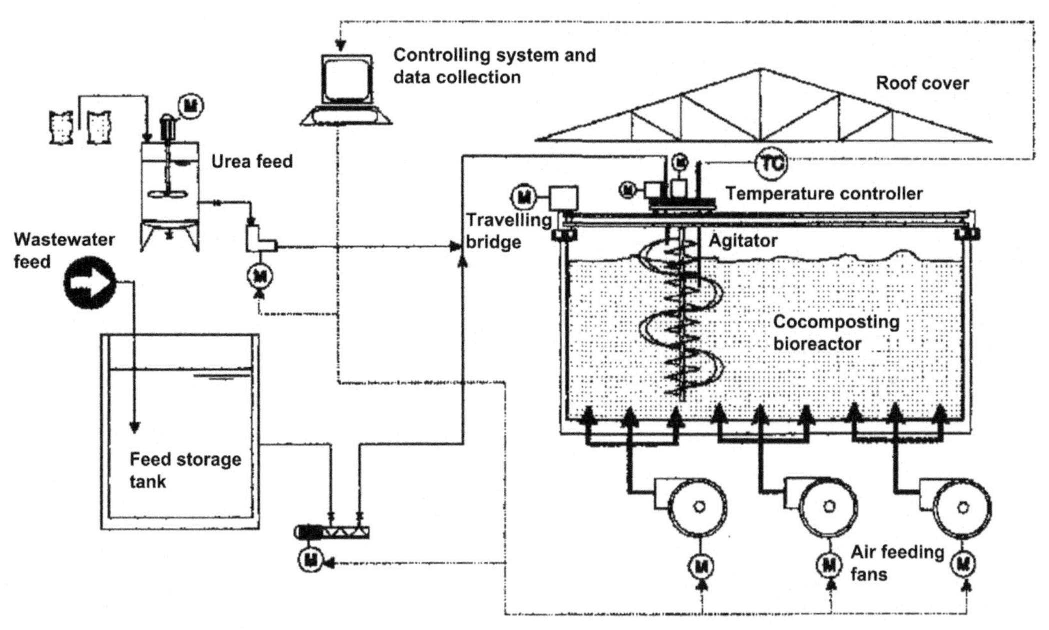

FIGURE 6.2 Schematic Diagram of a Cocomposting Reactor Installed in a Small-Scale Three-Phase Olive Mill in Crete Island, Greece (Vlyssides et al., 1996)

FIGURE 6.3 In Vessel Pilot-Scale Plant Installation for Cocomposting OMWW with Olive Kernel at Koutsouras Agricultural Cooperative Olive Mill

(A) In vessel cocomposting bioreactor and (B) the agitation and the OMWW feeding system.

the previous trial of (Vlyssides et al., 1989), was still low and below the requirements set for a functional composting unit. The OMWW was fed proportionally to temperature indications, so the moisture content was controlled from the temperature of the bulking material. The main reasons behind the low biodegradation yields were considered to be the absence of an independent temperature control system and the insufficient control of the carbon balance (Vlyssides et al., 1995, 1996).

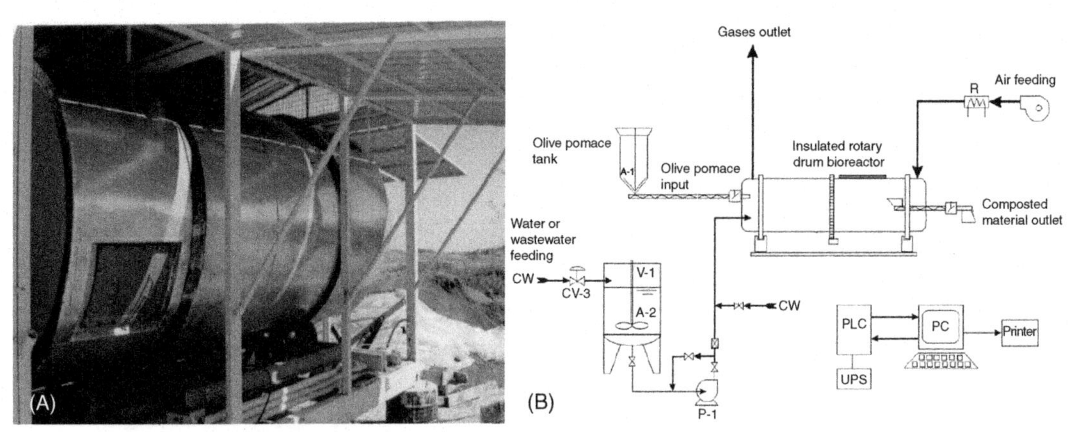

FIGURE 6.4 In Vessel Cocomposting of OMWW with Olive Kernel in a Rotary Drum Reactor Installed in the Industrial area of Nicosia, Cyprus

(A) The rotary drum bioreactor and (B) the schematic diagram of the process (Vlyssides et al., 2003).

A semicontinuous pilot scale composting unit was also constructed and installed in the industrial area of Nicosia, Cyprus in 1994. The unit of cocomposting was a horizontal closed rotary drum bioreactor with a volume of 5 m^3 (2.5 m length and 1.6 m diameter). The temperature of the system was controlled by regulating the air supply. The latter could be heated, if necessary, to provide heat during composting. Due to the fact that the system was closed, water and carbon balances were continuously monitored. Carbon balance was monitored through the generation of CO_2 and the generation of microbial heat. The automated system was able to control the moisture and the biodegradable carbon by feeding OMWW to the system and also by providing fresh Olive Kernel. The optimum conditions of cocomposting were found to be at 68°C, with a moisture content of 40% and a partial pressure of oxygen equal to 16%. The hydraulic retention time (HRT) was examined with different ratios of OMWW and Olive Kernel. It was found that ratios above 3 can be obtained with a HRT higher than 14 days. The unit was operated for 6 months. A photo of the installed unit and the schematic diagram of the process are shown in Fig. 6.4A–B, respectively (Vlyssides et al., 2003). The installed unit had the capability to increase the OMWW/Olive Kernel ratio up to 4 depending on the temperature and the oxygen partial pressure while the biodegradation rate (k) reached a high value equal to 0.7 d^{-1}. However, the high operational and capital cost of this facility as well as the elevated maturation time needed (around 8 months) due to residual toxicity of the substrate resulted in classifying this method as not feasible.

Results from all the above pilot- or industrial-scale units demonstrate that the sustainability of the biological processes either the cocomposting method of OMWW with Olive Kernel or the anaerobic digestion of OMWW are not dependent substantially on the control and automation of the unit. As we will see in the next section, the detoxification of the liquid wastes is a more drastic parameter for improving the efficiency of this method. The toxic compounds in this waste stream are primarily the high concentration of phenolic compounds and secondarily the lipid content. Two approaches have been applied for their effective treatment: (1) their extraction which is the most desirable one and (b) their oxidation with advanced oxidation processes (AOPs). Extracting them or applying a chemical pretreatment step prior to biological processes will result in removing/destroying toxic and difficult to dissimilate compounds, helping in this way the biological methods that follow (Mantzavinos and Kalogerakis, 2005).

6.5 ADVANCED OXIDATION PROCESSES (AOPS)

AOPs are aqueous phase chemical treatment procedures which are based on the intermediacy of highly reactive species, such as (primarily but not exclusively) hydroxyl radicals in the reactions resulting on the destruction of the target pollutant (Comninellis et al., 2008). AOPs are considered as promising detoxification technologies (Pera-Titus et al., 2004) and have been effectively used for a great variety of pollutants, such as pharmaceuticals, phenolic compounds, pesticides, coloring matters, and surfactants among others. Generally, they are used as pretreatment methods in biological treatment processes as they target to reduce the concentrations of toxic organic compounds that inhibit the biological activities (Stasinakis, 2008).

In Fig. 6.5, the most important AOPs are listed. In comparison with other AOPs, Fenton is a cost effective method with lower operational cost (Amaral-Silva et al., 2016) as there is no need for energy input to activate hydrogen peroxide. Other advantages of this method are the short reaction time and the use of easy to handle reagents (Bautista et al., 2007). A summary of AOPs which have been used for treating OMWW has been published by Oller et al. (2011). These processes include (Oller et al., 2011):

1. UV/ H_2O_2 combined with ultrafiltration;
2. Fenton and photo-Fenton (with or without coagulation);
3. electrochemical oxidation, photocatalysis, Fenton oxidation, ozonation;
4. zero-valent iron and H_2O_2;
5. solar photo-Fenton;
6. ozonation; and
7. wet-air oxidation.

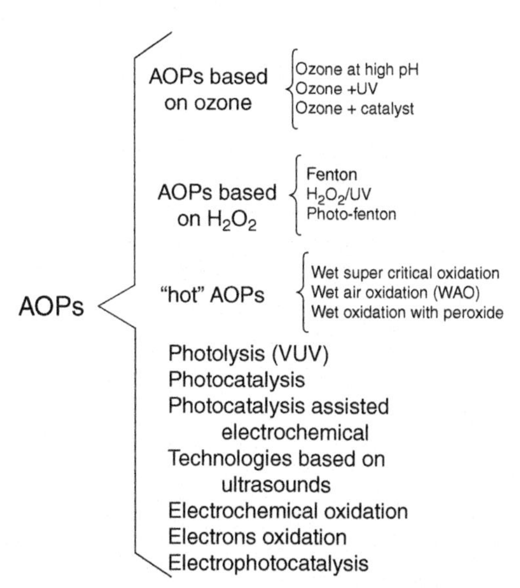

FIGURE 6.5 AOPs That Have Been Used for the Detoxification of OMWW (Oller et al., 2011)

6.6 FENTON OXIDATION

Fenton's history begins in 1894 when John F. Fenton reported at his published research the oxidation of tartaric acid by H_2O_2 in the presence of ferrous iron ions (Fenton, 1894). Until the late 1960s, Fenton was not generally applied (Barbusinski, 2009). However, the interest of many researchers was renewed in 1975, due to the published research of (Walling, 1975), who stated that Fenton's oxidation of organic compounds is carried out through the free radical phase and this statement remains valid until today. The Fenton is a homogenous catalytic oxidation process, which includes the reaction of hydrogen peroxide with ferrous iron. Products of this reaction are hydroxyl radical, with a high oxidation potential (2.80 V). When hydrogen peroxide (H_2O_2) and iron sulfate (II) ($FeSO_4$) is added in the OMWW under acidic conditions, flocculation of organic compounds is achieved (Lucas and Peres, 2009). The hydroxyl radical (•OH) is the second more powerful oxidant after fluorine and is able to nonselectively destroy most organic and organometallic contaminants until their complete mineralization (Babuponnusami and Muthukumar, 2014). The mechanism by which ferrous iron catalyzes hydrogen peroxide oxidations but also the identification of the participating intermediates is controversial (Enami et al., 2014). A review of the published research with various point of view in Fenton's reaction mechanisms is published by Barbusinski (2009).

The parameters which affect the Fenton process are mainly pH, temperature, Fe^{2+}, and H_2O_2 concentration. The pH of a system affects significantly the degradation of organic compounds. The ideal pH range in Fenton process has been found to be between 3 and 4 (Bautista et al., 2008). Particularly, for the chemical oxidation of OMWs, an optimum pH value equal to 3 has been published from several studies (Vlyssides et al., 2010; Dogruel et al., 2009; El-Gohary et al., 2009; Babuponnusami and Muthukumar, 2014). At pH higher than 3, Fe^{3+} starts precipitating as $Fe(OH)_3$ and H_2O_2 breaks down into O_2 and H_2O producing less hydroxyl radicals (Babuponnusami and Muthukumar, 2014). Besides, at high pH values, Fe(II) complexes are formed leading to a drop of Fe^{2+} concentration. On the other hand, at more acidic conditions ($pH < 3$), regeneration of Fe^{2+} is inhibited by the reaction of Fe^{3+} with H_2O_2 (Bautista et al., 2008). When the pH value is less than 2.5, the formation of $(Fe(II)\ (H_2O))^{2+}$ takes place while the reaction with hydrogen peroxide is comparatively slower. Thus less hydroxyl radicals are produced, and hence, the degradation efficiency reduces (Gogate and Pandit, 2004). Moreover, the scavenging of the hydroxyl radicals by hydrogen ions becomes significant at a very low pH. The reaction of Fe^{3+} with H_2O_2 is also suspended (Ranade and Bhandari, 2014). Although the rate of degradation is proportional to the concentration of ferrous ion, the extent of increase is sometimes observed to be marginal above a certain concentration of ferrous ion (Babuponnusami and Muthukumar, 2014).

Concentration of hydrogen peroxide plays a crucial role on determining the overall efficacy of the degradation process. As observed, the degradation of the pollutants increases with an increase in the dosage of H_2O_2 until it reaches an optimum value (Gogate and Pandit, 2004). A point of great attention is the investigation of the optimum dosage of H_2O_2 because in the case of a surplus, the remaining hydrogen peroxide can have critical negative effects in the produced effluent. It contributes to the COD of the wastewater, it has a hydroxyl radicals scavenging effect and it is harmful to most of the microorganisms used in biological oxidation methods that follow. Therefore, the total quantity of the H_2O_2 used in Fenton oxidation processes should be optimized in laboratory-scale studies (Gogate and Pandit, 2004). According to the literature, there is not any established temperature range as the optimum in Fenton process and few studies have worked on the effect of temperature on the rate of degradation (Babuponnusami and Muthukumar, 2014; Gogate and Pandit, 2004; Lucas and

Peres, 2009). According to Rivas et al. (2001), the degradation efficacy was not affected significantly from an increase in temperature from 10.9 to 40.9°C and Lin and Lo (1997) have observed that the optimum temperature value was 30°C.

6.7 LABORATORY-SCALE TRIALS ON OMWW DETOXIFICATION

In laboratory-scale trials, Vlyssides et al. (2003) have investigated the detoxification of OMWW by measuring the reduction in COD, BOD, and total phenolic content (TPC) using Fenton's reagents. The authors used different initial H_2O_2 concentrations by keeping the initial Fe^{2+} concentration constant at 0.4 g/L in an agitated, pH controlled 2 L batch reactor. The maximum reduction obtained in COD, BOD, and TPC was 60.6%, 31.9%, and 99.7%, respectively. All three maxima obtained at the highest added H_2O_2 concentration which was equal to 0.5% v/v in a H_2O_2 solution of 60%. In the case of TPC, apart from its reduction, the time of oxidation was also reduced significantly when the amount of added H_2O_2 increased. Oxidation of OMWW was completed around 40 min, when 0.5% v/v of H_2O_2 was used while the same OMWW quantity needed more than double time (approximately 100 h) when 0.3% v/v of H_2O_2 was added (Vlyssides et al., 2010). In the same study, the authors have measured the effectiveness of the detoxification process by using Fenton's reagents on an anaerobic digestion process by calculating the anaerobic sludge activity. The latter increased from 0 to 1.4 g-COD/g-VSS/d when 0.5% v/v of H_2O_2 was used in the detoxification step (Vlyssides et al., 2010). In another study, the sludge activity in an anaerobic digestion process was monitored in terms of the TPC of OMWW. When TPC reduced from 120 to 30 mg/L, the methane production from the anaerobic digester increased by 50% (Loutatidou, 2012).

6.8 OLIVE MILL COCOMPOSTING CASE STUDIES

After the completion of laboratory- and pilot-scale trials on the detoxification of the OMWW by using Fenton's reagents several industrial-scale facilities were designed from the Laboratory of Organic Chemical Technology and applied to real olive mill plants by using the laboratory-scale findings (Fig. 6.6).

6.8.1 ADVANCED BIOREFINERIES PROJECTS ON OLIVE MILL WASTES

In the final section of this chapter, two actual industrial scale projects for utilizing OMWs under an integrated approach are presented. The laboratory results on cocomposting and/or anaerobic digestion mentioned earlier were used as a guide for the installation and run of the two industrial-scale units. Both facilities are based in Greece and have successfully altered the OMWW disposal problem to advanced OMW biorefinery facilities valorizing OMWW with Olive Kernel. The core of this novel design is based on the cocomposting of Olive Kernel with detoxified OMWW as previously described in this chapter. However, prior to cocomposting, there are several novel steps for treating OMWW resulting on the production of a spectrum of added-value products, such as (1) pharmaceutical phenolic substances, (2) waste oil, (3) bioenergy, and finally (4) organo-humic fertilizer. The schematic diagram of Fig. 6.7 illustrates all the necessary steps for valorizing OMWW and Olive Kernel under an advanced biorefinery concept (Vlyssides et al., 2014, 2004).

FIGURE 6.6 Several Detoxification and Cocomposting Units of OMWW with Olive Kernel After Fenton Oxidation

(A) Fenton oxidation pilot-scale installation in Aspropyrgos, Attiki (2001), (B) full-scale fenton oxidation in Hania, Crete in a III-phase Olive Mill (2006), (C) industrial-scale cocomposting in a III-phase olive mill in Sitia, Crete (2008), (D) a cocomposting unit in Malia Heraklion, Crete in a II-phase Olive Mill (2009), (E) full-scale cocomposting unit in Katastari, Zakinthos in a III-phase Olive Mill (2009), and (F) a large scale cocomposting unit in Kalivia, Chalkidiki in a II-phase Olive Mill (2013).

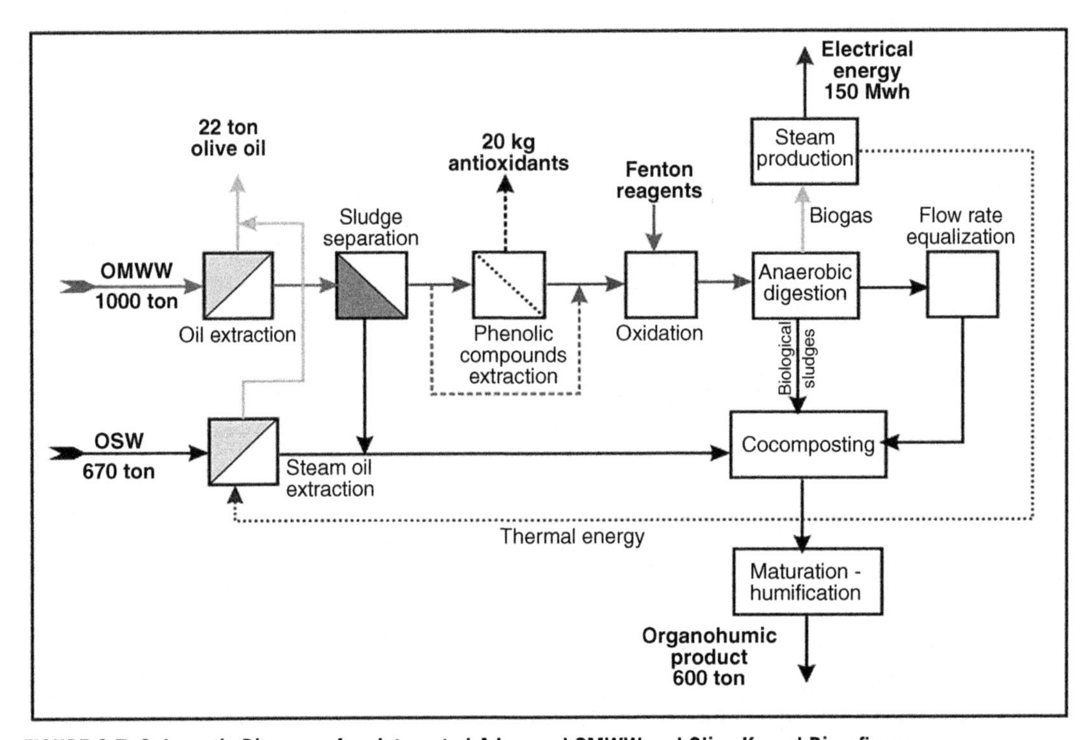

FIGURE 6.7 Schematic Diagram of an Integrated Advanced OMWW and Olive Kernel Biorefinery

Initially, the olive oil content of OMWW which is around 1–2% is recovered in a waste olive oil extraction unit using a salting out method. After oil removal, the sludge of OMWW is separated from liquid and it is mixed with Olive Kernel. The liquid content enters through an ion exchange resin where the phenolic compounds are extracted. With this method, up to 80% of the phenolic compounds can be extracted. The most important marketable antioxidant by-products found in the pulp of olives are tyrosol, hydroxyl-tyrosol, oleuropein, oleocein, and oleocanthal that can be used in the pharmaceutical industry (Karkoula et al., 2012). However, not all OMWW undergoes through phenolic extraction removal as there are market limitations in the absorbance of the final product. After the extraction of the phenolic compounds, OMWW are oxidized by using Fenton reagents where the OMWW is completely detoxified and enters into an anaerobic digestion unit. In anaerobic treatment, biological conversion of the organic matter to carbon dioxide and methane takes place. There are three outputs from the digester, the biogas that is used for energy recovery, the biological sludge which is fed to the cocomposting process, and the effluent which is also used as a feed and a moisture equalizer in the cocomposting process. The main raw material for the cocomposting process is Olive Kernel, which is mixed with sludges from OMWW for the production of an organohumic product. After cocomposting, the compost enters to the maturation/humification stage where the material is stabilized. At the end of the humification process, a stable organic soil conditioner has been produced (Vlyssides et al., 2004, 2014).

The earlier mentioned procedure can be implemented continuously and utilizes all the waste streams of an olive mill leaving zero wastes at the end. Initially, the pretreatment/detoxification units are necessary so as to allow the biological procedures to carry out with no inhibitory phenomena. Therefore, the oil extraction and phenolic compounds recovery processes not only produce two valuable products but also detoxify OMWW. Hence, the amount of Fenton reagents necessary to detoxify OMWW drops significantly, reducing in this way the production cost of the process. From 1000 ton of OMWW and 670 ton of Olive Kernel, 22 ton of olive oil can be extracted, 20 kg of antioxidants, energy production equivalent to 150 MWh and 600 ton of organic fertilizer. Market value of waste olive oil is around 0.9 €/L while the price of biogas is 0.21 €/kWh. Finally, the market price for compost is around 100–200 €/t (Vlyssides et al., 2014, 2004). On the other hand, the market of phenolic compounds is not well-established yet. Pure phenolic compounds can even reach several thousand dollars per gram. However, it is more likely for someone to produce concentrated phenolic extracts rather than pure compounds which can reach a market price of approximately 100–200 € per 100 mg (Schievano et al., 2015)

6.8.2 THE CASE OF AGIOS KONSTANTINOS

In 2013, the first integrated approach of OMWW treatment was implemented at a small olive mill in Agios Konstantinos, in central Greece, following the procedure described earlier. In Fig. 6.8, the constructed facilities of oil and phenolic compounds extraction units together with the anaerobic digester and the composting facility are illustrated. The construction of this unit was funded by ESPA 2007–14 projects, Organohumiki Thrakis P.C., NTUA, and NUTRIA SA. The project's code was 83-BET-2013 and it was entitled "Biorefinery using olive mill wastes" (Vlyssides, 2002). A UASB type of anaerobic reactor was constructed, the volume of which was 50 m^3. The reactor was designed to treat 10–20 m^3/d of detoxified OMWW by using Fenton's reagents and produce 160–220 m^3 CH_4/d. Also it was able to produce 100–200 L of waste oil per day and 3.6–7.2 ton/d of organohumic product.

FIGURE 6.8 Integrated Industrial-Scale Facility for Treating OMWW in Agios Konstantinos, Fthiotida

(A) Oil and phenolic compounds extraction unit, (B) anaerobic digestion unit after oxidation with Fenton's reagents, and (C) cocomposting unit.

6.8.3 THE CASE OF ORGANOHUMIKI THRAKIS P.C.

A second case of installing an advanced biorefinery strategy for valorizing OMWW and Olive Kernel in an integrated point of view was developed in the province of Thrakis by Organohumiki Thrakis P.C., in 2013 (Organohumiki Thrakis, 2015). Contrary to the previous case study of Agios Konstantinos, this facility was not installed inside an olive mill but it was built as a separate plant for treating OMWs derived from all the nearby olive mills. The plant at the moment can process 3000–4000 tons/year of OMWW and 2000–3000 tons/year of Olive Kernel. However maximum capacity of the plant will be reached in the coming years for 15,000 tons/year of OMWW and 6,000 tons/year of Olive Kernel. Fig. 6.9A shows a part of the facility in the province of Thraki where the preparation of the compost and the detoxification of the OMWW take place, while Fig. 6.9B illustrates the schematic diagram of the process that the company follows in order to produce an organic fertilizer from OMWW and Olive Kernel. At the imminent plans of the company is the construction of an anaerobic digester which will be able to process 35–40 m^3 per day of OMWW. The produced biogas will be used for the

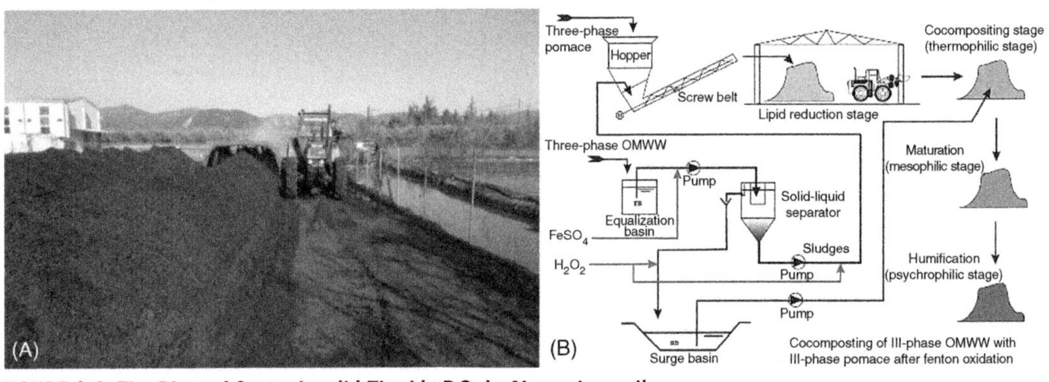

FIGURE 6.9 The Plant of Organohumiki Thrakis P.C. in Alexandroupolis

(A) Agitating cocomposted windrow piles and (B) process flow diagram of the cocomposting unit (Vlyssides et al., 2015)

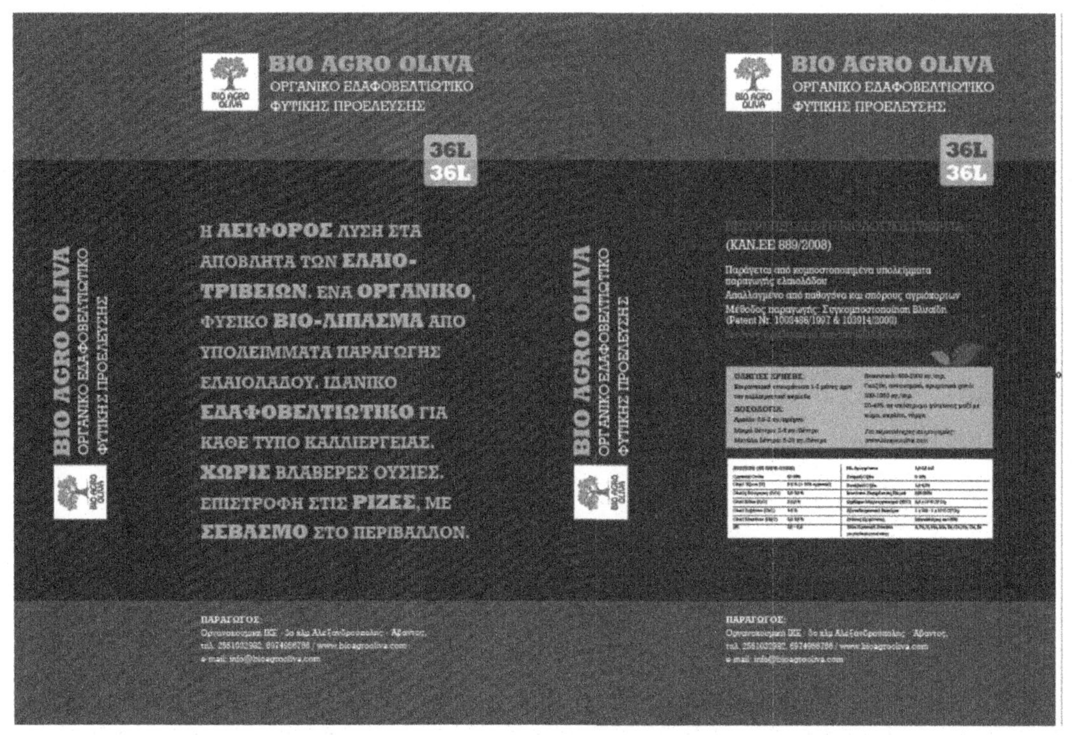

FIGURE 6.10 Compost Bag From Cocomposting Process of OMWW and OMSW Under the Brand Name BIO AGRO OLIVA Produced by Organohumiki Thrakis P.C

energy requirements of the plant. All the earlier information has been provided by our industrial partner Organohumiki Thrakis P.C. (Organohumiki Thrakis, 2015). At the end of the composting process, the compost is released to the market under the brand name BIO AGRO OLIVA (Fig. 6.10). The characteristics of this material can be seen in Table 6.1 (Organohumiki Thrakis, 2015).

6.9 ECONOMIC PERSPECTIVES OF OMWW AND OLIVE KERNEL BIOREFINERY

A cost estimation has been conducted by using economic data from the two plant facilities. The cost of raw materials is entirely attributed to the cost of Olive Kernel which has a fob (freight-on-board) price of around 20 €/ton. The transportation cost depends on the distance covered. An average transportation cost has been estimated to be around 22 €/ton. The transportation of the OMWW is more costly but it is usually paid entirely by the olive mill owners. In Fig. 6.11, four different production cost scenarios have been estimated (horizontal lines). The simplest scenario indicates that when compost is produced from raw material derived from a single olive mill, the production cost is around 110 €/ton. If the plant is a centralized installation unit for 10 olive mills, the production cost drops to 80 €/ton. With the coproduction of waste oil, the compost production process can decrease to 50 €/ton. Finally,

Table 6.1 Characteristics of the Soil Conditioner Produced from Organohumiki Thrakis P.C. (Personal Communication)

Parameter	Value
Water holding capacity	248 g/L
Germination index	235%
C/N ratio	17.9
Organic matter	85.5%
Ash	12.2%
Organic carbon	43.5%
Total nitrogen	2.42%
Total phosphorus	0.6–0.8%
Cation exchange capacity (CEC)	62 meq/100 g
Humic compounds	5–15%
pH	7.8

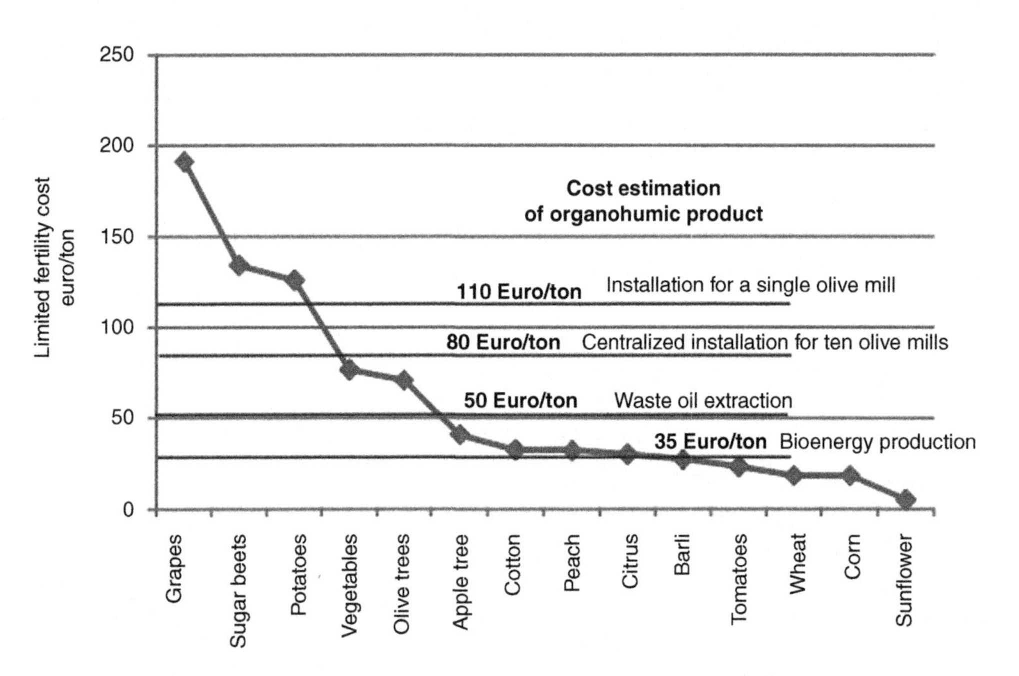

FIGURE 6.11 Production Cost of Humic Compost in Relation With the Capacity of the Plant and the Level of Applied Technology (Vlyssides, 2012)

if an anaerobic digestion unit is also installed for the production of energy, this cost is minimized to 35 €/ton (Vlyssides, 2012).

The points in Fig. 6.11, showed in *x* axis, illustrate the cost regarding the use of fertilizers for different types of cultivations. Each point above the horizontal lines indicates that a shift to the organohumic compost will be more beneficiary in terms of economics than the current fertilization method. For instance, organohumic product derived from an installation using a single olive mill can replace the type of fertilization used in grapes, sugar beets, and/or potatoes while organohumic product derived from a centralized unit that extracts oil and produces bioenergy can replace the fertilization type for most of the cultivations in terms of costing as shown in Fig. 6.11 (Vlyssides, 2012).

6.10 CONCLUSIONS

In this chapter, the implementation of the Fenton oxidation process as an effective and cost inexpensive pretreatment method for the successful detoxification of OMWW was illustrated. Several industrially developed case studies were described for treating OMWW and cocomposting it with Olive Kernel. Moreover, an integrated approach for producing a spectrum of added-value chemicals, bioenergy, and commodities was proposed. Under this perspective, we have underlined the opportunity to terminate the problematic treatment and disposal of OMWW by developing centralized facilities that receive and detoxify OMWW and cocompost it with Olive Kernel. It was also shown that in order to increase the sustainability of these facilities, added-value chemicals like phenolic compounds, marketable products like waste oil and bioenergy through anaerobic digestion need to be extracted and produced prior to cocomposting. Apart from the environmental benefits of this approach that leaves zero wastes there are also benefits in the socio-economic sector as the construction and development of these facilities will boost the local rural economy.

REFERENCES

Amaral-Silva, N., Martins, R.C., Castro-Silva, S., Quinta-Ferreira, R.M., 2016. Integration of traditional systems and AOP's technologies on the industrial treatment for olive mill wastewaters. Environ. Technol. 3330, 1–41.

Babuponnusami, A., Muthukumar, K., 2014. A review on Fenton and improvements to the Fenton process for wastewater treatment. J. Environ. Chem. Eng. 2, 557–572.

Barbusinski, K., 2009. Fenton reaction—controversy concerning the chemistry. Ecol. Chem. Eng. S 16, 347–358.

Bautista, P., Mohedano, A.F., Gilarranz, M.A., Casas, J.A., Rodriguez, J.J., 2007. Application of Fenton oxidation to cosmetic wastewaters treatment. J. Hazard. Mater. 143, 128–134.

Bautista, P., Mohedano, A.F., Casas, J.A., Zazo, J.A., Rodriguez, J.J., 2008. An overview of the application of Fenton oxidation to industrial wastewaters treatment. J. Chem. Technol. Biotechnol. 83, 1323–1338.

Christodoulou, V., Bampidis, V.A., Robinson, P.H., Israilides, C.J., Giouzelyiannis, A., Vlyssides, A., 2007. Nutritional and net energy value of fermented olive wastes in rations of lactating ewes. Czech J. Anim. Sci. 52, 456–462.

Christodoulou, V., Bampidis, V.A., Israilides, C.J., Robinson, P.H., Giouzelyiannis, A., Vlyssides, A., 2008. Nutritional value of fermented olive wastes in growing lamb rations. Anim. Feed Sci. Technol. 141, 375–383.

Comninellis, C., Kapalka, A., Malato, S., Parsons, S.A., Poulios, I., Mantzavinos, D., 2008. Advanced oxidation processes for water treatment: advances and trends for R&D. J. Chem. Technol. Biotechnol. 83, 769–776.

Cooperband, L., 2002. The Art and Science of Composting A Resource for Farmers and Compost Producers. University of Wisconsin-Madison, Center for Integrated Agricultural Systems, Madison, WI, United States, pp. 1–14.

Dareioti, M.A., Dokianakis, S.N., Stamatelatou, K., Zafiri, C., Kornaros, M., 2010. Exploitation of olive mill wastewater and liquid cow manure for biogas production. Waste Manag. 30, 1841–1848.

de Bertoldi, M., Vallini, G., Pera, A., 1983. The biology of composting: a review. Waste Manag. Res. 1, 157–176.

De Nobili, M., Petrussi, F., 1988. Humification index (HI) as evaluation of the stabilization degree during composting. J. Ferment. Technol. 66, 577–583.

DellaGreca, M., Monaco, P., Pinto, G., Pollio, A., Previtera, L., Temussi, F., 2001. Phytotoxicity of low-molecular-weight phenols from olive mill waste waters. Bull. Environ. Contam. Toxicol. 67, 352–359.

Dogruel, S., Olmez-Hanci, T., Kartal, Z., Arslan-Alaton, I., Orhon, D., 2009. Effect of Fenton's oxidation on the particle size distribution of organic carbon in olive mill wastewater. Water Res. 43, 3974–3983.

El-Gohary, F.A., Badawy, M.I., El-Khateeb, M.A., El-Kalliny, A.S., 2009. Integrated treatment of olive mill wastewater (OMW) by the combination of Fenton's reaction and anaerobic treatment. J. Hazard. Mater. 162, 1536–1541.

Enami, S., Sakamoto, Y., Colussi, A.J., 2014. Fenton chemistry at aqueous interfaces. In: Proceedings of the National Academy of Sciences of the United States of America.

Fenton, H.J.H., 1894. Oxidation of tartaric acid in presence of iron. J. Chem. Soc. Trans. 65, 899–910.

Fezzani, B., Cheikh, R. Ben, 2010. Two-phase anaerobic co-digestion of olive mill wastes in semi-continuous digesters at mesophilic temperature. Bioresour. Technol. 101, 1628–1634.

Filidei, S., Masciandaro, G., Ceccanti, B., 2003. Anaerobic digestion of olive oil mill effluents: evaluation of wastewater organic load and phytotoxicity reduction. Water Air. Soil Pollut. 145, 79–94.

Finstein, M.S., Miller, F.C., Strom, P.F., 1986. Monitoring and evaluating composting process performance. Water Pollut. Control Fed. 58, 272–278.

Gelegenis, J., Georgakakis, D., Angelidaki, I., Christopoulou, N., Goumenaki, M., 2007. Optimization of biogas production from olive-oil mill wastewater, by codigesting with diluted poultry-manure. Appl. Energy 84, 646–663.

Georgacakis, D., Kyritsis, S., Manios, B., Vlyssides, A.G., 1987. Economic optimization of energy production from olive oil wastewater Peza, Heraklion (Creta): a case study. In: First EEC Symposium on Energy from Agricultural Wastes, Italy.

Gogate, P.R., Pandit, A.B., 2004. A review of imperative technologies for wastewater treatment I: oxidation technologies at ambient conditions. Adv. Environ. Res. 8, 501–551.

Haug, R.T., 1980. Compost engineering; principles and practice. Technomic Publishing, Boca Raton, Florida.

Karkoula, E., Skantzari, A., Melliou, E., Magiatis, P., 2012. Direct measurement of oleocanthal and oleacein levels in olive oil by quantitative 1H NMR. Establishment of a new index for the characterization of extra virgin olive oils. J. Agric. Food Chem. 60, 11696–11703.

Lin, S.H., Lo, C.C., 1997. Fenton process for treatment of desizing wastewater. Water Res. 31, 2050–2056.

Loutatidou, S., 2012. Fenton's influence in activity of sludge from anaerobic digestion of olive mill waste water. (Thesis dissertation). Retrieved from Digital library of National Technical University of Athens.

Lucas, M.S., Peres, J.A., 2009. Removal of COD from olive mill wastewater by Fenton's reagent: kinetic study. J. Hazard. Mater. 168, 1253–1259.

MacGregor, S.T., Miller, F.C., Psarianos, K.M., Finstein, M.S., 1981. Composting process control based on interaction between microbial heat output and temperature. Appl. Environ. Microbiol. 41, 1321–1330.

Mantzavinos, D., Kalogerakis, N., 2005. Treatment of olive mill effluents: part I. Organic matter degradation by chemical and biological processes—an overview. Environ. Int. 31, 289–295.

Mantzavinos, D., Psillakis, E., 2004. Enhancement of biodegradability of industrial wastewaters by chemical oxidation pre-treatment. J. Chem. Technol. Biotechnol. 79, 431–454.

Muktadirul Bari Chowdhury, A.K.M., Akratos, C.S., Vayenas, D.V., Pavlou, S., 2013. Olive mill waste composting: a review. Int. Biodeterior. Biodegrad. 85, 108–119.

Niaounakis, M., Halvadakis, C., 2004. Olive Processing Waste Management, First. ed. Typothito-George Dardanos, Athens, Greece.

Oller, I., Malato, S., Sánchez-Pérez, J.A., 2011. Combination of advanced oxidation processes and biological treatments for wastewater decontamination-a review. Sci. Total Environ. 409, 4141–4166.

Organohumiki Thrakis P.C., 2015. Available from: www.bioagrooliva.com

Papadimitriou, E.K., Chatjipavlidis, I., Balis, C., 1997. Application of composting to olive mill wastewater treatment. Environ. Technol. 18, 101–107.

Paraskeva, P., Diamadopoulos, E., 2006. Technologies for olive mill wastewater (OMW) treatment: a review. J. Chem. Technol. Biotechnol. 81, 1475–1485.

Pera-Titus, M., García-Molina, V., Baños, M.A., Giménez, J., Esplugas, S., 2004. Degradation of chlorophenols by means of advanced oxidation processes: a general review. Appl. Catal. B Environ. 47, 219–256.

Ranade, V.V., Bhandari, V.M., 2014. Industrial Wastewater Treatment, Recycling And Reuse, Industrial Wastewater Treatment, Recycling And Reuse. Elsevier, Oxford.

Rincon, B., Borja, R., Martin, M.A., Martin, A., 2010. Kinetic study of the methanogenic step of a two-stage anaerobic digestion process treating olive mill solid residue. Chem. Eng. J. 160, 215–219.

Rivas, F.J., Beltrán, F.J., Frades, J., Buxeda, P., 2001. Oxidation of p-hydroxybenzoic acid by Fenton's reagent. Water Res. 35, 387–396.

Roig, A., Cayuela, M.L., Sánchez-Monedero, M.A., 2006. An overview on olive mill wastes and their valorisation methods. Waste Manag. 26, 960–969.

Rozzi, A., Malpei, F., 1996. Treatment and disposal of olive mill effluents. Int. Biodeterior. Biodegradation 38, 135–144.

Schievano, A., Adani, F., Buessing, L., Botto, A., Casoliba, E.N., Rossoni, M., Goldfarb, J.L., 2015. An integrated biorefinery concept for olive mill waste management: supercritical CO_2 extraction and energy recovery. Green Chem. 17, 2874–2887.

Stasinakis, A.S., 2008. Use of selected advanced oxidation processes (AOPs) for wastewater treatment—a mini review. Glob. Nest J. 10, 376–385.

Tomati, U., Galli, E., Pasetti, L., Volterra, E., 1995. Bioremediation of olive-mill wastewaters by composting. Waste Manag. Res. 13, 509–518.

Vasiliadou, I.A., Muktadirul Bari Chowdhury, A.K.M., Akratos, C.S., Tekerlekopoulou, A.G., Pavlou, S., Vayenas, D.V., 2015. Mathematical modeling of olive mill waste composting process. Waste Manag. 43, 61–71.

Vlyssides, A.G., 2000. Method of useful exploitation of liquid effluents of a high organic load by co-processing composting-topsoil formation with solid organic waste and agricultural by-products. GR Patent 1003486 filed February 26, 1997 and issued November 30, 2000.

Vlyssides, A.G., 2002. A method of processing oil-plant wastes and resulting organo-humic product. GR Patent 1003914, filed May 26, 2000 and issued June 25, 2002.

Vlyssides, A.G., 2012. Sustainable olive oil production. In: The Future of Eco-Innovation: The Role of Business Models in Green Transformation. OECD/European Commission/Nordic Innovation Joint Workshop, Copenhagen.

Vlyssides, A.G., Parlavanza, M., Balis, C., 1989. Co-composting as a system for handling of liquid wastes from olive oil mills. In: Third International Conference on Composting. Athens.

Vlyssides, A.G., Grivas, A., Kalergis, C., 1995. Co-composting as a new approach for treating olive oil process effluents: application in a pilot plant. In: Second International Symposium Wastewater Reclamation and Reuse. Iraklio (Crete).

Vlyssides, A.G., Bouranis, D.L., Loizidou, M., Karvouni, G., 1996. Study of a demonstration plant for the co-composting of olive-oil-processing wastewater and solid residue. Bioresour. Technol. 56, 187–193.

Vlyssides, A., Loizidou, M., Gimouhopoulos, K., Zorpas, K., 1998. Olive oil processing wastes production and their characteristics in relation to olive oil extraction methods. Fresenius Environ. Bull. 7, 308–313.

Vlyssides, A.G., Loizides, M., Karlis, P., Simonetis, I.S., 2002. Olive stone oil processing wastes production and their characteristics. Fresenius Environ. Bull. 12b, 1114–1118.

Vlyssides, A.G., Loizides, M., Barampouti, E.M., Mai, S., 2003. Treatment of high organic and toxic strength effluents using the method of co-composting with organic residuals. In: The Eighth European Biosolids and Organic Residuals Conference and Exhibition. Wakefield (UK).

Vlyssides, A., Loizides, M., Karlis, P., 2004. Integrated strategic approach for reusing olive oil extraction by-products. J. Clean. Prod. 12, 603–611.

Vlyssides, A.G., Mai, S., Barampouti, E.M., 2009. An integrated mathematical model for co-composting of agricultural solid wastes with industrial wastewater. Bioresour. Technol. 100, 4797–4806.

Vlyssides, A., Moutsatsou, A., Tsimas, S., Mai, S., Barampouti, E.M., 2010. Influence of oxidation pre-treatment on anaerobic digestion of olive mill wastewater. In: Environmental 2010 Situation and Perspectives for the European Union. Porto, Portugal, p. 8.

Vlyssides, A., Van der Smissen, N., Mai, S., Barampouti, E.M., 2014. Sustainable cultivation of olive trees by reusing olive mill wastes after effective co composting treatment processes. In: Building Organic Bridges. Johann Heinrich von Thünen-Institut.

Vlyssides, A., Barampouti, E.M., Mai, S., 2015. In: Sideris, J. (Ed.). In: Industrial Pollution, first ed., Athens, Greece.

Walling, C., 1975. Fenton's reagent revisited. Acc. Chem. Res. 8, 125–131.

CHAPTER

7

INTEGRATED BIOLOGICAL TREATMENT OF OLIVE MILL WASTE COMBINING AEROBIC BIOLOGICAL TREATMENT, CONSTRUCTED WETLANDS, AND COMPOSTING

Athanasia G. Tekerlekopoulou*, Christos S. Akratos*, Dimitrios V. Vayenas,†**

**Department of Environmental and Natural Resources Management, University of Patras, Agrinio, Greece; **Department of Chemical Engineering, University of Patras, Rio, Patras, Greece; †Institute of Chemical Engineering Sciences (ICE-HT), Platani, Patras, Greece*

7.1 INTRODUCTION

7.1.1 OLIVE MILL WASTES (OMW)

During the extraction of oil from olives, considerable volumes of solid wastes and wastewaters are generated that can have a major impact on land and aquatic environments due to their high organic loads and phytotoxicity. Two processes are currently used for the extraction of olive oil: the three-phase and the two-phase. The continuous centrifuge two-phase process generates a semisolid waste termed olive wet pomace or olive wet cake that is characterized by a high humidity level (62%) (Valta et al., 2015). The two by-products produced by the three-phase process are a solid residue (comprising olive pomace and olive leaves) with a humidity of about 55%, and large volumes of aqueous liquid known as olive mill wastewater (OMWW). The three-phase process usually yields 20% olive oil, 30% olive pomace waste, and 50% OMWW, which means that 4 times more waste is produced than the actual product. On the other hand, two-phase olive mills require a smaller amount of water for oil separation, and therefore produce smaller volumes (but more toxic) of wastes than three-phase systems. Regardless of these differences, both processes produce effluents with highly phytotoxic and antimicrobial properties, mainly due to the presence of poisonous, caustic crystalline compounds known as phenols.

7.1.2 CHARACTERISTICS OF OMW IN GREECE

Greece has about 2300 small-scale, rural, agroindustrial units that extract olive oil (Michailides et al., 2011). These are generally three-phase systems and produce significant quantities of solid wastes with outputs of 0.35 tons of olive pomace and 0.05 tons of olive leaves per tonne of olives

Olive Mill Waste. http://dx.doi.org/10.1016/B978-0-12-805314-0.00007-8

(Niaounakis and Halvadakis, 2004). The huge quantities of olive pomace and olive leaves produced within the short oil extraction season cause serious management problems in terms of volume and space. OMWW is formed from the water content of the olive fruit itself and the water used to wash and process them. Typically, the weight composition of OMWW is 83–96% water, 3.5–15% organics, and 0.5–2% mineral salts. The maximum chemical oxygen demand (COD) and biological oxygen demand (BOD) reach concentrations of 220 and 100 g/L, respectively (Paraskeva and Diamadopoulos, 2006). OMWW contains high concentrations of recalcitrant compounds such as lignins and tannins, which give it a characteristic dark color (52.3–180 g/L Pt-Co units) (Niaounakis and Halvadakis, 2004), but most importantly, it contains phenolic compounds and long-chain fatty acids which are toxic to microorganisms and plants. The phenolic compounds are present in the residue as a mixture of monomeric aromatics and as polymerized heterogeneous pigments (Stasinakis et al., 2008). The concentration of phenolics in OMWW varies greatly from 0.5 to 24 g/L. As an example of the scale of the environmental impact of OMWW, it should be noted that 10 million m^3/year of liquid effluents from three-phase systems corresponds to an equivalent load of the wastewater generated from about 20 million people (McNamara et al., 2008).

7.1.3 OMW TREATMENT METHODS

Due to the content of these effluents and potential environmental impacts, OMW should be treated before being discharged. Literature analysis concludes that all treatment processes developed for domestic and industrial wastes have been tested on OMW at some point, but that none of them appeared fully suitable to be adopted widely. The treatment methods investigated include physicochemical technologies, oxidation/advanced oxidation processes, biological (aerobic or anaerobic) technologies, as well as combinations of these.

Simple physical processes such as dilution, evaporation, sedimentation, filtration, and centrifugation, have been employed to treat OMWW, however none of these alone is able to reduce organic loads and toxicity to acceptable limits (Paraskeva and Diamadopoulos, 2006). The presence of hardly or nonbiodegradable dissolved organic pollutants in OMWW, mainly tannic acid, requires the employment of physicochemical treatments, which involve use of additional chemicals for their neutralization, flocculation, precipitation, adsorption, chemical oxidation, and ion exchange. Various different coagulants and polyelectrolytes have been investigated, such as alum, ferric, starch, chitosan, lime, anionic/cationic polyelectrolytes, etc. (Pelendridou et al., 2014). The viability of these technologies is questionable as they are relatively expensive (large quantities of chemicals are required), generate large volumes of sludge, and/or do not produce high-quality effluents. Therefore, these treatment processes must be combined with further purification phases.

Oxidation and advanced oxidation processes have been extensively tested to treat OMWW. These processes include ozonation, UV irradiation, photocatalysis, hydrogen peroxide/ferrous iron oxidation (the so-called Fenton's reagent), electrochemical oxidation, wet air oxidation, as well as various combinations of the earlier processes. Processes, such as electrochemical oxidation, Fenton oxidation, ozonation, and photocatalysis, can only achieve partial decontamination even after prolonged treatment times (Mantzavinos and Kalogerakis, 2005). High-temperature, high-pressure processes, such as wet air oxidation can achieve high COD removals in relatively short treatment times. However, one major drawback is that wet air oxidation is usually an expensive process to install and operate because of the intense conditions required.

There is little doubt that biological processes are the most environmentally friendly, reliable, and in most cases, cost effective methods of wastewater treatment. Synoptically, biological treatment employs the use of microorganisms to break down biodegradable chemical species and remove inorganic nutrients. Several studies have reported the biological treatment of OMWW and the potential reuse of the effluent. However, care needs to be taken in the selection of the microorganisms employed and their adaptation to treat OMW because the phenolic compounds contained in these wastes are inhibitory to microorganisms (Mantzavinos and Kalogerakis, 2005).

Bioreactor technology, which could be applied to remove organic pollutants from OMWW, is based on either aerobic or anaerobic bioreactor configuration. A variety of anaerobic bioreactor types have been tested for the treatment of OMWW (suspended biomass systems, upflow anaerobic sludge blanket, anaerobic baffled reactor, upflow and downflow anaerobic filters, and the fluidized bed). There are several potential advantages of this approach including the production of a useable biogas fuel (typical yields of 7–8 m^3/m^3 OMWW), low waste biomass generation (0.15–0.25 kg/kg COD removed) and the generation of an excellent soil conditioner and fertilizer (McNamara et al., 2008). Reductions in COD of 70–90% have been reported in some anaerobic processes (Paraskeva and Diamadopoulos, 2006). Results of single anaerobic treatment are not always satisfactory and some form of pretreatment, apart from a simple dilution and nutrient/alkalinity adjustment, is usually necessary. The presence of compounds toxic to methanogens in OMWW appears to be a significant problem in the anaerobic digestion of OMWW.

Aerobic bioreactor configurations, such as activated sludge or sequencing batch reactors have also been examined for the treatment of OMWW. Although the capital costs of aerobic processes are lower than anaerobic processes, the operating costs are substantially greater due to the process' oxygen requirements. Studies on activated sludge treatment after microorganism adaptation in laboratory-scale bioreactors report COD removal rates of 81–84% and phenol removal of about 90% (Paraskeva and Diamadopoulos, 2006).

Composting is another biological technology used for the disposal of OMW. Composting of OMW and by-products (mixed with a relatively large variety of agricultural and agroindustrial residues) provides efficient soil amendments and fertilizers, which could perform well in the market and complement and/or replace relevant chemical compounds. Of course, there are still issues that need addressing, and these are mostly related to possible adverse effects of the end product (e.g., high salinity, phytotoxicity), its fertilizing value and market price as compared with similar commercial compounds (Ouzounidou et al., 2010; Chowdhury et al., 2013, 2015). The composting process is presented analytically in Chapter 8.

Although constructed wetlands have been used to treat a variety of wastewaters, only few studies have been published on their treatment of OMWW (Sultana et al., 2015). The high organic load of OMWW makes constructed wetlands attractive only as a polishing treatment method. Thus, several prior-treatment technologies are employed. Constructed wetlands treat OMWW rather efficiently and have many advantages such as low construction and operation costs, and environmental and aesthetic benefits (Kapellakis et al., 2012; Grafias et al., 2010; Yalcuk et al., 2010; Herouvim et al., 2011; Michailides et al., 2015). However, constructed wetlands are a land-based technology and cannot be used everywhere. On sites with adequate space, the benefits of wetlands can result in substantial cost savings, especially for systems that have to operate over long periods of time.

Complete abatement of OMWW pollutants cannot be achieved by the adoption of a single process. Consequently, the effective management of OMWWs includes their pretreatment before the application of the selected process. Many hybrid systems for OMWW treatment have been developed and tested recently. These systems usually consist of a physicochemical method combined with a biological one, such as: ozonation/aerobic biological treatment, chemical oxidative procedure with $FeCl_3$/aerobic biological treatment, coagulation–flocculation/anaerobic biological process, electrochemical oxidation combined with constructed wetlands, electro-Fenton combined with anaerobic digestion or an aerobic biological process, and sequential treatment with adsorption, biological and photo-Fenton oxidation (Benitez et al., 1999; Aytar et al., 2013; Pelendridou et al., 2014). Each of these methods has both advantages and disadvantages. However, none of the proposed methods can be considered superior over the others in terms of its effectiveness and environmental and economic impact.

This chapter presents an integrated method including biological systems (an aerobic trickling filter, a constructed wetland, and composting) that offers a final solution for the complete treatment of OMW.

7.2 INTEGRATED BIOLOGICAL APPROACHES FOR OLIVE MILL WASTE TREATMENT

Although the problem of OMW treatment could be more or less solved technologically, it is still far from being solved realistically. In terms of both economics and convenience, existing OMWW treatment technologies have not yet been developed for small-scale enterprises. Consequently, the practices currently applied include land disposal, discharge into nearby rivers, lakes or seas, and storage/evaporation lagoons.

7.2.1 AEROBIC BIOLOGICAL TREATMENT OF OMW USING TRICKLING FILTERS

Literature analysis on the present situation of OMWW treatment shows that biological treatments (aerobic and anaerobic) with selected microorganisms are the most environmentally compatible OMWW treatment methods (Hamdi et al., 1992; Mantzavinos and Kalogerakis, 2005; Arvanitoyannis et al., 2007; Jalilnejad et al., 2011; González-González and Cuadros, 2015). Aerobic biological treatments may have higher removal efficiencies than anaerobic; however, the former are limited by the unacceptable cost of the continuously provided mechanical aeration (Benitez et al., 1997; Lafi et al., 2009).

In almost all cases, aerobic biological processes involve suspended growth systems with mixed microbial populations adapted from municipal sludge (Rozzi and Malpei, 1996; Benitez et al., 1997; Günay and Çetin, 2013). However, these processes have limited capability of biologically degrading polyphenols or substances that are generally toxic to biological treatments (Hamdi et al., 1992) and inhibit the efficiency of these processes. Limited studies have been conducted under aerobic conditions using single step processes (Mantzavinos and Kalogerakis, 2005; Paraskeva and Diamadopoulos, 2006; Arvanitoyannis et al., 2007; McNamara et al., 2008). Therefore, combined aerobic/anaerobic biological treatments (Hamdi et al., 1992; Jalilnejad et al., 2011; González-González and Cuadros, 2015) or chemical-electrochemical/aerobic methods (Benitez et al., 1999; Aytar et al., 2013) were recently tested for their ability to remove toxic compounds present from the wastewaters described earlier, showing satisfactory results. However, in terms of both economics and convenience, these OMWW treatment technologies have not yet been developed in small-scale enterprises.

Attached growth systems can be an alternative option for the treatment of OMWW as they are more advanced than suspended biomass processes. In particular, they maintain a high concentration of

FIGURE 7.1

The (A) laboratory-scale, (B) pilot-scale, and (C) full-scale trickling filters for OMWW treatment.

microorganisms resulting in high removal rates at relatively small hydraulic retention times. However, they have not been studied extensively for their ability to treat OMWW.

An environmentally friendly and cost effective technology for OMWW treatment is presented in this section by using a trickling filter with mixed, aerobically grown, indigenous cultures originating from olive pulp. This method combines the advantages of aerobic treatment in an attached growth process, which is known to provide very high biomass concentration and treat high COD levels. In addition, the operation of a trickling filter has the advantage of not requiring an external air supply since air is naturally convected through the filter under high wastewater recirculation rates (Vayenas, 2011). The development process of this innovative system is subsequently presented and describes all experiments performed in laboratory-, pilot-, and full-scale trickling filters in order to elaborate and refine the performance, effectiveness, and sustainability of such a system.

7.2.1.1 Laboratory and Pilot-Scale Trickling Filters

Initially, experiments in a laboratory-scale filter were conducted (at 28°C) under sequencing batch operation with recirculation (Fig. 7.1A). OMWW was obtained from an olive mill (three-phase centrifugal) located near Amfilochia city in Aitoloakarnania Prefecture (Western Greece). The original inoculum of microorganisms was obtained from olive pulp. This source was selected since both olives and OMWW contain a significant amount of phenolic compounds.

The laboratory-scale trickling filter was a Plexiglas tube, 160 cm high with a 9 cm internal diameter. The support media of the reactor comprised hollow plastic tubes (a readily available, inexpensive, electrological material), of 1.6 cm internal diameter, 3 cm length, 500 m^2/m^3 specific surface area (*As*),

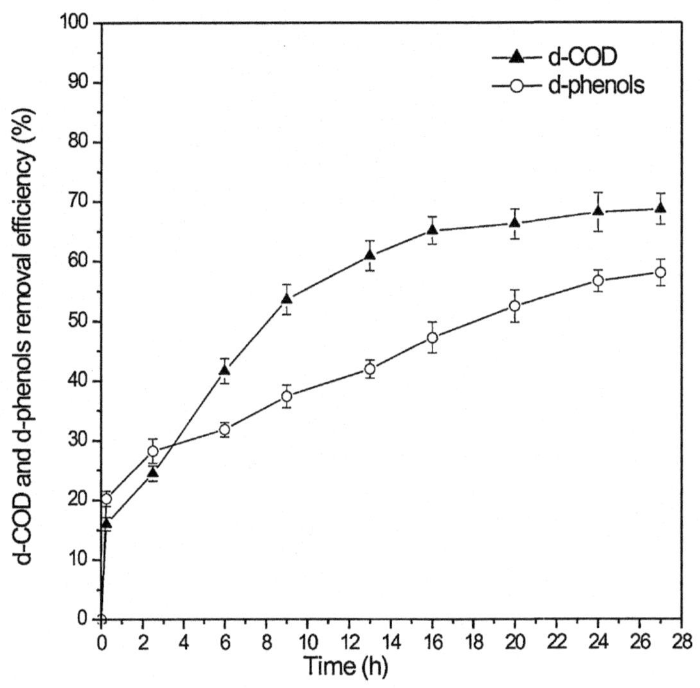

FIGURE 7.2 d-COD and d-Phenols Removal (%) in the Laboratory-Scale Filter

and 0.8 filter porosity (ε). The depth of support media in the filter was 143 cm. The pilot-scale filter was first operated as a batch reactor for a period of about 30 days to ensure attachment of the microorganism culture onto the support media and the development of a biofilm layer (start-up). The filter was then operated as a sequencing batch reactor with recirculation (draw-fill mode with recirculation). Recirculation was provided in order to obtain a completely mixed flow pattern within the bioreactor and sufficient exploitation of the filter. Dissolved oxygen was maintained above 4.5 mg/L throughout all the experiments. The average dissolved COD (d-COD) and dissolved phenolic (g syringic acid/L, d-phenols) contents of OMWW examined in the filter were 15.05 g/L and 2.9 g/L, respectively. Duration of the operating cycles was determined by the time that no further reduction of d-COD and dissolved-phenols (d-phenols) was observed. Filter backwash due to pore clogging from biomass growth was necessary and was performed at the end of each batch experiment using high upward water and air velocities. The sludge produced was subsequently treated in a composting unit (Chapter 8).

A series of operating cycles was performed, until a minimum period of 27 h was required for d-phenols and d-COD degradation of about 60 and 70%, respectively (Fig. 7.2). In addition, BOD measurements showed a BOD reduction of approximately 90%, which indicates that no further biological degradation was possible (Tziotzios et al., 2007). Based on the experiments conducted in the laboratory-scale filter, it can be concluded that the use of indigenous bacterial populations from olive pulp provides a certain advantage as it ensures durability under various operating conditions and provides high degradation rates of natural phenolic compounds, since indigenous populations are already acclimatized to high phenol concentrations. Additionally, the use of trickling filters under sequencing batch operation with recirculation reduces the operational cost of the reactor, since there is no need for an external oxygen supply.

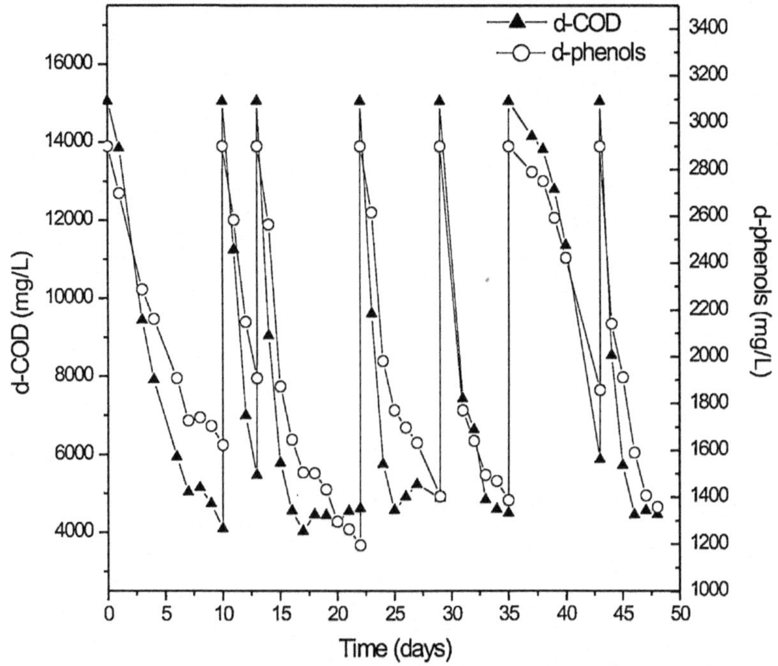

FIGURE 7.3 d-COD and d-Phenols Concentration in the Pilot-Scale Filter During Draw-Fill Operating Cycles With Recirculation

To validate and demonstrate the technology and generate scalable data that confirms the design and warrant of the system, a pilot plant was designed and operated within the site of an olive mill in Amfilochia, Western Greece (Fig. 7.1B). The pilot-scale filter consisted of a cylindrical polyethylene tank with a conical base, 1.5 m in height and 1.8 m of internal diameter. The support material was rippled plastic hollow tubes, 3.85 cm internal diameter and 4 cm in length, with specific surface area of 207.8 m^2/m^3 (*As*), and porosity 0.93 (ε). OMWW (500 L) was stored in a nearby tank and was continuously recirculated within the filter, from the top, by a rotary distributor. Inoculation of the filter was performed by using the mixed culture grown in the laboratory-scale filter. The pilot plant did not require an external oxygen supply since 11 ventilation ports were positioned at the base of the filter allowing air circulation. Sampling during the draw-fill cycles was performed on a daily basis. The d-phenols and d-COD concentrations recorded in the operating cycles are shown in Fig. 7.3. It was observed that after 3 days operation, d-COD removal reached about 70%, while d-phenols removal reached about 55% in 5 days operation, despite the low ambient temperature (16–20°C). These results indicated that an aerobic trickling filter is a very effective method for OMWW biodegradation.

Continuation of the earlier research led to the construction and operation of a full-scale trickling filter for the treatment of OMWW under real industrial conditions (Fig. 7.1C). d-COD and d-phenols removal, as well as several other parameters, were monitored throughout one olive harvest period (November, 2010–January, 2011) in order to evaluate the effectiveness of the proposed system.

FIGURE 7.4 The Full-Scale Trickling Filter for OMWW Treatment

(A) Fresh support media, (B) support media following several days of filter operation, (C) the rotating distributor, and (D) general view of the filter

7.2.1.2 Full-Scale Trickling Filter—A Case Study

The full-scale trickling filter was constructed on the site of the same olive mill in Amfilochia, Western Greece. On average this mill produces about 30 m^3/d of OMWW per harvest season (from November to January). Mean d-COD and d-phenol concentrations in the wastewater of this mill are about 45,000 and 9,000 mg/L, respectively. Wastewater temperature is about 35°C and its pH ranges between 4.5 and 5.0.

The full-scale OMWW treatment plant consisted of a trickling filter and a recirculation tank. The trickling filter was a metallic cylindrical tank of 1.8 m in diameter and 3 m in height. The porous media used was a random type high density polyethylene with a specific surface area of 188 m^2/m^3 (*As*), a specific weight of 47 kg/m^3, and filter porosity 95% (ε) (Figs. 7.1C and 7.4). The total height of the porous media inside the filter was 1.8 m. The filter was open to the atmosphere at the top, and a rotating distributor ensured uniform distribution of the wastewater at the top of the filter. In this application,

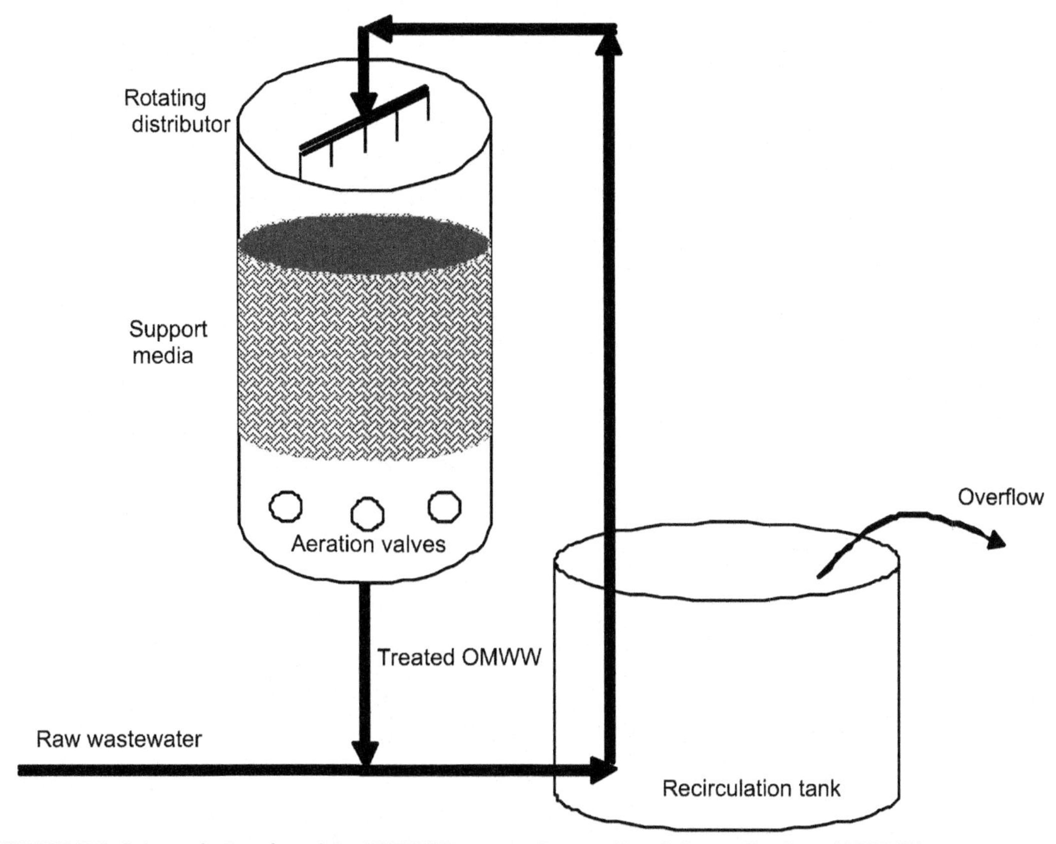

FIGURE 7.5 Schematic Drawing of the OMWW Treatment System (Greek Patent Number: 1007517)

only natural aeration took place, while mechanical aeration was used only for filter backwash. The driving force for air circulation through the filter is the temperature difference between the wastewater and the ambient air. Raw wastewater temperature was constant at 35°C. However, its temperature did not drop below 20°C within the recirculation tank even when olive oil production was interrupted (e.g., at night or during heavy rainfall), thus ensuring optimal temperature conditions.

Treated OMWW was routed to the bottom of a plastic cylindrical recirculation tank of 30 m^3 total volume, where it diluted the raw wastewater produced by the olive mill. The latest was also entered at the bottom of the same tank. A schematic drawing of the OMW treatment system is presented in Fig. 7.5. In order to ensure a hydraulic residence time of about 1 day, the size of the recirculation tank required was calculated to be 30 m^3 according to the mill's mean daily OMW production. However, for higher daily wastewater production, the hydraulic retention time was reduced (to about 12 h), and for lower daily production the hydraulic retention time was increased (up to 19 days at the end of the olive harvest period).

Inoculation of the filter was performed by using the mixed culture grown in the pilot-scale filter (Section 7.2.1.1). The filter was initially operated in batch mode with recirculation for 17 days. The mode was then changed to continuous operation with recirculation and filter efficiency reached a

Table 7.1 Statistics of Overall Influent and Effluent Concentrations and Removal Efficiencies in Each OMWW Treatment Unit

			d-COD (mg/L)		d-phenols (mg/L)		OP (mg/L)		NH_3 (mg/L)		TKN (mg/L)	
			Mean	SD	Mean	SD	Mean	SD	Mean	SD	Mean	SD
Full-scale trickling filter		*Influent*	42935	2294	9236	1084	146	40	3	1	909	226
		Effluent	21201	3819	4311	376	97	23	26	5	885	221
Pilot-scale constructed wetland	*Planted*	A_1	10192	4316	1923	586	47	24	149	44	280	47
		A_2	7932	3766	1545	549	38	26	122	34	168	38
		A_3	5325	2642	1088	395	21	18	72	32	117	31
		A_4	3580	2421	1014	302	12	10	36	26	99	81
	Planted	B_1	9215	3775	1861	568	47	20	160	53	252	55
		B_2	7893	4077	1462	584	28	26	84	57	196	56
		B_3	4967	3613	998	556	19	18	57	40	168	41
		B_4	3719	3226	657	359	11	8	32	28	97	75
	Unplanted	C_1	10342	4126	1814	635	47	27	138	54	420	59
		C_2	7438	3923	1359	582	29	20	92	41	210	47
		C_3	6266	3568	1230	549	23	22	80	45	170	41
		C_4	5214	3077	1025	509	15	16	52	49	145	95
Full-scale constructed wetland			1584	82	248	12	5.2	0.4	NM	NM	48	1.1

d-phenols, dissolved phenols; NM, not measured; OP, orthophosphate; TKN, total Kjeldahl nitrogen.
Herouvim, E., Akratos, C.S., Tekerlekopoulou, A.G., Vayenas, D.V., 2011. Treatment of olive mill wastewater in pilot-scale vertical flow constructed wetlands. Ecol. Eng. 37, 931–939; Michailides, M., Panagopoulos, P., Akratos, C.S., Tekerlekopoulou, A.G., Vayenas, D.V., 2011. A full-scale system for aerobic biological treatment of olive mill wastewater. J. Chem. Technol. Biotechnol. 86, 888–892; Michailides, M., Tatoulis, T., Sultana, M-Y., Tekerlekopoulou, A.G., Konstantinou, I., Akratos, C.S., Vayenas, D.V., 2015. Start-up of a free water surface constructed wetland for treating olive mill wastewater. Hemijskaindustrija 69, 577.

constant level in about 1 week. Recirculation provided dissolved oxygen concentrations of 1–1.5 mg/L at the filter outlet. During continuous operation with recirculation, an overflow system at the top of the recirculation tank (Fig. 7.5) routed the treated wastewater into a square concrete basin of 500 m^3. Backwash of the filter was necessary every 2 weeks to remove clogging due to biomass growth and suspended solids attached to the porous media. The sludge produced was subsequently treated in a composting unit adding olive leaves and olive pomace.

Under these operating conditions and without pretreatment, the trickling filter reduced d-COD and d-phenol concentrations by about 50–60%, for an organic load of 530 kg $COD/m^2/d$ (Table 7.1, Fig. 7.6, Michailides et al., 2011). Nutrients (nitrogen and phosphorus) were also removed at lower percentages (33 and 2.7%, respectively). Ammonia concentration in the trickling filter only rose from 3

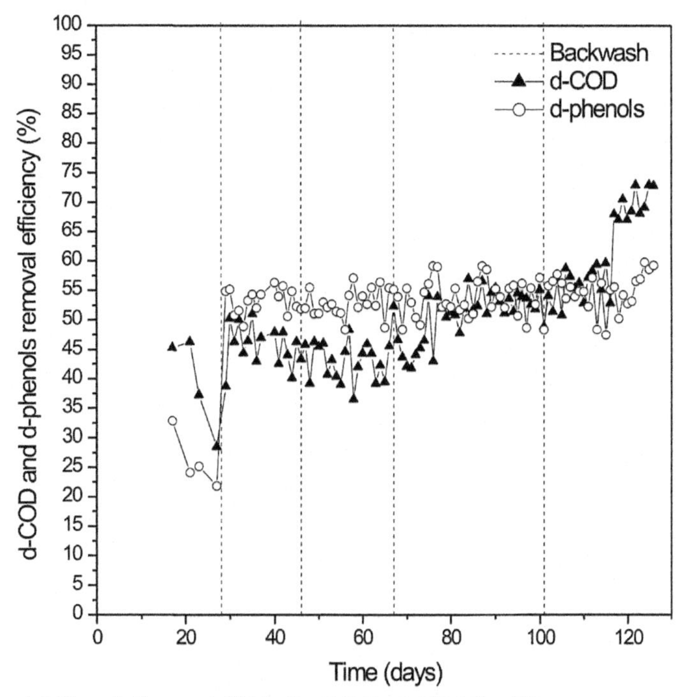

FIGURE 7.6 d-COD and d-Phenols Removal (%) in the Full-Scale Trickling Filter

to 26 mg/L because it was produced by organic nitrogen mineralization. One advantage of this trickling filter is that it can operate under high total solids concentrations. The influent concentration of total solids in the treatment system was 36 g/L while the effluent concentration reduced to about 25 g/L, indicating suspended solid filtration. It should be mentioned that activated sludge systems operate with total solids of between 2 and 4 g/L (Michailides et al., 2011).

Indigenous olive pulp microorganisms proved to be resistant to full-scale conditions, while concentrations of phosphorous and nitrogen in the fresh OMWW were sufficient for their growth; thus the addition of extra nutrients was not required. The treatment system employed might have failed to completely remove the OMWWs organic load. However, despite being an aerobic biological system, it did not use mechanical aeration and the hydraulic residence time was the lowest ever reported for biological systems treating OMWW (Michailides et al., 2011). The cost of the treatment process was limited to the energy required for wastewater recirculation and was about 0.10 €/m^3.

Consequently, it was concluded that this method could be easily applied as a pretreatment step to remove part of the OMWW's organic load and phenolic compounds at minimal cost. Additionally, combining this system with another technology, such as coagulation-flocculation or constructed wetlands, could drastically reduce OMWW toxicity. Therefore, experiments were conducted using either combinations of electrolytes/polyelectrolytes or pilot- and full-scale constructed wetlands (Sections 7.2.2 and 7.2.3, respectively) with very encouraging results.

7.2.2 THE COAGULATION-FLOCCULATION PROCESS AS A POSTTREATMENT STEP

Many research papers deal with the coagulation of OMWW, however, very little is known about the efficiency of the coagulation-flocculation method following biological treatment, used mainly as a pretreatment step (Pelendridou et al., 2014). Two inorganic materials [polyaluminum chloride (PAC) 18% of density 1.35 kg/L and pH 1.2 and aluminum sulfate (ASM) 1 g/L] and two organic polyelectrolytes [anionic polyacrylamide-Flocan 23-20 (AF) and anionic polyelectrolyte-Praestol 2240 (AP)] were examined for the treatment of OMWW by means of coagulation/flocculation. These materials are widely available at relatively low cost.

Using the biologically treated OMWW, coagulation/flocculation experiments were carried out in 1 L volume glass beakers. Samples of OMWW (400 mL) were placed into the beakers and experiments were performed without pH adjustment. The coagulant was then added in different concentrations (between 200 and 4000 mg/L) and the solution was mixed rapidly for 3 min to promote waste homogenization and neutralization of the suspended particles. Then the flocculant was added and the solution was mixed slowly for 10 min to allow agglomeration of the formed flocs. The solution was finally left to settle, and after 24 h, samples were collected from the supernatant phase for further analysis of COD, d-phenols, total solids, total suspended solids, and zeta potential (Pelendridou et al., 2014).

All the experiments clearly demonstrated that PAC coagulation (with AFas flocculant) is advantageous over ASM coagulation (with AP) as higher removal efficiencies were achieved for all coagulant concentrations tested from 200 to 4000 mg/L. According to the literature (Mert et al., 2010; Zagklis et al., 2013), the cost of coagulation-flocculation processes is in the range of 3.57–8.33 €/m^3 of treated OMWW. However, the cost of the hybrid treatment method presented here (aerobic biological treatment followed by coagulation/flocculation) was calculated as 0.36–2.03 €/m^3 and is among the lowest reported to date, thus suggesting that this hybrid method is both effective and low cost.

The combination of biological treatment and coagulation/flocculation showed very promising results with high total removal efficiencies (65.5, 66, 71, and 99.2% for d-COD, d-phenols, total solids, and total suspended solids, respectively) and low cost. Additionally, this combination is technically recommended because no pH correction is necessary before aerobic treatment as microorganisms usually prefer more alkaline values.

7.2.3 CONSTRUCTED WETLANDS AS A POSTTREATMENT STEP

Constructed wetlands are currently used not only for municipal but also for industrial and agroindustrial wastewater treatment (Stefanakis et al., 2014). Recently, constructed wetlands have also been used for OMWW treatment (Aktas et al., 2001; Bubba et al., 2004; Kapellakis et al., 2012; Grafias et al., 2010; Yalcuk et al., 2010; Herouvim et al., 2011; Gikas et al., 2013; Michailides et al., 2015) (Table 7.2). Due to the high pollutant loads of OMWW, all constructed wetlands applications included pretreatment stages, such as coagulation (Aktas et al., 2001), electrochemical oxidation (Grafias et al., 2010), and biological trickling filters (Herouvim et al., 2011). The only two available studies on OMWW treatment using constructed wetlands that lack a pretreatment stage (Kapellakis et al., 2012; Yalcuk et al., 2010) involved diluting the OMWW with tap water before introducing it into the constructed wetland system.

OMWW treatment by constructed wetlands appears to be sufficient when a horizontal subsurface system, a vertical flow system, or a combination of these is used. In some studies (Kapellakis et al., 2012; Michailides et al., 2015), free water surface wetland systems have demonstrated high removal efficiency for organic matter (86–95%), however the organic surface loads (5–57 $g/m^2/d$)

Table 7.2 Studies of Olive Mill Wastewater Treatment Using Constructed Wetlands

References	Constructed Wetland Type	Surface Area of Wetland (m^2)	Plant Species	Pretreatment	Surface Loads ($g/m^2/d$)					Removal Rates (%)				
					C	TKN	TP	TSS	d-phenols	C	TKN	TP	TSS	d-phenols
Aktas et al. (2001)	HSF	0.85	*Phragmites australis*	Coagulation	77.03	1.08	0.42	9.32	16.85	68.9	12.4	54.5	49.4	78.9
Bubba et al. (2004)	VF	0.24	—	Electrochemical oxidation	15	—	—	—	—	86	—	—	—	—
Herouvim et al. (2011)	VF	0.1256	*Typha latifolia, C. alternatifolius*	Dilution with tap water	114.71	—	OP: 2.74	—	—	72	—	OP: 95	—	—
Grafias et al. (2010)	VF		*P. australis*	Biological filter	6589	175	20.0	—	997	73	75	88	—	75
Yalcuk et al. (2010)	FWS		*P. australis*	Dilution	5–15	—	—	—	—	86	—	—	—	—
Gikas et al. (2013)	VF-HSF	1.0–2.9	*P. australis, T. latifolia*	Stabilization pond	2371	8	—	53	37.4	54	44	—	52	60

C, Organic matter; d-phenols, dissolved phenols; FWS, free water surface; HSF, horizontal subsurface load; OP, orthophosphate; TKN, total Kjeldahl nitrogen; TP, total phosphorus; TSS, total suspended solids; VF, vertical flow.

tested were lower than those applied in other constructed wetland applications. Horizontal subsurface constructed wetlands appear to be more efficient than free water surface wetland systems at treating OMWW. For example, Bubba et al. (2004) recorded removal efficiencies of 69% for COD, 12% for nitrogen, 55% for phosphorus, 50% for TSS, and 79% for phenols. It should be mentioned that these removal efficiencies were achieved with pollutant surface loads higher than those applied to free water surface wetland systems, but lower than those applied to vertical flow systems. The most efficient system of constructed wetlands for OMWW treatment appears to be the vertical flow system (Bubba et al., 2004; Grafias et al., 2010; Yalcuk et al., 2010; Herouvim et al., 2011) or a combination of vertical flow and free water surface constructed wetlands (Gikas et al., 2013). Vertical flow constructed wetlands show high removal efficiencies for all pollutants (72–86% for COD, 75% for nitrogen, 88–95% for phosphorus, 79% for phenols) even when exceptionally high pollutant surface loads were applied. It appears from the literature that vertical flow constructed wetlands are more efficient at organic matter and phenol removal, but attention should be paid to the presence of suspended solids, which can cause porous media to clog and thus damage the constructed wetlands system. Yalcuk et al. (2010) attribute vertical flow constructed wetlands treatment ability to the more efficient oxygen transport in the porous media. The higher oxygen concentration in vertical flow constructed wetlands results in increased organic matter oxidation and ammonia nitrification.

While the majority of published studies used common gravel as porous media, Yalcuk et al. (2010) examined the efficiency of zeolite in OMWW treatment. They report that the use of zeolite increased organic matter and ammonia removal, which is consistent with the results of other research groups (Stefanakis and Tsihrintzis, 2012a,b).

The case studies presented later used constructed wetlands units as a posttreatment stage for OMWW that has been treated by the biological trickling filter presented earlier. Initially pilot-scale experiments were conducted using vertical flow constructed wetlands (Herouvim et al., 2011), while a full-scale wetlands system was also constructed by modifying an evaporation lagoon into a free water surface constructed wetland (Michailides et al., 2015).

7.2.3.1 Pilot-scale units

Twelve pilot-scale vertical flow constructed wetland units were set up adjacent the oil mill mentioned in the previous section. Each unit was a plastic rectangular tank of 96 cm in length, 38.5 cm in width, and 31 cm in depth (Fig. 7.7A). Two identical series of four units (series A and B) were each planted with common reed (*Phragmites australis*) obtained from the neighboring area (6 stems per unit). The other four units formed a control series (series C). The four units in each series were placed at different heights (Fig. 7.7A) so the effluent of one stage became, via a siphon, the influent of the next stage. The OMWW was stored in a plastic tank (1 m^3) and fed into the first stage of the units with a pump, 3 times a day, twice a week (using a timer). To ensure the OMWW would flood the pilot-scale units and its level would remain 3–4 cm above the porous media, 13 L OMWW was added to each series in each loading (a total of 80 L per series per week). Vertical flow systems are usually loaded for 2 days and left to rest for the next 4. However, in this study the resting period was resolved to be longer (5 days) because OMWW influent concentrations are higher than those in municipal wastewater. The effluent of the last stage was stored in an 80 L plastic tank. Six vertical perforated plastic tubes were placed in each unit to ensure aeration of the porous medium. Perforated plastic tubes were also placed across and along the units to ensure the uniform distribution of the OMWW onto their surfaces. Three different types of vertical flow constructed wetlands were selected for the four stages of each series. The first stage (units A1,

FIGURE 7.7

The (A) Pilot-scale vertical flow and (B) full-scale free water surface constructed wetland units used for OMWW treatment.

B1, and C1) had a 7 cm deep drainage layer of cobbles (D_{50} = 60 mm). This layer was covered with 7 cm of medium gravel (D_{50} = 24.4 mm) and 17 cm of fine gravel (D_{50} = 6 mm). The second stage (units A2, B2, and C2) had a 9 cm deep drainage layer of cobbles (D_{50} = 60 mm), covered with a 6 cm deep layer of medium gravel (D_{50} = 24.4 mm), 9 cm of fine gravel (D_{50} = 6 mm), and then 6 cm of sand (D_{50} = 0.5 mm). Finally, the third (units A3, B3, and C3), and fourth stages (units A4, B4, and C4) had drainage layers of cobbles 5.5 cm deep (D_{50} = 60 mm). On top of the cobbles were placed layers of medium gravel (9.5 cm) (D_{50} = 24.4 mm), fine gravel (3.5 cm) (D_{50} = 6 mm), and sand (11.5 cm) (D_{50} = 0.5 mm). Igneous cobbles were obtained from a local river bed and the other porous media were obtained from a quarry and were of carbonate rock. The first stage contained the coarsest media gradation in order to retain suspended solids, while the next three stages had finer gradations in order to increase the filtration time of the OMWW. To reduce hydraulic residence time, the total thickness of the porous media in the units was lower than that proposed in the literature (Stefanakis and Tsihrintzis, 2009). In Greece, full-scale units usually comprise two or three vertical flow stages and a final horizontal subsurface flow stage (Tsihrintzis and Gikas, 2010). In this study however, the last horizontal subsurface stage was replaced with a vertical flow stage to reduce hydraulic residence times. During the experimental period, several measurements of OMWW retention time were taken, and on average the OMWW was retained in each stage for approximately 15 min, leading to a total retention time of 60 min in each series.

Although the pollutant loads applied were extremely high (i.e., 88–6589 $g/m^2/d$; 17–997 $g/m^2/d$; 3.0–175 $g/m^2/d$; 3.0–20.0 $g/m^2/d$, for COD, d-phenols, total Kjeldahl nitrogen and orthophosphate, respectively) the pilot-scale vertical flow constructed wetlands managed to achieve high removal rates for each pollutant. For the planted units (series A and B), COD removal rates ranged between 71% and 74%, total Kjeldahl nitrogen removal rates were approximately 75%, and orthophosphate removal rates were approximately 87% (Table 7.1). On the other hand, the unplanted units (series C) presented relatively lower removal rates for COD (63%), total Kjeldahl nitrogen (62%) and orthophosphate (84%) (Herouvim et al., 2011). As expected, the ambient temperature significantly affected all pollutant removal efficiencies, as the main removal mechanism exploited in vertical flow constructed wetlands is aerobic and anaerobic microbial oxidation (Greenway and Woolley, 1999; Steer et al., 2002; Vymazal, 2002), whereas bacterial activity increases when temperature rises. The research of Herouvim et al. (2011)

FIGURE 7.8 The Full-Scale Free Water Surface Constructed Wetland

(A) Land preparation, (B) view of cells 3, 4, 5, (C) view of cell 3, and (D) reed vegetation in cell 3.

is noteworthy due to the high phenol removal rates recorded, as the majority of studies concerning OMWW treatment with constructed wetlands do not examine phenol removal. Furthermore, Herouvim et al. (2011) proved that vertical flow constructed wetlands can tolerate and successfully treat wastewater with extremely high pollutant loads, as their system removed 4900 g/m^2/d, 130 g/m^2/d, 15 g/m^2/d, and 750 g/m^2/d of COD, total Kjeldahl nitrogen, orthophosphate, and phenols, respectively.

7.2.3.2 Full-Scale Unit

Based on the experimental results, olive mill's evaporation pond was transformed to a full-scale free water surface wetland was constructed and operated (Figs. 7.7B and 7.8) in order to assess its performance in real conditions. The full-scale free water surface wetland was divided into five cells by soil barriers. To avoid OMWW leakage, the base of the wetland was lined with clay. The wetland had an incline of 2% and cells 1, 2, 3, 4, and 5 covered a surface area of 400, 350, 700, 350, and 250 m^2, respectively (total area 2050 m^2). Cells 1 and 2, closest to the influent entry point, were kept unplanted because the high pollutant concentrations present here are toxic for plant growth. In March, cells 3, 4, and 5 were planted with common reeds (*P. australis*). Reed growth was rapid in the following 2 months. At the end of April, pretreated OMWW was introduced into the first cell for 1 day until all the wetland cells were flooded. The wetland was operated under batch conditions for 60 days (hydraulic residence time (HRT) of 60 days). The constructed wetland had an initial mean depth of 0.5 m thus, when flooded, the initial OMWW volumes within the wetland were 200, 175, 350, 175, and 125 m^3 for cells 1, 2, 3, 4, and 5, respectively (total OMWW volume of 1025 m^3).

Under real operation conditions the wetland received a maximum flow rate of 30 m^3/d, while the maximum hydraulic loading rate and organic loading rate was 14.6 $L/m^2/d$ and 285 gr $COD/m^2/d$, respectively. The wetland's minimum hydraulic retention time under continuous flow was 35 d. Due to high evapotranspiration rates, the level of the OMW in the wetland reduced to 0.2 m after 60 days. Liquid samples (38 samples per campaign) were collected in six monitoring campaigns from April to June from the influent, effluent, and various other points of each cell, in order to monitor operations over the entire wetland area. The constructed wetland operated for only 60 days, as during this period the OMWW level was minimal due to high evapotranspiration rates. In each monitoring campaign, the OMWW depth was also measured in order to estimate evapotranspiration values and correct measured pollutant concentrations. To calculate removal efficiencies, the pollutant concentration in cell 1 at day 0 was used as the influent value and the concentration in cell 5 at day 60 was used as the effluent value (Michailides et al., 2015).

Furthermore, to assess the exact contribution of photooxidation in pollutant removal, a series of irradiation experiments were performed. Photooxidation of centrifuged (4000 rpm for 20 min) wastewater samples taken from the fifth cell at day 0, was performed using a Suntest XLS+ solar light simulator from Atlas (Germany) equipped with a xenon arc lamp (2.2 kW) and special glass filters restricting the transmission of wavelengths below 290 nm. An average irradiation intensity of 765 W/m^2 (range between 300 and 800 nm) was maintained throughout the experiments and was measured by an internal radiometer. Chamber and black panel temperatures were regulated by the pressurized air cooling circuit and monitored using thermocouples supplied by the manufacturer. Irradiation experiments were performed using a cylindrical 250 mL Duran glass UV-reactor with a flat flange lid with three necks (Lenz, Germany; light absorbance $\lambda < 300$ nm) with a thermostatic jacket. A tap water cooling circuit was used to prevent heating of the solution and keep the temperature constant at $24 \pm 1°C$. Aliquots (10 mL) were collected at different time intervals, and the concentration of total phenols and COD was analyzed. To assess biological activity, a control experiment was also carried out under dark conditions. The same OMWW was placed in a vessel and aliquots were collected at the same time intervals as the aliquots for the photooxidation experiments. During this experiment (data not shown), no alterations in COD or phenol concentrations were observed, leading to the conclusion that no significant microbial activity took place. Equivalent days of sunlight were determining according to the following equation based on the OECD guidelines (OECD, 2002; Oliver et al., 2013) for the testing of chemicals:

$$\text{Days of sunlight} = \frac{h \times r}{0.75 \times 12}$$

where h represents hours of irradiation by the xenon lamp of the Sun test simulator, r is the ratio of intensity of the xenon radiation to that of sunlight during the studied period, 0.75 is the correction factor for diurnal variation of natural sunlight, and 12 is the conversion of hours to days.

Following the 60-day retention time in the wetland, the pretreated OMWW in each cell was homogenized and pollutant concentrations were found to have reduced from 27400, 4800, 191, and 770 mg/L for COD, phenols, orthophosphate, and total Kjeldahl nitrogen, respectively (day 0), to 1584, 248, 5.2, and 18 mg/L for COD, phenols, orthophosphate, and total Kjeldahl nitrogen, respectively (corrected values for day 60). Therefore, the free water surface wetland succeeded in removing above 95% of all the pollutants. Nevertheless, pollutant concentrations measured in the effluent (3960, 656, 13, and 45 mg/L for COD, phenols, orthophosphate, and total Kjeldahl nitrogen, respectively) remained far above legislation limits for disposal into a water body (EU Directive 1991/271/EEC).

Although pollutant removal efficiencies were extremely high in the free water surface constructed wetlands, it seems that photooxidation has only a limited contribution to this efficiency. The irradiation experiments resulted in COD and phenol removal rates of 18% and 31%, respectively, in contrast to the full-scale wetland that removed 81% of COD and 86% of phenols (Michailides et al., 2015).

7.2.4 COMPOSTING

Apart from the large volumes of OMWW produced in a three-phase olive mill, large quantities of solid wastes are also produced. In a three-phase olive mill, OMW mainly comprises olive pomace and olive leaves. Both by-products can serve as composting materials and can produce a high quality soil amendment. Several experiments using OMW as compost materials have been carried out in pilot and full-scale units (Chowdhury et al., 2014), in order to examine the effect of several important parameters such as pH, particle size, porosity, total pile volume, initial moisture and nutritional balance, moisture content, temperature, and aeration (Chowdhury et al., 2013). The composting process is presented in detail in Chapter 8.

7.3 CONCLUSIONS

Biological systems are gaining ground against physicochemical systems concerning the treatment of wastewaters with high pollutant loads. However, biological systems alone cannot achieve compete biodegradation and therefore hybrid systems should be applied. This work presents integrated hybrid treatment systems that comprise either a trickling filter with coagulation/flocculation as a posttreatment step or a trickling filter and a constructed wetland. The latter hybrid system is of special significance because both the trickling filter and the constructed wetland are straightforward to construct and simple to operate. This is the first time that an effective and at the same time low-cost method was applied within a small-scale olive mill plant. This hybrid treatment system is cost effective and can be installed on the grounds of any small-scale, rural, olive mill. The overall removal efficiency of this specific system concerning organic and phenol loads exceeds 90%, while its operation cost is limited to approximately 0.10 €/m^3. In this integrated approach, OMW is also treated by composting which leads to the production of a high quality soil amendment (Chapter 8). The hybrid system offers an integrated approach that retains solid and liquid waste streams within the area of an olive mill plant and diminishes uncontrolled disposal and its associated environmental impact. In addition, quality of life in areas close to OMW disposal sites would improve significantly, as aerobic processes eliminate bad odors. The proposed system provides a viable solution to the thousands of olive mills situated within the Mediterranean basin and could drastically alleviate pollution from OMW.

REFERENCES

Aktas, E.S., Imre, S., Ersoy, L., 2001. Characterization and lime treatment of olive mill wastewater. Water Res. 35, 2336–2340.

Arvanitoyannis, I.S., Kassaveti, A., Stefanatos, S., 2007. Olive oil waste treatment: a comparative and critical presentation of methods, advantages & disadvantages. Crit. Rev. Food Sci. Nutr. 47, 187–229.

Aytar, P., Gedikli, S., Sam, M., Farizoğlu, B., Çabuk, A., 2013. Sequential treatment of olive oil mill wastewater with adsorption and biological and photo-Fenton oxidation. Environ. Sci. Pollut. Res. 20, 3060–3067.

Benitez, J., Beltran-Heredia, J., Torregrosa, J., Acero, J.L., Cercas, V., 1997. Aerobic degradation of olive mill wastewaters. Appl. Microbiol. Biotechnol. 47, 185–188.

Benitez, F.J., Beltran-Heredia, J., Torregrosa, J., Acero, J.L., 1999. Treatment of olive mill wastewaters by ozonation, aerobic degradation and the combination of both treatments. J. Chem. Technol. Biotechnol. 74, 639–646.

Bubba, M.D., Checchini, L., Pifferi, C., Zanieri, L., Lepri, L., 2004. Olive mill wastewater treatment by a pilot-scale subsurface horizontal flow (SSF-h) constructed wetland. Ann. Chim. 94, 875–887.

Chowdhury, A.K.M.M.B., Akratos, C.S., Vayenas, D.V., Pavlou, S., 2013. Olive mill waste composting: a review. Int. Biodeterior. Biodegradation 85, 108–119.

Chowdhury, A.K.M.M.B., Michailides, M.K., Akratos, C.S., Tekerlekopoulou, A.G., Pavlou, S., Vayenas, D.V., 2014. Composting of three phase olive mill solid waste using different bulking agents. Int. Biodeterior. Biodegradation 91, 66–73.

Chowdhury, A.K.M.M.B., Konstantinou, F., Damati, A., Akratos, C.S., Vlastos, D., Tekerlekopoulou, A.G., Pavlou, S., Vayenas, D.V., 2015. Is physicochemical evaluation enough to characterize olive mill waste compost as soil amendment? The case of genotoxicity and cytotoxicity evaluation. J. Clean. Prod. 93, 94–102.

EU Directive, 1991. EU Directive 91/271/EEC, Concerning Urban Waste Water Treatment.

Gikas,G.D., Tsakmakis, I.D., Tsihrintzis, V.A., 2013. Treatment of olive mill wastewater in pilot-scale natural systems. Proceedings of the Eighth International Conference of EWRA Water Resources Management in an Interdisciplinary and Changing Context, Porto, Portugal, 26–29 June 2013, paper #232, pp. 1207–1216.

González-González, A., Cuadros, F., 2015. Effect of aerobic pretreatment on anaerobic digestion of olive mill wastewater (OMWW): an ecoefficient treatment. Food Bioprod. Process. 95, 339–345.

Grafias, P., Xekoukoulotakis, N.P., Mantzavinos, D., Diamadopoulos, E., 2010. Pilot treatment of olive pomace leachate by vertical-flow constructed wetland and electrochemical oxidation: an efficient hybrid process. Water Res. 44, 2773–2780.

Greenway, M., Woolley, A., 1999. Constructed wetlands in Queensland: performance efficiency and nutrient bioaccumulation. Ecol. Eng. 12, 39–55.

Günay, A., Çetin, M., 2013. Determination of aerobic biodegradation kinetics of olive oil mill wastewater. Int. Biodeterior. Biodegradation 85, 237–242.

Hamdi, M., Garcia, J.L., Ellouz, R., 1992. Integrated biological process for olive mill wastewater treatment. Bioprocess Eng. 8, 79–84.

Herouvim, E., Akratos, C.S., Tekerlekopoulou, A.G., Vayenas, D.V., 2011. Treatment of olive mill wastewater in pilot-scale vertical flow constructed wetlands. Ecol. Eng. 37, 931–939.

Jalilnejad, E., Mogharei, A., Vahabzadeh, F., 2011. Aerobic pretreatment of olive oil mill wastewater using Ralstonia eutropha. Environ. Technol. 32, 1085–1093.

Kapellakis, I.E., Paranychianakis, N.V., Tsagarakis, K.P., Angelakis, A.N., 2012. Treatment of olive mill wastewater with constructed wetlands. Water (Switzerland) 4, 260–271.

Lafi, W.K., Shannak, B., Al-Shannag, M., Al-Anber, Z., Al-Hasan, M., 2009. Treatment of olive mill wastewater by combined advanced oxidation and biodegradation. Sep. Purif. Technol. 70, 141–146.

Mantzavinos, D., Kalogerakis, N., 2005. Treatment of olive mill effluents. Part I: organic matter degradation by chemical and biological processes—an overview. Environ. Int. 31, 289–295.

McNamara, C.J., Anastasiou, C., O'Flaherty, V., Mitchell, R., 2008. Bioremediation of olive mill wastewater. Int. Biodeterior. Biodegradation 61, 127–134.

Mert, B.K., Yonar, T., Kilic, Y.M., Kestioğlu, K., 2010. Pre-treatment studies on olive oil mill effluent using physicochemical, Fenton and Fenton-like oxidation processes. J. Hazard. Mater. 174, 122–128.

Michailides, M., Panagopoulos, P., Akratos, C.S., Tekerlekopoulou, A.G., Vayenas, D.V., 2011. A full-scale system for aerobic biological treatment of olive mill wastewater. J. Chem. Technol. Biotechnol. 86, 888–892.

Michailides, M., Tatoulis, T., Sultana, M.-Y., Tekerlekopoulou, A.G., Konstantinou, I., Akratos, C.S., Vayenas, D.V., 2015. Start-up of a free water surface constructed wetland for treating olive mill wastewater. Hemijskaindustrija 69, 577.

Niaounakis, M., Halvadakis, C.P., 2004. Olive-Mill Waste Management–Literature Review and Patent Survey. Typothito-George Dardanos.

OECD, 2002. Guidelines for the testing of chemicals. Phototransformation of chemicals on soil surfaces.

Oliver, R.G., Wallace, D.F., Earll, M., 2013. Variation in chlorotoluron photodegradation rates as a result of seasonal changes in the composition of natural waters. Pest Manage. Sci. 69, 120–125.

Ouzounidou, G., Zervakis, G.I., Gaitis, F., 2010. Raw and microbiologically detoxified olive mill waste and their impact on plant growth. Terr. Aquat. Environ. Toxicol. 4 (Special Issue 1), 21–38.

Paraskeva, P., Diamadopoulos, E., 2006. Technologies for olive mill wastewater (OMW) treatment: a review. J. Chem. Technol. Biotechnol. 81, 1475–1485.

Pelendridou, K., Michailides, M.K., Zagklis, D.P., Tekerlekopoulou, A.G., Paraskeva, C.A., Vayenas, D.V., 2014. Treatment of olive mill wastewater using a coagulation–flocculation process either as a single step or as post-treatment after aerobic biological treatment. J. Chem. Technol. Biotechnol. 89, 1866–1874.

Rozzi, A., Malpei, F., 1996. Treatment and disposal of olive mill effluents. Int. Biodeterior. Biodegradation 38, 135–144.

Stasinakis, A.S., Elia, I., Petalas, A.V., Halvadakis, C.P., 2008. Removal of total phenols from olive-mill wastewater using an agricultural byproduct olive pomace. J. Hazard. Mater. 160, 408–413.

Steer, D., Fraser, L., Boddy, J., Seibert, B., 2002. Efficiency of small constructed wetlands for subsurface treatment of single-family domestic effluent. Ecol. Eng. 18, 429–440.

Stefanakis, A.I., Tsihrintzis, V.A., 2009. Performance of pilot-scale vertical flow constructed wetlands treating simulated municipal wastewater: effect of various design parameters. Desalination 248, 753–770.

Stefanakis, A.I., Tsihrintzis, V.A., 2012a. Effects of loading, resting period, temperature, porous media, vegetation and aeration on performance of pilot-scale vertical flow constructed wetlands. Chem. Eng. J. 181–182, 416–430.

Stefanakis, A.I., Tsihrintzis, V.A., 2012b. Use of zeolite and bauxite as filter media treating the effluent of vertical flow constructed wetlands. Micropor. Mesopor. Mater. 155, 106–116.

Stefanakis, A.I., Akratos, C.S., Tsihrintzis, V.A., 2014. Vertical Flow Constructed Wetlands: Eco-Engineering Systems for Wastewater and Sludge Treatment, First ed. Elsevier, Burlington, USA.

Sultana, M.-Y., Akratos, C.S., Vayenas, D.V., Pavlou, S., 2015. Constructed wetlands in the treatment of agro-industrial wastewater: a review. Hem. Ind. 62, 127–142.

Tsihrintzis, V.A., Gikas, G.D., 2010. Constructed wetlands for wastewater and activated sludge treatment in north Greece: a review. Water Sci. Technol. 61, 2653–2672.

Tziotzios, G., Michailakis, S., Vayenas, D.V., 2007. Aerobic biological treatment of olive mill wastewater by olive pulp bacteria. Int. Biodeterior. Biodegradation 60, 209–214.

Valta, K., Aggeli, E., Papadaskalopoulou, C., Panaretou, V., Sotiropoulos, A., Malamis, D., Moustakas, K., Haralambous, K.J., 2015. Adding value to olive oil production through waste and wastewater treatment and valorisation: the case of Greece. Waste Biomass Valor. 6 (5), 913–925.

Vayenas, D.V., 2011. Attached growth biological systems in the treatment of potable water and wastewater. Moo-Young, M. (Ed.), Comprehensive Biotechnology, vol. 6, second ed. Elsevier, Oxford, pp. 371–383.

Vymazal, J., 2002. The use of sub-surface constructed wetlands for wastewater treatment in the Czech Republic: 10 years experience. Ecol. Eng. 18, 633–646.
Yalcuk, A., Pakdil, N.B., Turan, S.Y., 2010. Performance evaluation on the treatment of olive mill waste water in vertical subsurface flow, constructed wetlands. Desalination 262, 209–214.
Zagklis, D.P., Arvaniti, E.C., Papadakis, V.G., Paraskeva, C.A., 2013. Sustainability analysis and benchmarking of olive mill wastewater treatment methods. J. Chem. Technol. Biotechnol. 88, 742–750.

CHAPTER

8 COCOMPOSTING OF OLIVE MILL WASTE FOR THE PRODUCTION OF SOIL AMENDMENTS

Christos S. Akratos*, Athanasia G. Tekerlekopoulou*, Ioanna A. Vasiliadou, Dimitrios V. Vayenas[†,‡]**

**Department of Environmental and Natural Resources Management, University of Patras, Agrinio, Greece; **Department of Engineering and Architecture, University of Trieste, Trieste, Italy; [†]Department of Chemical Engineering, University of Patras, Rio, Patras, Greece; [‡]Institute of Chemical Engineering Sciences (ICE-HT), Platani, Patras, Greece*

8.1 INTRODUCTION

Composting is nature's way of recycling. It biodegrades organic wastes, such as food waste, leaves, grass trimmings, paper, wood, feathers, and crop residues, turning them to a valuable organic fertilizer. Composting is a natural biological process when carried out under controlled aerobic conditions. In this process, various microorganisms break organic matter down into simpler substances. Composting usually requires a long time period, the duration of which can be classified to three phases: (1) the initial activation phase, (2) a thermophilic phase, and (3) a mesophilic phase (Ryckeboer et al., 2003). Composting is an exothermic process, as the microbes responsible for organic matter degradation, generate heat (Cooperband, 2002). Studies have shown that particular groups of microbes, such as *Aspergillus niger* and *Aspergillus terreus, Geotrichum candidum, Lentinula edodes*, strains of *Penicillium, Pleurotus ostreatus,* and *Azotobacter chroococcum*, can efficiently degrade low-molecular weighted compounds (Chowdhury et al., 2013).

Composting has gained appeal in recent decades as agricultural soils continue to lose their fertility and the use of soil amendments is frequently necessary to replenish organic matter and nutrient content. Compost fulfills both these requirements and has a low production cost, too. Additionally, it can enhance soil water capacity, cation exchange, reduce pesticide numbers, and increase microbial activity (Rynk, 1992; Cooperband, 2002). The olive oil extraction industry is an important activity in the Mediterranean region but can lead to serious environmental problems by producing vast amounts of wastes (by-products) within a short production period. These by-products are olive kernel and olive mill wastewater (OMWW) for three-phase systems, and wet olive pomace for two-phase systems. Three-phase olive mills separate oil from by-products using a three-phase centrifuge (decanter), whereas two-phase olive mills use a two-phase centrifuge to separate oil from three-phase OMWW, which is a mixture of wastewater and olive kernel. The production rate of olive oil is about 1.4–1.8

Olive Mill Waste. http://dx.doi.org/10.1016/B978-0-12-805314-0.00008-X

million tons per year in the Mediterranean, resulting in 30 million m^3 of by-products and 20 million tons of olive kernel (Chowdhury et al., 2013).

In both two-phase and three-phase olive mills significant quantities of solid by-products are produced during the olive oil extraction process. Wet olive pomace is lignocellulosic humid husk material (Alfano et al., 2008), with a high water (56.6–74.5%), phenols (0.62–2.39%), and lipids content (Ranalli et al., 2002; Alburquerque et al., 2004). On the other hand, three-phase olive mills produce significant quantities of solid wastes with outputs of 0.35 tons of olive kernel and 0.05 tons of leaves per ton of olives. The huge quantities of olive kernel and leaves produced within the short oil extraction season cause serious management problems in terms of volume and space. By-products from both two-phase and three-phase olive mills have been used for compost production in the past, usually resulting in a high quality soil amendment. Several experiments using OMW as compost materials have been carried out in different reactor types. At these studies, the effect of several critical factors in the composting process have examined, including pH, particle size, porosity, total pile volume, initial moisture and nutritional balance, moisture content, temperature, and aeration (Chowdhury et al., 2013).

Temperature is a key factor for the composting process, as it governs all the biological processes occurring during composting (Cooperband, 2002). All OMW used in composting experiments have produced composts that successfully reached the thermophilic phase, however the duration of and maximum temperatures observed in this phase presented substantial variations, fluctuating from 16 to 250 days and 43–70°C, respectively. These variations could be attributed to several factors including: (1) the different types of olive mill waste, (2) the bulking agent applied, (3) the aeration system tested, and (4) the volume of the compost pile. According to the literature, compost temperature evolution is likely governed by several actions:

- Bulking agents that increase compost porosity and oxygen availability, optimize microbial activities and hence temperature increase (Cayuela et al., 2006; Sánchez-Arias et al., 2008; Chowdhury et al., 2013).
- Forced aeration systems that lead to a shorter composting process (Cegarra et al., 2006).
- Moisture content exceeding 55% could shorten composting duration (Chowdhury et al., 2013).
- The volume of the compost pile. Small piles can easily maintain aerobic conditions, however, larger piles can achieve higher temperatures (Cooperband, 2002).

A bibliography review revealed that compost derived from OMW is of exceptional quality (Sánchez-Arias et al., 2008; Altieri et al., 2011; Chowdhury et al., 2013). According to Chowdhury et al. (2013), the factors indicating this high quality are:

- The C/N ratio, which is one of the most important quality characteristics, ranges from 25 to 35, and in some cases exceeds 40%.
- All other quality characteristics (i.e., nutrient content, germination index) of olive mill waste compost define it as an excellent quality soil amendment.

The majority of published experiments using olive mill waste for composting have been conducted in pilot-scale units. These units provide the requisite control of parameters to study the dynamics of composting in a systematic manner. In addition, composting research at pilot-scale is critical for the development of optimized full-scale plants. Mathematical models are also useful tools for the scale-up and prediction of the operation of composting units. Mathematical models offer the potential to reduce, or even replace, the need for physical experimentation when exploring new material and/or process options. This chapter is divided into three sections. The first section reviews all research results on OMW composting from both two-phase and three-phase olive mills. The second section presents two case studies on composting OMW (pilot-scale and full-scale unit) using different bulking agents. Finally, the third section introduces an integrated detailed mathematical model that can accurately simulate and predict the composting process.

8.2 FACTORS AFFECTING OMW COMPOSTING

8.2.1 COMPOST TEMPERATURE AND DURATION

One of the most important factors governing the compost process is temperature as it affects microbial activity and thus organic matter degradation. As shown in Fig. 8.1, temperature characterizes the three distinguished compost phases. The initial activation phase takes place as temperature rises from ambient values to 45°C, and is followed by the thermophilic phase where temperature remains above 45°C (temperatures above 55°C are necessary to kill pathogens). The maturation phase commences when temperature drops below 45°C (Chowdhury et al., 2013). The evolution of temperature throughout the compost process is controlled by a series of parameters including pile volume, aeration strategy, compost porosity, and moisture content.

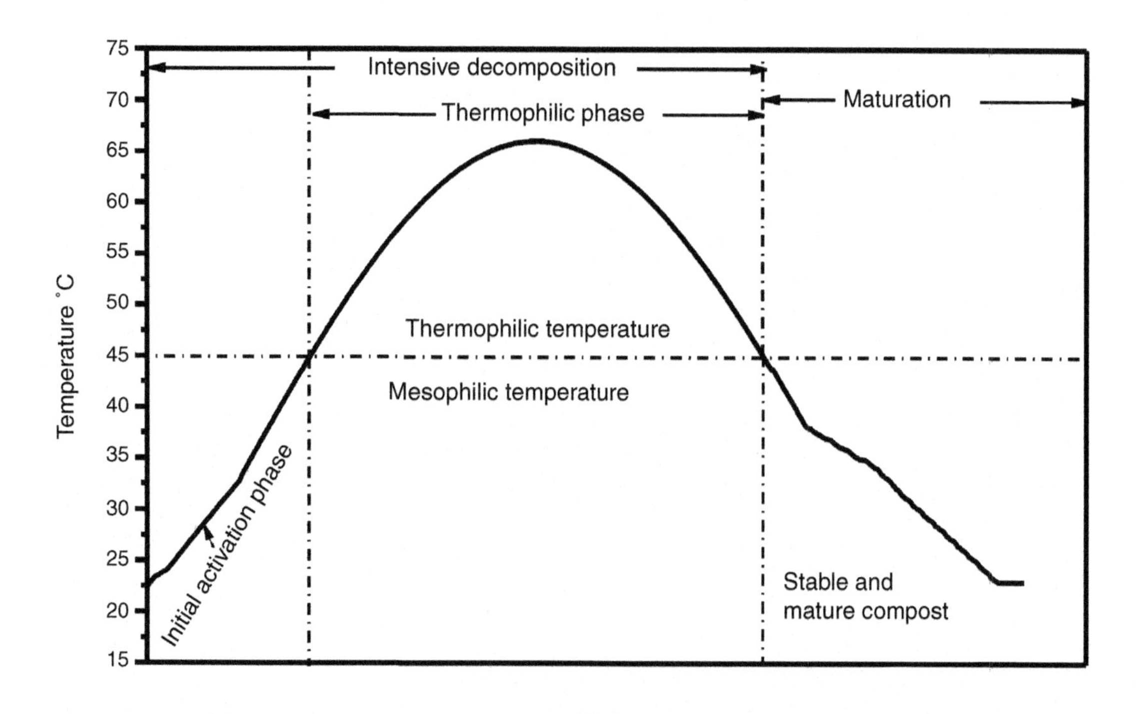

FIGURE 8.1 The Different Phases of the Composting Process

During the process, care should be taken to maintain temperature values below 70°C, as microbial activity is limited at very high temperatures (Bernal et al., 2009). A review of OMW composting experiments and applications in the literature reveals that temperature values vary from 43 to 70°C (Table 8.1), and the duration of the thermophilic phase also fluctuates significantly (from 33 to 365 days) (Chowdhury et al., 2013). These differences in temperature values and thermophilic phase durations can most probably be attributed to the different operational parameters (e.g., OMW origin, bulking agents, compost pile volume, and aeration strategy) tested in each experiment (Miller, 1992; Paredes et al., 2002; Cayuela et al., 2006; Cegarra et al., 2006). The effect of these parameters on the evolution of temperature during composting of OMW can be summarized as follows:

- Turning and the addition of bulking agents prolong the thermophilic phase (Alfano et al., 2008; Tortosa et al., 2012).
- Higher compost porosity values enhance oxygen diffusion within the compost pile. Therefore, using bulking agents of higher porosity improves the compost process (Chowdhury et al., 2013).
- Organic matter content influences temperature evolution as materials with high organic matter content lead to a prolonged thermophilic phase (Tortosa et al., 2012).
- Aeration strategies affect compost temperature as increased oxygen contents (through forced aeration systems) shorten the composting process (Cegarra et al., 2006).

8.2.2 ORGANIC MATTER DEGRADATION

Organic matter degradation in composting is exclusively a biological process, as different types of microorganisms degrade the organic matter. The types of microorganisms responsible for organic matter degradation depend on the compost's temperature. Bacteria are active during the initial activation phase, fungi are active throughout the entire process, and *Actinomycetes* sp. dominate the maturation phase (Bernal et al., 2009). The origin of the composted materials also plays an important role on organic matter degradation. Therefore, materials containing simple carbohydrates, fats, and amino acids are degraded easily and quickly, while materials containing cellulose, hemicelluloses, and lignin are not fully degraded (Haug, 1993). During the process, organic matter content decreases by 55–68% (Chowdhury et al., 2013). Relatively low organic matter loss occurs when composting materials contain mostly lignin, hemicelluloses and cellulose. In order to increase organic matter loss during composting, the thermophilic phase should be prolonged (Tomati et al., 1995; Paredes et al., 1996; Sánchez-Monedero et al., 1999; Paredes et al., 2002).

Water soluble carbon is produced during the composting process via organic matter degradation (Said-Pullicino and Gigliotti, 2007). Specifically, in OMW composting, the concentration of water soluble carbon increases due to the transformation of lignocellulosic substrates to soluble compounds of low-molecular weight (e.g., phenols and sugars) (Gigliotti et al., 2012). Water soluble carbon content is considered an indicator of compost maturity (below 1% in mature composts) (Alburquerque et al., 2006b; Gigliotti et al., 2012). Another indicator of compost maturity related to water soluble carbon is the ratio of hydrophobic to hydrophilic fraction, as values greater than one indicate a mature compost (Said-Pullicino and Gigliotti, 2007). Another factor affecting the quality of the final compost is its content of humic and fulvic acids. Increased humic and fulvic acid contents indicate a high quality soil amendment (Bernal et al., 2009). The majority of OMW composting research presents final humic acid and fulvic acid values of 18.6% and 12.8%, respectively, while composts with contents above 7% are characterized as high quality soil amendments (Chowdhury et al., 2013).

Table 8.1 The Main Characteristics of Olive Mill Waste Compost (Chowdhury et al., 2013)

Suitable bulking agents	Yard trimmings, grape stalks, poultry manure, wheat straw, cotton gin waste, agricultural by-products, sheep manure, wool waste, cow manure, municipal solid waste, winery distilled waste, domestic sewage sludge, olive tree prunings, cereal straw, sesame bark, wood chips, rice by-products, bean straw, gypsum, ammonium sulfate, urea, maize waste
Aeration strategy	Mechanical turning, forced aeration, forced ventilation
Moisture (%)	40–68
Max. temperature (°C)	43–70
Thermophilic phase duration (days)	16–250
Total composting duration (days)	33–365
pH	
Initial	4.7–8.6
Final	5.4–9.5
Electrical conductivity (ds/m)	
Initial	0.2–1080
Final	0.01–568
Organic matter	
Initial	65–93
Final	47–88
Total nitrogen (%)	
Initial	0.7–13
Final	0.7–12
Total phosphorous (%)	
Initial	0.1–0.4
Final	0.1–1.6
C/N	
Initial	13–83
Final	11–53
Germination index (%)	
Initial	5–80
Final	66–201

8.2.3 C/N RATIO

The carbon to nitrogen (C/N) ratio is significant in composting because microorganisms need a good balance of carbon and nitrogen (ranging from 25 to 35) in order to remain active. High C/N ratios can lead to prolonged composting duration and low C/N ratios enhance nitrogen loss. The C/N ratio can be regulated by selecting the most suitable combination of compost materials and added bulking agents to ensure a final ratio within the optimum range. As shown in Table 8.1, a wide variety of bulking agents has been assessed in OMW compost experiments and applications. These agents can either lower the C/N ratio (e.g., green materials, animal wastes) or increase it (e.g., woody materials, dead leaves) (Rynk, 1992). During the OMW composting process, a decrease in C/N ratio is observed due to the release of organic matter content, however, compost can be characterized as mature only when the C/N ratio is below 20 and nitrogen content is above 3% (Chowdhury et al., 2013).

8.2.4 COMPOST PHYTOTOXICITY AND MICROBIAL ACTIVITY INDICATORS

Before applying compost to agricultural cultivations, its potential phytotoxicity should be evaluated. This is especially relevant for composts deriving from OMW as their high phenolic compound contents could reduce their quality. Chowdhury et al. (2013) state that, during OMW composting, phenolic content is reduced by 93% after the maturation phase. The key microorganisms for phenol degradation in OMW composting are fungi as they can break down polyphenols during the thermophilic phase using enzymes (e.g., polyphenoloxidase, laccases, peroxidases, and tyrosinases).

The possible phytotoxic effects of final composts are evaluated using a variety of tests and indicators. The most commonly used test for phytotoxicity evaluation is the determination of the germination index according to Zucconi et al. (1981), which can also be used as an indicator of compost maturity. Germination index values above 80% indicate the absence of phytotoxicity (Zucconi et al., 1981). Composts deriving from OMW show germination index values below 80% in the initial composting phase (Chowdhury et al., 2013). However, after their maturation phases the majority of OMW composts transform from phytotoxic materials to quality soil amendments, as final germination index values of up to 201% are recorded (Zorpas and Costa, 2010). Compost maturity is also assessed by various respirometric tests that evaluate microbial activity. The most common tests are the dynamic respiration index (DR4) and the static respiration index at 4 days (AT4) (Bernal et al., 1998; Lasaridi et al., 2006). The AT4 test is used by the European Commission (EC, 2001) to assess compost stability, which is achieved when the AT4 indicator is below 10 g O_2/kg dry matter. OMW composts have AT4 values well below 10 g O_2/kg dry matter, ranging from 3.94 to 4.8 g O_2/kg dry matter (Komilis and Tziouvaras, 2009; Michailides et al., 2011; Chowdhury et al., 2013).

8.2.5 PHYSICOCHEMICAL PARAMETERS

As compost will be applied on agricultural land, its physicochemical parameters (i.e., pH, electrical conductivity, and mineral concentrations) should be within a specific range in order to prevent any adverse effects on soil quality. While OMW composts have initially acidic pH values, final pH values are neutral thus ideal for agricultural applications (Baeta-Hall et al., 2005; Gigliotti et al., 2012). On the other hand, initial acidic pH values favor microbial activity during composting (Said-Pullicino and Gigliotti, 2007) and enhance lignin and cellulose degradation (Paredes et al., 1999). Electrical conductivity values should be lower than 4000 mS/cm to be safe for agricultural usage (Lasaridi et al., 2006). OMW composts present a wide variety of electrical conductivity values ranging from 0.01 dS/m (Abid and Sayadi, 2006) to 9.2 dS/m (Hachicha et al., 2009a). Although values tend to increase during composting (as the compost mass decreases) in some OMW compost experiments, electrical conductivity decreased during composting due to mineral leaching (Paredes et al., 2002; Abid and Sayadi, 2006;

Alburquerque et al., 2006a; Hachicha et al., 2009b; Makni et al., 2010; Gigliotti et al., 2012). Finally, high levels of mineral nutrients (e.g., phosphorus, potassium, and sodium) are essential to improve the quality of composts intended for agricultural purposes. Many studies report that composts based on OMW have high mineral nutrient contents (phosphorus: 0.1–3%; potassium: 0.12–4.4%; sodium: 0.05–4.1%) (Sánchez-Arias et al., 2008; Hachicha et al., 2009b; Michailides et al., 2011; Chowdhury et al., 2013).

8.3 CASE STUDIES OF OLIVE MILL COMPOSTING

Considering the above research data, it can be concluded that the composting of OMW leads to mature, pathogen-free compost that benefits crop production. To date, great effort has been made to optimize the factors that could improve the composting process and increase the quality of the final product. However, few studies have examined the effect of composting duration on the quality of the final product and the use of various bulking agents in composting OMW. Recently, a series of parallel pilot-scale experiments for composting olive kernel, olive leaves, and rice husk have been performed to define the best material ratios, optimum humidity, and minimum composting duration (Michailides et al., 2011; Chowdhury et al., 2014; Chowdhury et al., 2015).

8.3.1 PILOT-SCALE COMPOSTING UNITS

To examine OMW (i.e., olive kernel, olive leaves) composting, four experiments were conducted using various materials as bulking agents: rice husk, sawdust, wood shavings, and chromium treated reed plants (Chowdhury et al., 2014; Chowdhury et al., 2015). In addition, two different humidifying agents were used (i.e., water and OMW water). Olive kernel, olive leaves, and OMWW were collected from a three-phase olive mill located in Amfilochia, Western Greece. Rice husk was obtained from the Agrino rice mill in Agrinio, Western Greece, and wood shavings and sawdust were obtained from a local wood processing industry. Finally, chromium treated reed plants (*Phragmites australis*) were collected by harvesting reed biomass from pilot-scale constructed wetlands used for hexavalent chromium removal. The experimental units comprised plastic trapezoidal bins 1.26 m in length, 0.68 m in width, 0.73 m in depth, and with a total volume of 0.62 m^3 (Fig. 8.2), while several bulking agents were used

FIGURE 8.2 The Pilot-Scale Olive Mill Waste Composting Units

Table 8.2 Description of Pilot-Scale Experimental Set-Ups (Chowdhury et al., 2014, 2015; Sultana et al., 2015)

Composting materials: ratios per volume	*Olive kernel*	1–3
	Olive leaves	1–3
	Rice husk	1–2
	Wood shavings	0.5
	Sawdust	0.5
	Cr treated reed plants	0.25–0.5
Initial compost mass (kg)	49–204	
Final compost mass (kg)	73–144	
Initial compost volume (L)	252–435	
Final compost volume (L)	160–281	

(i.e., rice husk, sawdust, wood savings, and reed biomass) in different mixing ratios (Table 8.2). The main objectives of the experiments were:

1. to examine the effects of different olive kernel and olive leaves ratios on compost quality,
2. to control total composting duration using a compost cooling process,
3. to examine the optimum mixing ratio of rice husk,
4. to examine the effect of prolonged composting (not controlled by cooling) on final compost quality, and finally
5. to determine potential toxicity to human lymphocytes (genotoxicity and cytotoxicity) when using different bulking agents.

In almost all the conducted pilot-scale experiments, composts quickly reached the thermophilic phase (at day 5), while maximum compost temperature values (56°C) were obtained in day 12 (Fig. 8.3). These results are consistent with those of previous composting experiments. However, satisfactory temperature evolution was not achieved when compost moisture content exceeded 65% and additional water was added before reaching the thermophilic phase. Moisture content greatly affected the process as initial contents exceeding 60% decreased thermophilic phase duration. Nevertheless, it is essential not to add excessive water to the compost prior or during the thermophilic phase. Concerning the effect of bulking agents on compost temperature, it was found that the excessive use of bulking agents with high porosity (i.e., rice husk) resulted in a shorter thermophilic phase.

Table 8.3 presents the physicochemical parameters of the final compost. The main outcomes of these compost experiments were:

1. the short composting period (maximum 120 days) achieved, which is one of the shortest reported,
2. the high nutrient content (i.e., nitrogen, phosphorus, potassium, and sodium) of the final compost that makes it suitable as a soil amendment, and
3. the high germination index recorded indicating that the final compost product can be beneficial for crop cultivations.

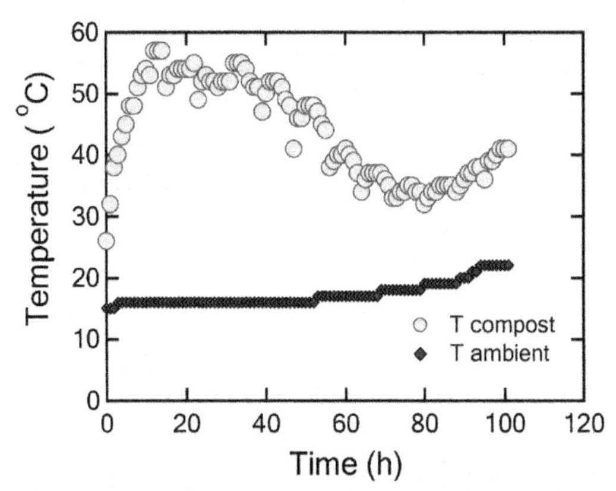

FIGURE 8.3 Temperature Profile for the Experimental Compost Mixture Containing Olive Kernel, Olive Leaves and Rice Husk

Table 8.3 Physicochemical Characteristics of Mature Compost Deriving From Pilot-Scale and Full-Scale Experiments (Michailides et al., 2011; Chowdhury et al., 2014, 2015; Sultana et al., 2015)

Physicochemical Characteristics	Pilot-Scale	Full-Scale
pH	6.83–8.15	7.58
EC (mS/cm)	587–1050	1.21
Dry matter (%)	44–53	45.0
Volatile solids (%)	83–97	94.3
Organic matter (%)	76–92	83.7
Total organic carbon (%)	44–53	48.5
Phosphorous (%)	0.13–0.19	0.17
Total Kjeldhal nitrogen (%)	3.1–4.1	1.79
C/N	12–17	27.1
Potassium (K) (%)	0.018–0.047	0.62
Sodium (Na) (%)	0.08–0.20	0.23
Germination index (%)	85–164	198

Nevertheless, it is important to ensure that the final compost has matured fully before application, otherwise it can lead to cytotoxic or genotoxic effects.

In all the experiments compost pH values were initially slightly acidic (ranging from 6.74 to 7.58), then reached neutral and slightly basic values (ranging from 7.58 to 8.87) after the maturation phase. These values indicate aerobic microbial decomposition of organic matter. In addition, low electrical conductivity values were reported in both initial activation phase (ranging from 336 to 663 μS/cm) and maturation phase (ranging from 548 to 1050 μS/cm). Although initial organic matter contents varied due to the different bulking agents and mixing ratios tested, all the final composts presented an approximate 7% decrease of inorganic matter. Among the examined bulking agents, only the excessive use of fresh olive leaves lead to insufficient organic matter decomposition, while the use of rice husk enhanced microbial activity and organic matter decomposition. The total Kjeldahl nitrogen content of the final compost appears to be affected by:

1. rice husk ratio, as higher rice husk ratios lead to lower total Kjeldahl nitrogen contents, and
2. the use of OMWW as a wetting agent which increased total Kjeldahl nitrogen values.

Lastly, C/N ratios appear to decrease with increased use of fresh olive leaves. However, this effect was not statistically significant.

The pilot-scale experiments showed that the compost process reduced water soluble phenol by 90%, thus reducing the potential phytotoxic effect of the final product. The use of rice husk as a bulking agent facilitates water soluble phenol reduction. The absence of phytotoxicity in OMW composts was also proved by the recorded germination index values, ranging from 101% to 122% in mature compost. Besides, germination index values fall below 100% only when high olive leaf contents applied. Finally, OMW composts contain all high mineral nutrient concentrations (Table 8.3).

In order to evaluate if physicochemical analysis and standard phytotoxicity tests are sufficient to characterize compost, a series of genotoxic and cytotoxic evaluation tests were conducted using human lymphocytes. Specifically, genotoxicity and cytotoxicity were evaluated through micronuclei (MN) and an in vitro cytokinesis block micronucleus (CBMN) assay. The genotoxic and cytotoxic evaluation was performed in compost samples taken at the beginning of compost process, the end of thermophilic phase, and the end of maturation phase, while three different concentrations of compost [i.e., 1, 2, and 5% (v/v)] were used. Results revealed that from all the initial composting mixtures, only the ones containing either excessive amounts of rice husk or sawdust presented genotoxic effects. Similar results have also been reported by Chatterjee et al. (2013). Concerning cytoxocity, the initial compost mixtures containing either excessive amounts of olive kernel and rice husk or sawdust presented cytotoxic effects, while the use of reed biomass with high Cr content did not present genotoxic or cytotoxic effects. On the other hand, all compost samples taken at the end of the thermophilic phase presented both genotoxic and cytotoxic effects. This result can be attributed to the activity of thermophilic microorganisms that tend to degrade complex organic compounds and form toxic intermediates. The appearance of genotoxic and cytotoxic effects at the end of the thermophilic phase was also evaluated and verified in several other studies (Kreja and Seidel, 2002; Axelsson et al., 2006; Gandolfi et al., 2010). Samples of compost taken after the maturation phase did not present genotoxic or cytotoxic effects, with the exception of those containing excessive olive kernel and rice husk. From the above genotoxic and cytotoxic evaluation, it is evident that future compost evaluations should include these types of tests, in order to safely characterize compost products as soil amendments (Chowdhury et al., 2014).

Based on the pilot-scale results, research was extended to the development of full-scale cost effective systems that could be applied in olive mill plants. Therefore, a full-scale composting unit was constructed and operated allowing the evaluation of environmental conditions effect on the composting process.

8.3.2 FULL-SCALE COMPOSTING UNITS

The full-scale composting unit was built adjacent an olive mill in Amfilochia, Western Greece (Fig. 8.4). It was a covered rectangular windrow of 8 m in length, 1 m in width, and 0.5 m in depth, while only

FIGURE 8.4

The full-scale olive mill waste composting unit showing the (A) basic compost materials, (B) aeration system, and (C) final compost product.

olive kernel and olive leaves were used as compost materials. Compost remained in the windrow for 60 days. During this first period, the pile was turned approximately once each week or when compost temperature decreased. The turning mechanism (for aeration) consisted of a parallel steel frame onto which was mounted a rotating cylinder with three endless screws (augers) set in opposite directions forming two "V"s. Tap water was added to the windrow when necessary in order to keep moisture levels high (above 50%). After this 60-day period, the compost was removed from the windrow to mature. During the maturation period (60 days) the compost was piled in a covered area. Although the compost pile was open to the air and ambient temperatures were not high, compost temperature evolution was rapid, as it reached the thermophilic phase in just 3 days. The total duration of the compost process was 120 days, including the maturation phase, which is one of the lowest reported in full-scale units (Chowdhury et al., 2013). Compost temperature presented spatial variations during the process that were attributed to the ambient temperature effect, as temperatures in the center of the windrow were higher than those measured at its borders.

The physicochemical characteristics of the final compost (Table 8.2) were found to be in compliance with current legislation limits for soil amendments. Additionally, its nutrient content was high thus benefiting plant growth. Germination index experiments using cress (*Lepidium sativum*) seeds revealed that the mature compost presented germination indices of approximately 200%, indicating that the current compost had excellent characteristics.

One effective method of evaluating the effect of compost on soil fertility is to study the growth of one plant species (such as lettuce) after applying compost. With this aim, mature compost was added to the surface of four pots into which 10-day old lettuce plants (*Lactuca sativa)* had been planted (Fig. 8.5). The weights of the compost added to each different pot were 100, 200, 300, and 400 g. These masses correspond to 31.8, 63.7, 95.6, and 127.4 kg/ha, respectively. Three replicates were used for each pot. The lettuce plants were placed in a green house and each pot was irrigated with 1 L of water per day. After 20 days each plant was removed from its pot and its leaves and roots were weighed. The addition of mature compost greatly improved the percentage yield of lettuce plants, as the addition of 100 g (i.e., equivalent to 31.5 tons/ha) of compost increased lettuce leaf yield by 145%.

8.4 MATHEMATICAL MODELING OF THE OMW COMPOSTING PROCESS

One of the main problems hindering compost production is the difficulty to control factors that could improve the quality of the final compost. Mathematical modeling has been widely utilized in science and engineering to improve understanding of the behavior of systems, predict system performance, and aid the solution of practical design problems. For this reason, an integrated model for the composting of OMW was developed. This model can be used as a guide to design and assess the conditions under which good quality compost may be expected when using a wide variety of compost mixtures.

To date, great effort has been made to optimize the factors that could improve the composting process and increase the quality of the final compost. These include C/N ratio, addition of chemicals, bulking materials and aeration strategy, temperature, moisture content, and oxygen availability. Mathematical models are an effective tool for predicting the operation of the composting procedure and may be used as a guide in designing and assessing the conditions under which good compost quality may be expected. Relatively few modeling studies have been conducted to simulate the composting process of OMW (Paredes et al., 2002; Garcia-Gomez et al., 2003; Sánchez-Arias et al., 2008; Alburquerque

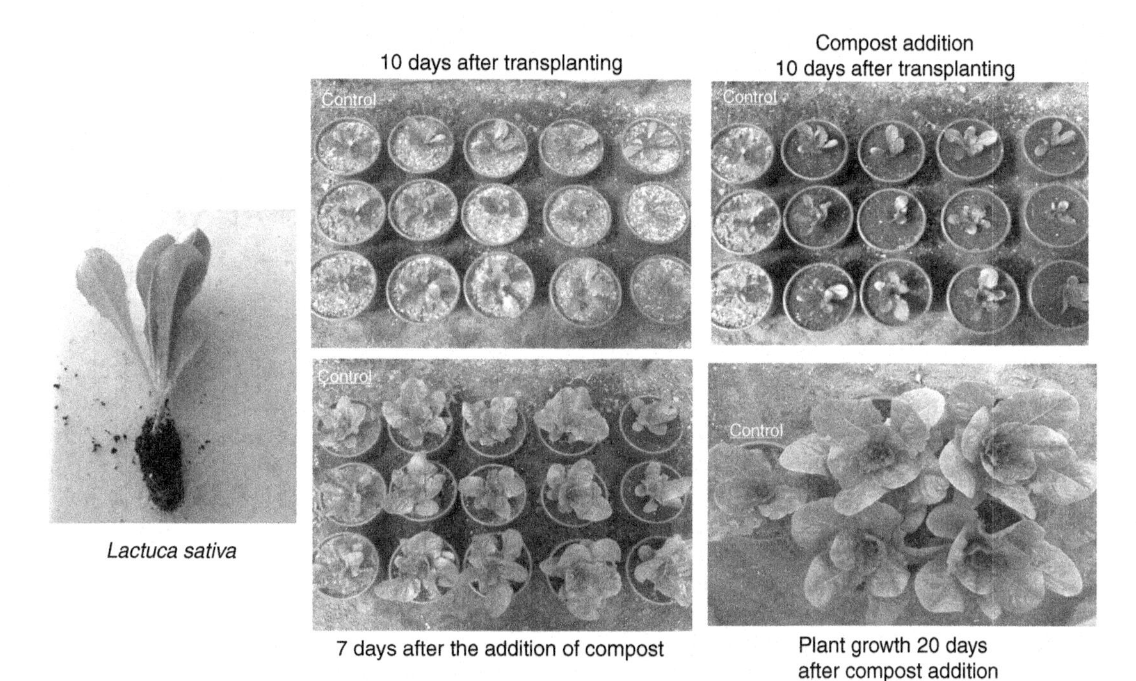

FIGURE 8.5 Photographs of the *Lactuca sativa* (lettuce) Phytotoxicity Experiment

et al., 2009; Vlyssides et al., 2009; Serramiá et al., 2010). Almost all of these studies present simplified modeling approaches, including only organic matter change without describing the main physicochemical and biological mechanisms involved in the composting process.

In this section, an integrated mathematical model capable of describing both physicochemical and biological processes in the composting of OMW is presented. Some experimental data presented in Section 8.3.1 were used to calibrate and validate the model. The multicomponent model simulates the composting of three-phase olive mill solid waste (olive kernel) with olive leaves and other materials, such as rice husk, wood shavings, sawdust, and chromium treated reed plants as bulking agents in different initial ratios. Specifically, the composting materials C1, C2, C3, C4, C5, and C6 consisted of olive kernel/olive leaves/rice husk/wood shavings, sawdust, and chromium treated reed plants with ratios of 3.0/2.0/1.0/0/0/0, 3.5/2.0/0.5/0/0/0, 3.5/2.0/0/0.5/0/0, 3.5/2.0/0/0/0.5/0, 3.5/2.0/0/0/0/0.5, and 3.75/2.0/0/0/0/0.25, respectively (Vasiliadou et al., 2015). Sensitivity analysis with respect to the model outputs was also performed by varying the optimized set of kinetic parameters.

8.4.1 MODEL DEVELOPMENT

The modeling system includes heat transfer, organic substrate degradation, oxygen consumption, carbon dioxide production, water content change, and biological processes (Vasiliadou et al., 2015), while the total volume of the composting material consists of three parts: insoluble particulate matter, a gaseous phase, and a liquid phase. First-order kinetics were used to describe the hydrolysis of insoluble organic matter (S_s) [Eq. 8.1], which lead to the formation of biomass. The constant of the hydrolysis rate is a function of temperature and thus the first-order coefficient, K_h, was adjusted by a temperature correction function ($F_{hydro}(T) = 0.0182 \cdot T$).

$$\frac{dS_s}{dt} = -K_h S_s F_{hydro}(T) \tag{8.1}$$

The balance of soluble organic substrate (S_L) is given by the following equation [Eq. 8.2]:

$$\frac{dS_L}{dt} = K_h S_s F_{hydro}(T) - \frac{1}{Y_{X/S_L}} \mu X + Y_{S_L/X} k_d X \tag{8.2}$$

This equation includes the production of soluble organic substrate due to hydrolysis and its consumption for microbial growth, as well as the conversion of dead biomass into soluble organic substrate. Microbial biomass growth was modeled with a double-substrate limitation by hydrolyzed available organic substrate and oxygen using Monod kinetics according to the following equation [Eq. 8.3].

$$\frac{dX}{dt} = \frac{r_{max} S_L}{K_L \varepsilon_w V_C + S_L} \frac{O_2}{K_{O_2} \varepsilon_w V_C + O_2} F_T F_{inh} F_{moist} X - k_d X \tag{8.3}$$

The coefficients F_T, F_{inh}, and F_{moist} are functions expressing the dependence of microbial biomass growth on temperature, phenolic fraction presence, and moisture content.

It is important to understand the way in which heat transfer occurs within composting systems so that the process can continue to be refined and improved. Assuming that no heat is lost through radiation, the energy balance analysis of a composting pile is defined by Eq. 8.4:

$$\frac{d(mcT)}{dt} = G_a(H_i - H_0) - UA(T - T_a) + h_{fg} G_a \left(H_s(T_a) - H_s(T)\right) + \frac{H_c}{Y_{X/S_L}} \mu X \tag{8.4}$$

This equation describes the heat inlet due to dry air and the heat loss of dry exit gas (first term on the right-hand side), the heat loss by conduction from the surface (second term), the heat due to the latent heat of evaporation of water (third term), and finally the heat produced from biomass growth.

During composting, the microorganisms in the compost material consume oxygen (O_2) while feeding on organic matter. Microorganisms also break down organic matter and produce carbon dioxide. The oxygen consumption and carbon dioxide production can be estimated by the following expressions [Eq. 8.5 and Eq. 8.6, respectively]:

$$\frac{d(O_2)}{dt} = G_a(X_{O_2,in} - X_{O_2,exit}) - \frac{Y_{O_2/S_L}}{Y_{X/S_L}} \mu X \tag{8.5}$$

$$\frac{d(CO_2)}{dt} = G_a(X_{CO_2,in} - X_{CO_2,exit}) + \frac{Y_{O_2/S_L}}{Y_{X/S_L}} \mu X \tag{8.6}$$

The first terms on the right-hand side describe the incoming and outgoing oxygen and carbon dioxide flows, while the second terms describe the oxygen consumption and carbon dioxide production through biological reactions.

The water balance in the composting material changes with time during organic matter degradation. Taking into account the biological and physical processes, the water balance is represented by the following equation [Eq. 8.7]:

$$\frac{d(\mathrm{H_2O})}{dt} = G_a\left(H_s(T_a) - H_s(T)\right) + \frac{Y_{\mathrm{H_2O}/S_L}}{Y_{X/S_L}}\mu X - \frac{1}{Y_{\mathrm{H_2O}/X}}\mu X + Y_{\mathrm{H_2O}/X\mathrm{dead}} k_d X + W_{\mathrm{added}} \tag{8.7}$$

This equation includes an aeration process that moves water out with the hot humid air stream (first term on the right-hand side), the water formed during the oxidation of soluble substrate (second term), the water needed for the formation of new cells (third term), and the amount of water that returns due to microbial death (fourth term). The amount of water externally added into the compost material is represented by the last term while the amount of water needed for hydrolysis was assumed to be negligible.

The model calculates mass balances and forms of nitrogen [Eq. 8.8] and phosphorus [Eq. 8.9] by assuming that these nutrients enter the system through hydrolysis of insoluble organic matter and are then consumed for biomass growth (Vlyssides et al., 2009).

$$\frac{dN_L}{dt} = N_{S_S X} K_h S_s F_{\mathrm{hydro}}(T) - N_{N/X}\mu X + N_{N/X} k_d X \tag{8.8}$$

$$\frac{dP_L}{dt} = P_{S_S X} K_h S_s F_{\mathrm{hydro}}(T) - P_{P/X}\mu X + P_{P/X} k_d X \tag{8.9}$$

Eqs. 8.8 and 8.9 involve the production of soluble nitrogen and phosphorous due to the hydrolysis process (first terms on the right-hand side), the consumption of available soluble nitrogen and phosphorous for biomass growth (second terms), and finally the conversion of dead biomass to soluble nitrogen and phosphorous.

8.4.2 MODEL EVALUATION AND RESULTS

The model was evaluated by using the experimental results of six controlled composting experiments previously conducted by Chowdhury et al. (2014) (Section 8.3.1). Four of these experiments were used to estimate the model's parameters (calibration sets: C1, C2, C4, C5), and the remaining two were used for model validation (sets: C3, C6). Finally, sensitivity analysis with respect to the model outputs was performed by varying the optimized set of kinetic parameters. Fig. 8.6 presents the kinetic parameters that most affect the model's outputs, according to their sensitivity coefficients. Results revealed that oxygen simulations were more sensitive to the input parameters of the model than those of water, temperature, and insoluble organic matter. Moreover, the insoluble carbon variable (data not shown) was not influenced by any of the parameters with the exception of the hydrolysis rate constant which had a sensitivity coefficient of 5.1×10^{-3}. Fig. 8.7 shows an example of a dataset with the experimental (symbols) and simulated (lines) profiles of temperature, water, insoluble carbon, nitrogen, phosphorous, and compost pile volume for two mixtures containing olive kernel/olive leaves/rice husk and olive pomace/olive leaves/sawdust. All model simulations were in good agreement with the experimental results, according to the Nash and Sutcliff index (Nash and Sutcliffe, 1970) that was evaluated for each variable (Fig. 8.7). As shown in Fig. 8.8, expected values (E) were estimated for both calibration and validation data sets. Results indicated that the insoluble organic matter and the total compost volume

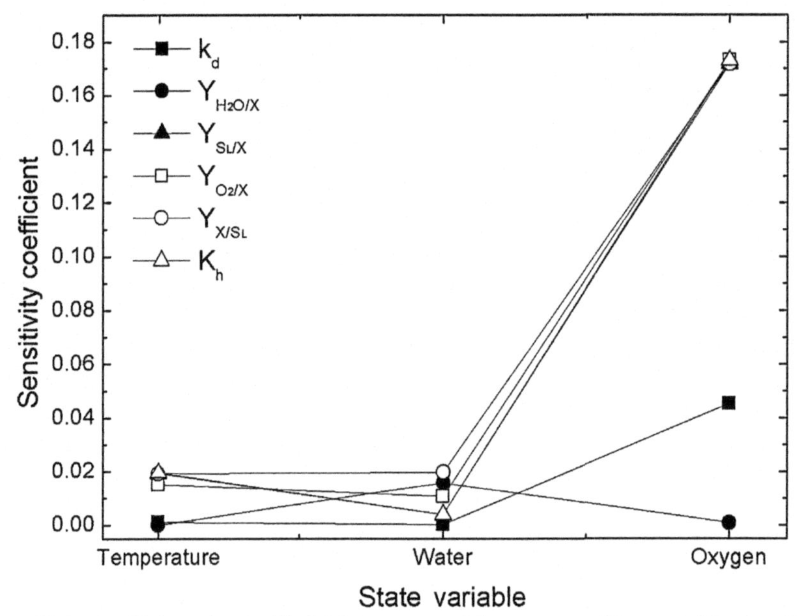

FIGURE 8.6 Sensitivity Coefficients for the Model Parameters of Temperature, Water, and Oxygen Model Outputs

k_d, death rate of microorganisms (1/d); $Y_{H_2O/X}$, metabolic yield of water (kg-X/kg-H_2O); $Y_{S_L/X}$, biomass yield on soluble substrate (kg-S_L/kg-X); Y_{O_2/S_L}, metabolic consumption of oxygen (kg-O_2/kg-S_L); Y_{X/S_L}, cell yield per unit substrate consumed (kg-X/kg-S_L); K_h, hydrolysis constant (1/d).

values were better predicted by the model than other variables. Only water content was not simulated satisfactorily, as its simulation involves many kinetic parameters. The mathematical model presented here can accurately describe the most significant mechanisms and reactions involved in the composting of OMW. As different composting mixtures were used for the model's evaluation, it can be concluded that this model could be successfully used to predict the behavior of a wider variety of composting mixtures.

8.5 CONCLUSIONS

OMW composting has proved to be an efficient valorization method as its final product is characterized as an excellent soil amendment. Composting OMW transforms a hazardous waste into an excellent quality final product and could increase the income and viability of small-scale olive mill plants. In addition, the cocomposting of several other agroindustrial by-products makes composting an ideal method for the valorization of these by-products while simultaneously providing significant environmental benefit. The curse of OMW turns into a blessing. From the analysis of the previous sections it becomes clear that during OMW composting, several operational parameters affect the quality of the final compost. For this reason, the following suggestions are proposed to ensure a stable composting process:

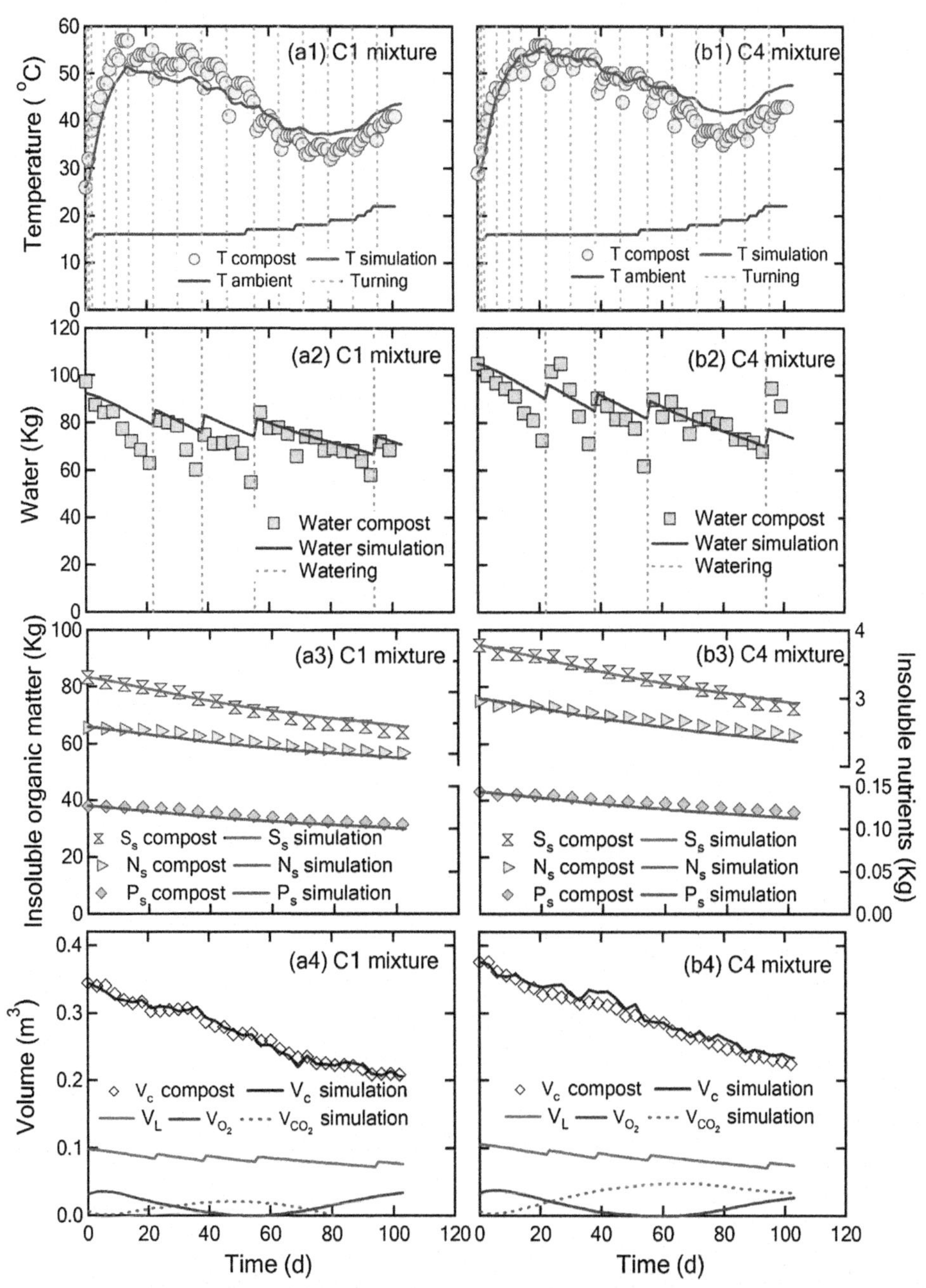

FIGURE 8.7 Experimental (*symbols*) and predicted profiles (*lines*) of two compost mixtures containing olive kernel/olive leaves/rice husk (C1) and olive kernel/olive leaves/sawdust (C4)

(a1) and (b1), temperature profiles; (a2) and (b2), water content profiles; (a3) and (b3), insoluble organic matter, insoluble nitrogen and phosphorous profiles; (a4) and (b4), profiles of volume of total compost, water, oxygen, and carbon dioxide.

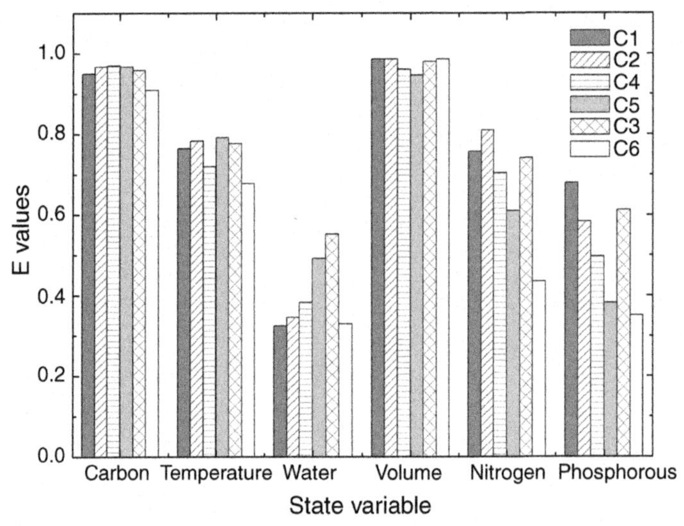

FIGURE 8.8 The Nash and Sutcliff Index (E) Values Calculated to Calibrate and Validate the Data Set for All Variables Included in the Mathematical Model

- Aeration type has no significant effect on the composting process, as long as oxygen concentrations are between 15% and 20%.
- Suitable bulking agents should be selected to achieve compost porosity values between 35% and 50% and a final C/N ratio between 10 and 20.
- Composting should be at least 90 days in duration.
- Genotoxic and cytotoxic effects should be evaluated before composts are used on agricultural crops.

Since composting is long-term process, the development of detailed mathematical models could alleviate the experimental effort, and provide immediate answers and fast responses. This chapter presents a very precise and detailed model that considers all biological and physicochemical processes taking place during the distinct composting phases. It is a very useful tool for engineers and composting unit operators that can be used for the design of efficient composting units, to provide guidelines for proper operation, and to guarantee a high quality final product.

NOMENCLATURE

A	reactor surface area (m^2)
c	specific heat of the composting material (kJ/kg °C)
$\mathbf{CO_2}$	carbon dioxide mass (kg)
$\mathbf{F_{moist}}$	function used to express the effect of moisture on the growth of biomass
$\mathbf{F_{inh}}$	function used to express the effect of phenolic organic matter content on biomass growth
$F_{hydro}(T)$	function used to express the effect of temperature on hydrolysis
$\mathbf{F_T}$	function used to express the effect of temperature on biomass growth
$\mathbf{G_a}$	mass flow rate of air (kg-dry air/d)
$\mathbf{H_2O}$	water mass (kg)

$\mathbf{H_c}$	heat of combustion of the substrate (kJ/kg-SL)
$\mathbf{H_i}$	inlet gas enthalpies (kJ/kg)
$\mathbf{H_o}$	exit gas enthalpies (kJ/kg)
$\mathbf{H_s}$	saturated humidity (kg-H_2O/kg-dry air)
$\mathbf{h_{fg}}$	latent heat of evaporation of water (kJ/kg)
$\mathbf{k_d}$	death rate of biomass (1/d)
$\mathbf{K_h}$	hydrolysis constant (1/d)
$\mathbf{K_L}$	saturation constant for soluble organic matter (kg/m^3 liquid)
K_{O_2}	saturation constant for oxygen (kg/m^3 liquid)
m	mass of the composting material (kg)
$\mathbf{N_L}$	mass of soluble nitrogen (kg-NL)
N_{S_SX}	nitrogen content of particulate matter (kg-N/kg- SS)
$N_{N/X}$	nitrogen content of biomass (kg-N/kg-X)
$\mathbf{O_2}$	oxygen mass (kg)
$P_{P/X}$	phosphorous content of biomass (kg-P/kg-X)
P_{S_SX}	phosphorous content of particulate matter (kg-P/kg- SS)
P_L	mass of soluble phosphorous (kg-PL)
r_{max}	maximum growth rate (1/d)
S_L	mass of soluble organic matter (kg-SL)
S_S	mass of insoluble organic matter (kg-SS)
t	time (d)
T	temperature of the composting material (°C)
T_a	ambient temperature (°C)
U	overall heat transfer coefficient (kW/m^{2}°C = 3600 kJ/h m^{2}°C)
V_C	volume of the compost (m^3)
$\mathbf{W_{added}}$	water added during the process (kg-H_2O/d)
X	biomass (kg)
$\mathbf{X_{O_2}}$	concentration of oxygen (kg-O_2/kg-dry air)
$\mathbf{X_{CO_2}}$	concentration of carbon dioxide (kg-CO_2/kg-dry air)
$\mathbf{Y_{X/S_L}}$	cell yield per unit substrate consumed (kg-X/kg-S_L)
$\mathbf{Y_{S_L/X}}$	biomass yield on soluble substrate (kg-SL/kg-X)
$\mathbf{Y_{H_2O/S_L}}$	metabolic yield of water (kg-H_2O/kg-SL)
$\mathbf{Y_{H_2O/X}}$	metabolic yield of water (kg-X/kg-H_2O)
$\mathbf{Y_{H_2O/Xdead}}$	metabolic yield of water of dead biomass (kg-H_2O/kg-X)
$\mathbf{Y_{O_2/S_L}}$	metabolic consumption of oxygen (kg-O_2/kg-S_L)
ε_w	water fraction related to moisture (m^3 H_2O/ m^3 compost)
μ	specific growth rate of biomass (1/d)

REFERENCES

Abid, N., Sayadi, S., 2006. Detrimental effect of olive mill wastewater on the composting process of agricultural wastes. Waste Manag. 26, 1099–1107.

Alburquerque, J.A., Gonzalvez, J., Garcia, D., Cegarra, J., 2004. Agrochemical characterization of "alperujo", a solid byproduct of the two-phase centrifugation method for olive oil extraction. Bioresour. Technol. 91, 195–200.

Alburquerque, J.A., Gonzalvez, J., García, D., Cegarra, J., 2006a. Effects of bulking agent on the composting of "alperujo", the solid by-product of the two-phase centrifugation method for olive oil extraction. Process Biochem. 41, 127–132.

Alburquerque, J.A., Gonzalvez, J., García, D., Cegarra, J., 2006b. Measuring detoxification and maturity in compost made from "alperujo", the solid by-product of extracting olive oil by the two-phase centrifugation system. Chemosphere 64, 470–477.

Alburquerque, J.A., Gonzálvez, J., Tortosa, G., Baddi, G.A., Cegarra, J., 2009. Evaluation of "alperujo" composting based on organic matter degradation, humification and compost quality. Biodegradation 20, 257–270.

Alfano, G., Belli, C., Lustrato, G., Ranalli, G., 2008. Pile composting of two-phase centrifuged olive husk residues: technical solutions and quality of cured compost. Bioresour. Technol. 99, 4694–4701.

Altieri, R., Esposito, A., Nair, T., 2011. Novel static composting method for bioremediation of olive mill waste. Int. Biodeterior. Biodegradation 65, 786–789.

Axelsson, V., Pikkarainen, K., Forsby, A., 2006. Glutathione intensifies gliotoxin induced cytotoxicity in human neuroblastoma SH-SY5Y cells. Cell Biol. Toxicol. 22, 127–136.

Baeta-Hall, L., Sàagua, C.M., Bartolomeu, M.L., Anselmo, A.M., Rosa, M.F., 2005. Biodegradation of olive oil husks in composting aerated piles. Bioresour. Technol. 96, 69–78.

Bernal, M.P., Alburquerque, J.A., Moral, R., 2009. Composting of animal manures and chemical criteria for compost maturity assessment. A review. Bioresour. Technol. 100, 5444–5453.

Bernal, M.P., Paredes, C., Sanchez-Monedero, M.A., Cegarra, J., 1998. Maturity and stability parameters of compost prepared with a wide range of organic wastes. Bioresour. Technol. 63, 91–99.

Cayuela, M.L., Sánchez-Monedero, M.A., Roig, A., 2006. Evaluation of two different aeration systems for composting two-phase olive mill wastes. Process Biochem. 41, 616–623.

Cegarra, J., Alburquerque, J.A., Gonzálvez, J., Tortosa, G., Chaw, D., 2006. Effects of the forced ventilation on composting of a solid olive-mill by-product ("alperujo") managed by mechanical turning. Waste Manag. 26, 1377–1383.

Chatterjee, N., Eom, H.-J., Jung, S.-H., Kim, J.-S., Choi, J., 2013. Toxic potentiality of bio-oils, from biomass pyrolysis, in cultured cells and Caenorhabditis elegans. Environ. Toxicol. 29 (12), 1409–1419.

Chowdhury, A.K.M.M.B., Akratos, C.S., Vayenas, D.V., Pavlou, S., 2013. Olive mill waste composting: a review. Int. Biodeterior. Biodegradation 85, 108–119.

Chowdhury, A.K.M.M.B., Konstantinou, F., Damati, A., Akratos, C.S., Vlastos, D., Tekerlekopoulou, A.G., Pavlou, S., Vayenas, D.V., 2015. Is physicochemical evaluation enough to characterize olive mill waste compost as soil amendment? The case of genotoxicity and cytotoxicity evaluation. J. Clean. Prod. 93, 94–102.

Chowdhury, A.K.M.M.B., Michailides, M.K., Akratos, C.S., Tekerlekopoulou, A.G., Pavlou, S., Vayenas, D.V., 2014. Composting of three phase olive mill solid waste using different bulking agents. Int. Biodeterior. Biodegradation 91, 66–73.

Cooperband, L., 2002. The Art and Science of Composting: A Resource for Farmers and Compost Producers. University of Wisconsin-Madison, Madison, WI.

EC (European Commission), 2001. Biological Treatment of Biowaste. Working document, 2nd draft.

Gandolfi, I., Matteo, S., Franzetti, A., Fontanarosa, E., Santagostino, A., Bestetti, G., 2010. Influence of compost amendment on microbial community and ecotoxicity of hydrocarbon-contaminated soils. Bioresour. Technol. 101, 568–575.

Garcia-Gomez, A., Roig, A., Bernal, M.P., 2003. Composting of the solid fraction of olive mill wastewater with OL: organic matter degradation and biological activity. Bioresour. Technol. 86, 59–64.

Gigliotti, G., Proietti, P., Said-Pullicino, D., Nasini, L., Pezzolla, D., Rosati, L., Porceddu, P.R., 2012. Co-composting of olive husks with high moisture contents: organic matter dynamics and compost quality. Int. Biodeterior. Biodegradation 67, 8–14.

Hachicha, R., Hachicha, S., Trabelsi, I., Woodward, S., Mechichi, T., 2009a. Evolution of the fatty fraction during co-composting of olive oil industry wastes with animal manure: maturity assessment of the end product. Chemosphere 75, 1382–1386.

Hachicha, S., Cegarra, J., Sellami, F., Hachicha, R., Drira, N., Medhioub, K., Ammar, E., 2009b. Elimination of polyphenols toxicity from olive mill wastewater sludge by its co-composting with sesame bark. J. Hazard. Mater. 161, 1131–1139.

Haug, R.T., 1993. The Practical Handbook of Compost Engineering. Lewis Publishers, Florida.
Komilis, D.P., Tziouvaras, I.S., 2009. A statistical analysis to assess the maturity and stability of six composts. Waste Manag. 29, 1504–1513.
Kreja, L., Seidel, H.-J., 2002. Evaluation of the genotoxic potential of some microbial volatile organic compounds (MVOC) with the comet assay, the micronucleus assay and the HPRT gene mutation assay. Mutat. Res. 513, 143–150.
Lasaridi, K., Protopapa, I., Kotsou, M., Pilidis, G., Manios, T., Kyriacou, A., 2006. Quality assessment of composts in the Greek market: the need for standards and quality assurance. J. Environ. Manag. 80, 58–65.
Makni, H., Ayed, L., Ben Khedher, M., Bakhrouf, A., 2010. Evaluation of the maturity of organic waste composts. Waste Manag. Res. 28, 489.
Michailides, M., Christou, G., Akratos, C.S., Tekerlekopoulou, A.G., Vayenas, D.V., 2011. Composting of olive leaves and pomace from a three-phase olive mill plant. Int. Biodeterior. Biodegradation 65, 560–564.
Miller, F.C., 1992. Composting as a process based on the control of ecologically selective factors. In: Metting, F.B.J. (Ed.), Soil Microbial Ecology, Applications in Agricultural and Environmental Management. Marcel Dekker Inc, New York, pp. 515–544.
Nash, J.E., Sutcliffe, J.V., 1970. River flow forecasting through conceptual models. J. Hydrol. 10, 282–290.
Paredes, C., Bernal, M.P., Cegarra, J., Roig, A., 2002. Bio-degradation of olive mill wastewater sludge by its co-composting with agricultural wastes. Bioresour. Technol. 85, 1–8.
Paredes, C., Bernal, M.P., Roig, A., Cegarra, J., Sánchez-Monedero, M.A., 1996. Influence of the bulking agent on the degradation of olive-mill wastewater sludge during composting. Int. Biodeterior. Biodegradation 38, 205–210.
Paredes, C., Cegarra, J., Roig, A., Sanchez-Monedero, M.A., Bernal, M.P., 1999. Characterisation of olive mill wastewater (alpechin) and its sludge for agricultural purposes. Bioresour. Technol. 67, 111–115.
Ranalli, G., Principi, P., Zucchi, M., Da Borso, F., Catalano, L., Sorlini, C., 2002. Pile composting of two-phase centrifuged olive husks: bioindicators of the process. In: Insam, H., Riddech, N., Klammer, S. (Eds.), Microbiology of Composting. Springer-Verlag, Berlin Heidelberg, pp. 165–176.
Ryckeboer, J., Mergaert, J., Coosemans, J., Deprins, K., Swings, J., 2003. Microbiological aspects of biowaste during composting in a monitored compost bin. J. Appl. Microbiol. 94, 127–137.
Rynk, R., 1992. On-Farm Composting Handbook. Northeast Regional Agricultural Engineering Service (NRAES) Pub.
Said-Pullicino, D., Gigliotti, G., 2007. Oxidative biodegradation of dissolved organic matter during composting. Chemosphere 68, 1030–1040.
Sánchez-Arias, V., Fernández, F., Villasenor, J., Rodríguez, L., 2008. Enhancing the co-composting of olive mill wastes and sewage sludge by the addition of an industrial waste. Bioresour. Technol. 99, 6346–6353.
Sánchez-Monedero, M.A., Roig, A., Cegarra, J., Bernal, M.P., 1999. Relationships between water-soluble carbohydrate and phenol fractions and the humification indices of different organic wastes during composting. Bioresour. Technol. 70, 193–201.
Serramiá, N., Sánchez-Monedero, M.A., Fernández-Hernández, A., García-Ortiz Civantos, C., Roig, A., 2010. Contribution of the lignocellulosic fraction of two-phase olive-mill wastes to the degradation and humification of the organic matter during composting. Waste Manag. 30, 1939–1947.
Sultana, M., Chowdhury, A.K.M.M.B., Michailides, M.K., Akratos, C.S., Tekerlekopoulou, A.G., Vayenas, D.V., 2015. Integrated Cr(VI) removal using constructed wetlands and composting. J. Hazard. Mater. 281, 106–113.
Tomati, U., Galli, E., Pasetti, L., Volterra, E., 1995. Bioremediation of olive-mill wastewaters by composting. Waste Manag. Res. 13, 509–518.
Tortosa, G., Alburquerque, J.A., Ait-Baddi, G., Cegarra, J., 2012. The production of commercial organic amendments and fertilisers by composting of two-phase olive mill waste ("alperujo"). J. Clean. Prod. 26, 48–55.

Vasiliadou, I.A., Muktadirul Bari Chowdhury, A.K., Akratos, C.S., Tekerlekopoulou, A.G., Pavlou, S., Vayenas, D.V., 2015. Mathematical modeling of olive mill waste composting process. Waste Manag. 43, 61–71.

Vlyssides, A., Mai, S., Barampouti, E.M., 2009. An integrated mathematical model for co-composting of agricultural solid wastes with industrial wastewater. Bioresour. Technol. 100, 4797–4806.

Zorpas, A.A., Costa, N.C., 2010. Combination of Fenton oxidation and composting for the treatment of the olive solid residue and the olive mill wastewater from the olive oil industry in Cyprus. Bioresour. Technol. 101, 7984–7987.

Zucconi, F., Forte, M., Monaco, A., De Bertoldi, M., 1981. Biological evaluation of compost maturity. BioCycle 22, 27–29.

CHAPTER

THE USE OF OLIVE MILL WASTE TO PROMOTE PHYTOREMEDIATION

9

Tania Pardo*,, Pilar Bernal*, Rafael Clemente***

**Spanish National Research Council (CSIC). Campus Universitario de Espinardo, Murcia, Spain; **Spanish National Research Council (CSIC). Av. De Vigo, Santiago de Compostela, Spain*

9.1 OLIVE OIL INDUSTRY BY-PRODUCTS AND MAIN CHARACTERISTICS

The olive oil industry is a significant socioeconomic sector for many countries, especially in the Mediterranean region, where olive (*Olea europaea* L.) cultivation is a centuries old tradition. Nowadays, the product is widely consumed around the world. About 3.32 millions of tons of olive oil are produced annually worldwide, with 72% produced in Europe (FAOSTAT, 2013). The European countries that are the main contributors in the production of olive oil are Spain (41.6%), Italy (17.2%), and Greece (10.5%), the three countries generating 96.4% of the total annual European production. Other relevant Mediterranean basin countries are Turkey (6.2%), Syria (6.0), Tunisia (5.8%), and Morocco (3.6%). Olive oil production is not restricted to the Mediterranean basin, and other smaller producers without previous tradition in olive cultivation are steadily increasing around in Asia, Africa, and America (with a 14.5%, 11.7%, and 1.5% of worldwide production, respectively; FAOSTAT, 2013).

The extraction of the olive oil from the olives is based on a mechanical process. Three different systems can be used industrially: the traditional press-cake system (practically obsolete); the three-phase decanter system; and the two-phase centrifugation system, the latter being the most commonly used currently in Spain (Alburquerque et al., 2004; Morillo et al., 2009). In the three-phase decanter system, the oil, tissue water, pulp, and seed of the olive are separated in a continuous process. This technology requires the addition of a high amount of water in the different steps of the process (Paredes et al., 1999), which implies the generation of large quantities of wastewater (1200 kg per ton of processed olive; Azbar et al., 2004). Then, the three-phase technology produces two by-products: the olive mill wastewater (OMWW or alpechín) composed of the olive tissue water plus the process water; and olive kernel (or orujo), which is the solid by-product that consists mainly of the olive pulp and olive stones (Brunetti et al., 2005; Morillo et al., 2009), and from which further oil is chemically extracted. The two-phase centrifugation system is considered the most ecologically friendly techniques, because it saves >80% of the water (the process requires only a small amount of water) and >20% of the energy required by the three-phase system. Then, using this two-phase technology, only a semisolid sludge is produced as by-product called wet olive pomace, olive wet husk, or alperujo (800 kg per ton of processed olive; Azbar et al., 2004). Residual oil from the solid olive mill waste (OMW) can be

Olive Mill Waste. http://dx.doi.org/10.1016/B978-0-12-805314-0.00009-1

obtained by a second centrifugation, and the resulting solid by-product can be also further chemically treated for the extraction of the remaining oil (Alburquerque et al., 2004).

The characteristics of the by-products from the olive oil industry may vary depending on the fruit variety and maturity, pedoclimatic conditions of the culture, olive storage system, and the production method. All these by-products are dark-colored and contain a high amount of organic matter. The OMWW contains most of the water-soluble chemical species present in the olive fruit, an elevated load of low degradable organic matter (COD/BOD5 ratio between 2.5 and 5; Azbar et al., 2004) including proteins, lipids, polysaccharides, and aromatic and polyphenolic compounds (up to 80 g/L), acidic pH, and high electrical conductivity (Morillo et al., 2009; Alburquerque et al., 2004; Paredes et al., 1999). The wet olive pomace is a dense sludge, with lower moisture content than OMWW, but higher than that of olive kernel from the three-phase system. This material is slightly acidic and presents high concentrations of organic matter (containing lignin, hemicelluloses and cellulose, and also proteins, lipids, and a small proportion of phenolic compounds; Morillo et al., 2009; Alburquerque et al., 2004). The dark color of the OMWs and their content of phenolic compounds are considered the main factors responsible for their associated pollution problems.

Phenolic compounds are originally synthesized by the olive plants, but can be also formed during oil extraction process (Morillo et al., 2009). The majority of these compounds remains in the olive mill by-products, and only a small part is extracted in the oil (Rodis et al., 2002). Such compounds are considered responsible for the phytotoxic character and antimicrobial characteristics of OMWs (Hachicha et al., 2009; Pierantozzi et al., 2012). Direct application to soil of OMWW and wet olive pomace has been reported to cause inhibition of seed germination and seedling growth on *Lactuca sativa*, lethality on *Artemia salina*, and genotoxicity effect (inhibition of mitosis) on *Allium cepa* (Pierantozzi et al., 2012), and a decrease in phosphatase and fluorescein diacetate hydrolase activities in the soil (Piotrowska et al., 2006).

9.1.1 OMWs COMPOSTING

The transformation of olive mill by-products by composting has been widely recommended before their application to soil. During the composting process, the labile organic compounds are degraded by bacteria, fungi, and other microorganisms, this leading to an organic-matter rich material, free of phytotoxicity and potentially beneficial for plant growth due to the elevated concentration of macro and micronutrients, the absence of toxic elements, and low salinity (Alburquerque et al., 2006; Hachicha et al., 2009). However, olive mill by-products have certain properties inadequate for composting, such as excessive moisture content (>90% in OMWW) and the presence of noneasily degradable compounds and fats and polyphenols that provide antimicrobial characteristics (Ramos-Cormenzana et al., 1995). In addition, its dense and sticky physical texture (especially wet olive pomace) makes it difficult to maintain aerobic conditions inside the material for composting. Cocomposting procedures have been developed for olive mill by-products with several agricultural wastes: olive tree leaves, straw, cotton gin waste, grape stalks, animal manures, etc (Alburquerque et al., 2006; Cayuela et al., 2004; García-Gomez et al., 2003; Paredes et al., 2002).

Composts from OMWs have demonstrated to be adequate organic fertilizers for horticultural crops (Alburquerque et al., 2006), olive trees (Cayuela et al., 2004) and also as part of the substrate or growing media for ornamental plants culture (García-Gomez et al., 2002). Nowadays, most olive oil producers have established composting systems for the treatment of their by-products. In Spain, legislation for fertilizer products includes compost from wet olive pomace as an organic amendment, and specifies the physicochemical requirements of the product. The majority of the compost produced in the olive oil industry comes from the cocomposting of wet olive pomace with olive leaves, animal manures, or straw.

9.1.2 BIOCHAR PRODUCTION

Recently, much attention has been paid for the production and use of biochar from different organic wastes. Biochar is a C-rich organic material produced during slow exothermic decomposition of biomass at temperatures ≤700°C under zero oxygen or low oxygen conditions (Lehman and Joseph, 2009). Pyrolysis is the major anaerobic thermochemical conversion process leading to three product phases: noncondensable gas and condensable vapors; liquids (biooil and tars); and solid (char or ash). Biochar is highly recalcitrant due to its high aromaticity and can sequester C in soil for a very long period of time.

Feedstock characteristics and pyrolysis conditions (e.g., maximum exposure temperature) affect the physical and chemical characteristics of biochars (Gaskin et al., 2008; Novak et al., 2009; Singh et al., 2010). Olive wastes, mainly olive tree cuttings or pruning and olive stones have been satisfactorily processed through pyrolysis obtaining biochar (Andrados et al., 2015; Alburquerque et al., 2014). However, other waste materials from the olive mill industry may require predrying due to their high moisture content (Table 9.1). To avoid the costly predrying step, wet pyrolysis, often called hydrothermal carbonization, is considered as an efficient technology to carbonize moist biomass. In this process, biomass is heated in a high pressure reactor at temperatures below 350°C. The reaction products are different gases and a mixture of solid (referred to as hydrochar) and liquid (containing the solvent used in the hydrothermal carbonization reaction) phases easy to separate mechanically (Benavente et al., 2015). Hydrochars from wet olive pomace have a moisture content between 27.1% and 59.5% (Benavente et al., 2015), greater than the values found for biochar (Table 9.1), with a composition of 1.95–3.35% ash; 63.3–73.0% C, 7.96–8.33% H, 1.12–1.72% N, 14.8–24.1% O, and <0.1% S. Olive kernel was air-dried for 30 days before being transformed in biochar by pyrolysis at 3 different temperatures and heating rates (Hmid et al., 2014). Average C concentrations of these biochars were greater than those found in the hydrochar by Benavente et al. (2015); however, greater ash content was found in the biochar than in the hydrochar (Hmid et al., 2014; Benavente et al., 2015).

In general, biochar obtained from OMWs and olive trees have low concentration of N (Table 9.1), with the lower values in biochar from olive stones (Alburquerque et al., 2014; Zabaniotou et al., 2015), which could limit their fertilizer value. According to Olmo et al. (2014), biochar can provoke positive effects on plant biomass production, due to the increase of fine root proliferation, which facilitated water and nutrient acquisition. Then, positive effects of biochar are related to their high concentration of highly recalcitrant C, which showed the highest concentration in biochar from olive stones (Alburquerque et al., 2014; Zabaniotou et al., 2015). The recalcitrant nature of biochar is responsible for its low mineralization and thus nutrient release in soil (Olmo et al., 2014). But their alkaline nature and liming potential, associated to their elevated Ca and Mg concentration (Alburquerque et al., 2014), and their sorption (metal retention) capacity can make them useful for remediation strategies.

Based on strengths, weaknesses opportunity, and threats (SWOT) analysis, Zabaniotou et al. (2015) demonstrated that pyrolysis of agro-residues targeting biochar production can fulfil the aim of closing the loop in agriculture and circular economy objectives, applicable in Mediterranean countries. In a three-phase olive mill system, the wastes—tree pruning, olive kernels—can be converted via pyrolysis

Table 9.1 Chemical Composition of Olive Mill Wastewater (OMWW, Fresh Weight), Wet Olive Pomace, Composts and Biochar From Olive Mill Waste (OMW) and Olive Tree Residues (Dry Weight)

Parameters	OMWW[a]	Wet Olive Pomace[a]	Composts[a]	Biochar[b]
Dry matter (%)	6.33–7.19	49.6–71.4	—	98.1–98.6
pH	4.2–5.2	4.9–6.8	5.4–9.5	8.6–11.5
EC (dS/m)	5.5–12.0	0.98–5.2	1.2–7.3	0.28–2.15
Organic matter (g/kg)	46.5–62.1	848–976	260–900	(1.58–39)[e]
TOC (g/kg)	34.2–39.8	495–539	110–580	483–933
TN (g/kg)	0.61–2.10	7.0–18.5	11–54	4.1–13
C/N	52.3–54.3	28.2–72.9	9–36	62.4–227
P (g/kg)	0.16–0.31	0.7–2.2	1–30	0.08–8.34
K (g/kg)	1.97–8.97	7.7–29.7	6–44	1.6–5.4
Na (g/kg)	0.11–0.42	0.5–1.6	2–41	—
Ca (g/kg)	0.20–0.64	1.7–9.2	7–72	86.0
Mg (g/kg)	0.04–0.22	0.7–3.8	0.9–57	2.0
Fe (mgkg)	18.3–120	78–1462	100–4100	—
Cu (mg/kg)	1.5–6	12–29	1.4–79	—
Mn (mg/kg)	1.1–12	5–39	13–131	—
Zn (mg/kg)	2.4–12	10–37	38–138[d]	—
Phenols (%)	0.98–10.7	0.5–2.4	0.1–3.8[c]	—

—, Not provided.
[a]Alburquerque et al. (2004); Morillo et al. (2009); Chowdhury et al. (2013).
[b]Andrados et al. (2015); Alburquerque et al. (2013, 2014); Hmid et al. (2014); Zabaniotou et al. (2015)
[c]Water soluble fraction
[d]Alburquerque, personal communication
[e]Ash content (%)
Adapted from Clemente, R., Pardo, T., Madejón, P., Madejón, E., Bernal, M.P., 2015. Food byproducts as amendments in trace elements contaminated soils. Food Res. Int. 73, 176–189.

into a liquid fuel, a gas fuel, and biochar; the liquid and gas fuels supplied the required energy for olive mill operation and olive kernel drying, with a surplus that could be sold and provide an extra income. The obtained biochar was returned directly to the cultivated land.

9.2 RESTORATION OF TRACE ELEMENTS CONTAMINATED SOILS

The term "trace element" (TE) is generally applied to elements found in small concentrations in different environmental compartments, and is also frequently used to refer to heavy metals and metalloids that can cause environmental and toxicological problems (Alloway, 2013). Some TEs are essential for the development of living organisms (like Co, Cu, Fe, Mn, Ni, and Zn), whereas others (As, Cd, Hg, and Pb) do not have any known physiological function (Adriano, 2001). In the soil, TEs are usually present in relatively low concentrations and even those with a defined biological function can be toxic to living organisms when present above certain concentrations and depending on their availability.

9.2.1 SOIL CONTAMINATION WITH TRACE ELEMENTS: SOURCES, ASSOCIATED PROBLEMS, AND REMEDIATION STRATEGIES

TE contamination of soils has received great attention due to the potential risk it poses to human health and because contamination of agricultural soils by these pollutants may threaten food safety and disturb the ecosystem (Lee et al., 2013). The presence of TEs in the soil may arise from different origins: lithogenic, coming from the lithosphere (bedrock); pedogenic, coming also from the lithosphere but after the processes of soil formation; and anthropogenic, arising from human activities. Lithogenic TEs are relatively immobile in the soil, but are subject to changes under certain conditions (like the influence of microorganisms and plant root exudates) that make them more mobile species (Kabata-Pendias and Adriano, 1995). Also, rock minerals weathering produces local accumulation of TEs in the soil. However, this accumulation does not usually exceed toxicity thresholds. In addition, volcanic emissions, fires, dust storms, and other phenomena can naturally introduce TEs on the ecosystem (Naidu et al., 2001).

There are also numerous anthropogenic causes that can lead to high concentrations of TEs in soils, which can exceed those of natural origin. Human activities have led to the modification of the natural cycles of TEs, so that the contributions to the soils far outweigh the losses by volatilization, leaching, or withdrawn by the crops. Operations arising from mining activities (ore extraction, processing, waste disposal, and transport of semiproducts) release metals and other TEs that adversely affect the environment in which the activity occurs. The abandonment of the mine once the exploitation process is completed, as well as accidents caused by mining activities contribute to the spread of contamination and represent a significant risk of TEs contamination (Conesa et al., 2012).

Industrial activities are also a source of pollution, and deposition of contaminated particles is an important gateway of TEs in soils. Similarly, the application of phosphate fertilizers and different amendments is the main source of soil pollution from agricultural activities (Adriano 2001; Kabata-Pendias and Pendias, 2001).

9.2.2 CONTAMINATED SOILS ASSOCIATED PROBLEMS AND RECOVERY

Soil contamination is one of the main environmental problems on a global scale. Only in Europe, it is estimated that the number of potentially contaminated sites exceeds 2.5 million and this number is expected to increase up to a 50% in 2025. More than 35% of the contaminated soils are affected by TEs (Mench et al., 2010), whose recovery represents a challenging task as a consequence of their toxicity, resistance (they cannot be degraded by microorganisms, unlike organic contaminants), and relative immobility in the soil (Kabata-Pendias and Pendias, 2001). Total TE concentrations considered to represent pollution vary according to the element, legislative body or country and soil use. However, they do not necessarily indicate soil toxicity, and bioavailable concentrations (the fraction able to interact with living organisms) are often more revealing on this regard (Clemente et al., 2015).

The selection of the appropriate technique or combination of techniques to be used for recovering a contaminated soil will depend primarily on the level of cleaning desired, the duration of the cleaning, the concentration and chemical forms in which the contaminants are found, the characteristics of the affected area, and the cost of operations (Adriano, 2001). The complete removal or the sealing of the contaminated soil would be necessary only when the decontamination and/or stabilization of the site do no warrant short or long term effects. As an alternative to the removal and sealing, a number of techniques based on the application of chemical and physical processes, have been developed (Evanko and Dzombak, 1997). However, most of the techniques based on physical or physicochemical processes, such as electroosmosis, physicochemical metal extraction, and acid wash, or in situ immobilization by vitrification, are rather drastic, precise from expensive equipment and are only suitable for ex-situ treatments or the decontamination of small areas. In addition, these techniques reduce soil biological activity, affect negatively soil structure and functionality, and its productivity is drastically reduced, so that soils reclaimed in this way are not suitable for cultivation or plant growth (Bernal et al., 2007).

As a consequence, most of the research dealing with the recovery of TEs contaminated soils has focused on the development of alternative methods that can be applied "in situ," which require less soil management and are able to eliminate or immobilize the TEs, and are at the same time cost effective and respectful with the environment (Adriano et al., 2004). Among these techniques, assisted natural attenuation and bioremediation have received special attention because both can be considered "soft" techniques, that is, with low cost, low demand for infrastructure, and easy to execute, and they mean an environmentally friendly alternative to the destructive physicochemical techniques described earlier (Bernal et al., 2007).

Natural attenuation refers to the use of unenhanced natural processes to immobilize the pollutants and prevent the spread of contamination (Bernal et al., 2007). Regarding TEs, the ultimate goal of these techniques is the reduction of their mobility or availability, and their stabilization in the soil through adsorption on soil minerals surface and soil organic matter, formation of insoluble compounds (precipitation), assimilation by soil microorganisms and plants, and occasionally volatilization (Hg, and methylated compounds of Hg, As, and Se) (Adriano et al., 2004). Assisted natural attenuation basically enhances these natural processes through the addition of soil amendments and the use of plants (Adriano et al., 2004).

Bioremediation techniques are based on the action of microorganisms and/or plants on soil TEs. Microorganisms may influence TEs mobility and availability through oxidation or reduction mechanisms, methylation, chelation, biosorption, and intracellular accumulation (Srivastava et al., 2011; Qian et al., 2012). Phytotechnologies (and more precisely phytoremediation) are based on the effects that plants and their associated microorganisms exert on soil contaminants. These techniques help recovering the basic processes that define ecosystems functionality and sustainability (Mench et al., 2010).

9.2.3 PHYTOREMEDIATION OF CONTAMINATED SOILS

Phytotechnologies take into account the mechanisms or strategies naturally developed by the plants to protect themselves from contamination (exclusion, regulation of absorption and transport to the shoot, and accumulation) (Dickinson et al., 2009; Conesa et al., 2012). Excluder plants maintain very low levels of TEs in their aerial part, while indicator plants concentration of TEs reflects the concentration of the soil, thanks to the regulation of the absorption and transport to the aerial part. Accumulator plants store TEs preferably in the aerial part due to an efficient transport from the root, resulting in a concentration ratio between shoot and root greater than one (Kidd et al., 2009).

Phytostabilization and phytoimmobilization techniques follow the containment of the contaminants in the soil (Robinson et al., 2009). Phytostabilization is based on the mechanical action of roots, which avoids erosion and soil particles transport, and reduces leaching due to increased evapotranspiration (Bolan et al., 2011). Phytoimmobilization involves the absorption, precipitation, and accumulation of the pollutants in the root zone, which is carried out in collaboration with the microorganisms associated to plant roots (Wenzel et al., 1999). Both techniques aim at creating a vegetation cover to prevent the incorporation of the TEs to the food chain, avoiding TEs leaching to groundwater and dissemination and its associated negative impact on surrounding ecosystems, and facilitating the retention of particles that can be directly ingested or inhaled (Mendez and Maier, 2008; Vangronsveld et al., 2009). Excluder and indicator plants are the most appropriate ones to be used in these techniques.

Phytoextraction is based on the use of plants able to tolerate and accumulate high concentrations of TEs in the aerial part (accumulator plants), so that this biomass can be harvested and removed from the contaminated site (Kidd et al., 2009). Most of the studies related to phytoextraction have been based on the identification and use of hyperaccumulator species, which have the ability to accumulate TEs as Cd, Mn, Ni, and Zn at very high concentrations (0.1–10% of the leaf mass) in the aerial part (Kidd et al., 2009). These techniques have frequently too many drawbacks to constitute a real alternative to the existing methods of recovery, notably the low biomass that hyperaccumulator plants develop, their low adaptation capacity to different environments, and their selectiveness for just one or two elements, which makes them not effective in multielement contaminated soils (Wenzel et al., 1999; Bhargava et al., 2012). Induced phytoextraction makes use of tolerant species of high biomass too which hyperaccumulation is induced when plants reach their maximum yield, through the addition of substances that facilitate TEs solubilization, like chelating agents (Kidd et al., 2009). This is frequently not a suitable option as TEs infiltration and movement to downwards soil horizons may contaminate groundwater (Kahn et al., 2000).

Phytovolatilization techniques use plants that absorb soil contaminants and turn them into volatile species (Wenzel et al., 1999); this can be used in the decontamination of soils contaminated with organic compounds and/or TEs able to from volatile species. Finally, in phyto- and rhizodegradation processes, plants absorb the pollutants and degrade them to nontoxic metabolites in the aerial part or at root level, respectively. This technique could be only applied to organic pollutants as TEs cannot degrade.

9.2.4 THE USE OF ORGANIC WASTES AS SOIL AMENDMENTS IN RESTORATION PROCEDURES

The improvement of the soil physicochemical properties is usually needed when the aim of the restoration of TEs contaminated soils is the reestablishment of a vegetation cover (Clemente et al., 2015). Contaminated soils usually have low fertility, low organic matter content, and sometimes very acidic pH and/or high salinity, and the establishment of a vegetation cover on these soils is very difficult unless organic and/or inorganic amendments are applied (Bernal et al., 2007).

Soil amendments (especially organic) facilitate the establishment of plants in degraded and contaminated soils as they improve soil physicochemical properties, fertility and microbial activity and diversity (Bolan et al., 2011; Renella et al., 2008). This is of relevance, as soil microorganisms play a significant role in phytoremediation because they affect not only the mobility of TEs, but also the development of plants (Kidd et al., 2009). In addition, amendments can directly affect TEs mobility and availability through the modification of soil physicochemical and biological conditions (pH, redox conditions, nutrients supply, etc.; Bernal et al., 2007).

The addition of different types of residual organic materials to the soil as a source of organic matter is a common practice to improve soil properties and might be an environmentally friendly and cost effective approach to restore extensive areas with a moderate level of contamination (Dickinson et al., 2009). The use of waste materials with this aim would mean an additional environmental benefit, as an alternative for their recycling and reutilization and a way to reduce waste disposal and accumulation (Clemente et al., 2015; Belaqziz et al., 2016). In this sense, recent changes in legislation have forced waste managers to face the development of alternative strategies and techniques to divert biodegradable waste materials from landfill. As a consequence, large amounts of low grade composts are produced and there is an urgent need to find alternative uses for these products (Farrell and Jones, 2010). But potential benefits from the application of waste-derived amendments to contaminated soils have to be balanced against the possible risks due to undesired pollutants (like TEs) and pathogenic organisms that can be present in these materials and may accumulate in the soil (Karaka, 2004).

9.3 THE USE OF FRESH AND PROCESSED OMWs IN PHYTOREMEDIATION

The suitability of the different OMWs as soil amendments, both fresh and processed, in the remediation of TEs contaminated soils has been evaluated in terms of their influence on soil physicochemical (pH, electrical conductivity, TEs and nutrients solubility, and availability) and/or biologic properties (microbial diversity and development, plant growth, TEs uptake, ecotoxicity, etc.) under different experimental scales. Most of the investigations have been focused on the evaluation of wet olive pomace (fresh or composted), as the two-phase system is currently the most frequent oil extraction method in many oil producing countries. The use of fresh versus processed OMWs (compost and biochar) has been described in detail in this chapter, focusing on key aspects regarding soil remediation.

9.3.1 TRACE ELEMENTS SOLUBILITY AND FRACTIONATION IN OMWs TREATED SOILS

The influence of OMWs on TE solubility and fractionation in contaminated soil has been evaluated under laboratory (Madrid and Díaz-Barrientos, 1998; Burgos et al., 2010; Alburquerque et al., 2011; Pardo et al., 2011), greenhouse (Nogales and Benítez, 2006; de la Fuente et al., 2011; Pardo et al., 2014a, 2016), and field conditions (Clemente et al., 2007b; Pardo et al., 2014c,d). Different results for the different wastes have been observed, mainly associated to their application fresh or processed (Romero et al., 2005; Nogales and Benítez, 2006; de la Fuente et al., 2011).

Numerous studies have reported an increase of TEs solubility in soils after the addition of fresh OMWs, enhancing leaching, availability, and toxicity to plants (Nogales and Benítez, 2006; de la Fuente et al., 2008, 2011). This behavior has been related to the slightly acidic nature, but especially to the elevated concentration of soluble organic compounds (de la Fuente et al., 2008; Piotrowska et al., 2006). Soluble phenols can chelate heavy metals, blocking their sorption and promoting leaching through the formation of soluble metal complexes (Madrid and Díaz-Barrientos, 1998). Congruently, Cu chelation by the phenolic compounds of wet olive pomace increased Cu solubility in a calcareous soil indirectly affected by mining activities in a 2-year experiment (Clemente et al., 2007b). However,

Rodríguez et al. (2016) observed a decrease of EDTA-extractable concentrations of Pb and Zn in contaminated soils amended with wet olive pomace due to a pH effect, but non extractable-Cu remained constant or slightly above that of untreated soils, probably due to the formation of soluble organic complexes. However, other authors (Burgos et al., 2010) suggested that the organic matter supplied by wet olive pomace may cause the formation of stable organic complexes with some metals reducing their solubility in soil, as lower concentrations of $CaCl_2$-extractable Cu in treated soils than in nonamended contaminated soils were found.

Besides the chelation effect, the soluble phenolic compounds of OMWs can highly influence redox processes in soils during their degradation and alter TEs speciation (Clemente et al., 2007b; de la Fuente et al., 2008). Oxidation and polymerization of phenolic compounds in soil can be accelerated nonenzymatically by iron and manganese oxides, resulting in the release of reduced elements: Mn^{2+} and Fe^{2+}, which are highly soluble (Huang, 1990). In agreement with this, less-oxidizing conditions (reduction in the redox potential) have been described by de la Fuente et al. (2008) in a metal-contaminated soil 14 days after wet olive pomace addition, showing a 70% decrease in the concentration of water-soluble phenols. The increase in the solubility of Fe, Mn, and other elements associated with Fe and Mn oxides, such as Zn and Pb, after the application of fresh OMWs to contaminated soils has been associated to this redox effect by several studies: Romero et al. (2005) showed an enhancement of water-extractable Zn and Pb soil fractions after the application of wet olive pomace to a mine soil from a tailing in an incubation experiment; Piotrowska et al. (2006, 2011) also reported elevated extractable concentrations of Fe and Mn immediately after the application of different rates of OMWW; Clemente et al. (2007b) reported a significant initial increase of Fe and Mn availability in a contaminated soil amended with wet olive pomace, but after 1.5 year under field conditions, only available Mn concentrations remained higher than in nonamended soil. It is known that Fe^{2+} is quickly oxidized to Fe^{3+} in soil, but Mn^{2+} oxidation is quite slow, so soluble Mn species can persist longer in the soil (Clemente et al., 2007b).

Moreover, in addition to changing soil redox conditions, the mineralization process of the organic compounds present in OMWs leads to the release of inorganic compounds, while in calcareous soils the CO_2 produced during organic matter mineralization can alter the carbonate/bicarbonate equilibrium, affecting metal precipitation (Clemente et al., 2006; de la Fuente et al., 2008).

When the OMWs are processed, most of their physicochemical characteristics change (Table 9.1), and their influence on TEs solubility and availability is consequently different (de la Fuente et al., 2011). Composted OMW have an alkaline pH and a higher stability of their organic matter than fresh materials and, therefore, present higher buffering and complexing capacity. For this reason, the use of composted OMWs in contaminated soils generally implies TEs immobilization (de la Fuente et al., 2011; Pardo et al., 2011, 2014a; Moreno-Jiménez et al., 2013). For example, the ability of wet olive pomace compost to decrease the readily available fractions of metals has been often associated to a soil pH increase (Fornes et al., 2009; Alburquerque et al., 2011). However, the richness in humic substances is considered the main responsible of the effects of composted OMW on TEs behavior through chelation, adsorption, or retention processes. Nogales and Benítez (2006) observed that the application of olive pomace compost and vermicompost to a contaminated soil decreased water-soluble and DTPA-extractable fractions of Pb and Zn related to metal chelation and adsorption by the humified organic matter supplied. Pardo et al. (2016) found a decrease of the dissolved concentrations of Cd, Cu, Mn, Pb, and Zn and the modification of the chemical species distribution of some TEs present in soil pore water of a mine tailing after the addition of olive pomace-compost (combined with a "red mud" derivate), organometallic forms being the dominant species for Pb and Cu. The role of the humic substances of OMWs compost was demonstrated by Clemente and Bernal (2006), who observed Pb and Zn immobilization after the application of humic acids isolated from a compost prepared with the solid fraction of OMWW to an acid contaminated soil, showing the significance of the provision of exchange sites by this organic matter fraction.

Mineralization rates between 3% and 5% of the organic-C coming from olive pomace compost have been described in contaminated soils (Alburquerque et al., 2011; Pardo et al., 2011) versus 15–39% of the raw waste (Romero et al., 2005; de la Fuente et al., 2008). The phenolic compounds present in OMWs are degraded during composting (Hachicha et al., 2009), but a small resistant fraction may be later oxidized in the soil provoking mobilization of TEs. Significant increases in the $CaCl_2$- and DTPA-extractable Mn concentrations were found up to 1 year after olive pomace compost addition to two soils with different contamination degree and pH by Clemente et al. (2012) and Pardo et al. (2014d) in two field experiments. However, these authors and also de la Fuente et al. (2011) observed that the increase in soil Mn solubility did not result in Mn enrichment of the leaves and fruits of the established vegetation.

Given the well-known affinity of Cu for soil organic matter, a mobilization of this element after OMWs compost addition to contaminated soils has been reported by numerous studies (Burgos et al., 2010; Pardo et al., 2011; Moreno-Jiménez et al., 2013). However, Pardo et al. (2014a) found, in a column experiment, that complexation of Cu by the organic matter provided by olive pomace compost reduced its leaching through the soil profile. Moreover, the application of composted OMWs to contaminated soils often implies the mobilization of As (Pardo et al., 2011, 2014a,d, 2016; Clemente et al., 2012; Moreno-Jiménez et al., 2013; Beesley et al., 2014). The pH increase caused by compost may be responsible of the solubilization of As, but also the phosphates and soluble organic compounds supplied with the compost can compete with As for adsorption sites in the soil contributing to its mobilization (Fitz and Wenzel, 2002). This effect could question the use of this type of compost in the remediation of trace elements contaminated soils, and implies that the dose applied must be carefully studied to avoid As (and/or any other toxic elements) solubilization in the soil (Clemente et al., 2012; Moreno-Jiménez et al., 2013).

The use of biochar from OMWs as amendments in contaminated soils has been little studied. Even though the effect of biochar addition depends on the characteristics of the soil and of the feedstock material of biochar (Beesley et al., 2014), its application to soil can generally enhance the cation exchange capacity of the soil and increase soil pH due to its high surface area and porous structure and its elevated pH (Alburquerque et al., 2013). In agreement with this, Hmid et al. (2015) observed a significant reduction of $Ca(NO_3)_2$-extractable concentrations of Cd, Pb, and Zn (27–77%) with increasing rates (5–15%) of olive kernel biochar added to a contaminated soil after 90 days of stabilization.

9.3.2 NUTRIENTS AVAILABILITY

The addition of OMWs to contaminated soils usually increases total and water soluble concentrations of organic C (TOC, water soluble organic C) and the availability of N, K, and P (from compost) (Romero et al., 2005; Clemente et al., 2007a; Moreno-Jiménez et al., 2013; Curaqueo et al., 2014). For example, Clemente et al. (2012) and Pardo et al. (2014d) observed high concentrations of TOC, TN, $NaHCO_3$-extractable P, and $NaNO_3$-extractable K in two contaminated soils with different physicochemical characteristics 2 years after compost addition to field experimental plots, reaching levels around 10-fold and 5-fold for K and P, respectively, in the treated soils compared to control soils. This improvement is of great importance for the stimulation of biogeochemical nutrient cycles in these typically poor soils and their fertility (Pardo et al., 2014b,c).

Kinetic models developed for C mineralization in contaminated soils after OMWs application showed that a high proportion of easily mineralizable organic matter is provided by raw materials (de la Fuente et al., 2008), while composts supply mostly organic compounds with a slow mineralization rate (Pardo et al., 2011). Therefore, the organic matter of compost provides a gradual release of C compounds, remaining longer in the soil and supposing a long-term reservoir/pool of C (Romero et al., 2005; Alburquerque et al., 2011). However, the high amount of phenolic compounds and highly-resistant ligno-cellulosic compounds of fresh OMWs possess toxic and antimicrobial properties that lead to a slow microbial degradation of the organic matter in soil despite its easily degradable nature (Romero et al., 2005). This effect was reported by de la Fuente et al. (2011), who observed a lower mineralization of water soluble organic C of wet olive pomace than that provided by olive pomace-compost although the same amount of this C fraction of each amendment was added to soil. This difference was associated to the reduction of toxic compounds in the composting process. Besides the relevance of organic matter for soil fertility, the increase in the soil organic matter content by OMWs may affect the soil aggregation and structure (López-Piñeiro et al., 2008). Mahmoud et al. (2010, 2012) showed an increase of soil aggregates stability and soil hydrophobicity, and a reduction of the effective diffusion coefficient into aggregates and the drainable porosity with the regular application of OMWW for 5–15 years.

In OMWs, N is mainly present in organic forms, and its mineralization in soil by microbial degradation is relatively low, independently of if they are applied raw or as compost (Brunetti et al., 2005; Alburquerque et al., 2006). It has been reported that the phenolic compounds of fresh OMWs are nitrification inhibitors and provoke a delay in the conversion of ammonium into nitrate in soil (Aguilar-Carrillo et al., 2009). In agreement with this, Di Serio et al. (2008) observed that spreading of OMWW in high amounts in a noncontaminated soil decreased the soil nitrifying community while increased the soil denitrifying populations. Nevertheless, the application of compost from OMWs to contaminated soils stimulates nitrification, which is usually inhibited in this type of soils (Alburquerque et al., 2011; Pardo et al., 2014a,c). The release of nitrates in soils amended with composted OMWs means an increase in N availability, but may also imply a significant risk of nitrate leaching due to its high mobility in soil (Pardo et al., 2014a).

The large content of K in raw and composted OMWs provokes a significant increase of K availability when added to soil (Walker and Bernal, 2008; Alburquerque et al., 2011; Clemente et al., 2012). This increment is usually reflected in leaf K concentrations, which have been found to be higher in plants grown in OMWs treated soils than in those receiving other or no amendments (Martínez-Fernández and Walker, 2012). Pardo et al. (2014a) observed in a mine soil amended with olive pomace-compost that the soluble K provided was mainly retained in the exchange complex of soil, although a high proportion leached down the soil profile. Therefore, it is remarkable that due to the leaching risk of nitrate and K in soils treated with composted OMWs, the doses applied must be carefully adjusted.

9.3.3 SOIL BIOLOGICAL PROPERTIES

The improvement of soil conditions, together with the reduction of the contaminants availability and the supply of essential nutrients by OMWs addition allow the stimulation of soil microbial communities in TEs contaminated soils (Romero et al., 2005; Clemente et al., 2007a,b; de la Fuente et al., 2011;

Pardo et al., 2014b,c,d). The stress to which the microorganisms are exposed to is reduced, and consequently their growth, activity, and functional diversity increased (Pardo et al., 2014b,c).

The application of fresh OMWs to soil can cause a short-term negative response of the soil microbiota (even in noncontaminated soils) mainly due to the toxicity associated to their richness in phenolic and other potentially toxic organic compounds (Piotrowska et al., 2006; Aguilar-Carrillo et al., 2009; de la Fuente et al., 2011). Di Bene et al. (2013) observed a significant reduction of microbial biomass C (B_C) and higher values of metabolic quotient (qCO_2, an indicator of stress conditions for soil microorganisms) in two agricultural soils immediately after OMWW application with respect to untreated soils. Similarly, Romero et al. (2005) found values of dehydrogenase (an intracellular enzyme indicator of overall microbial activity) and hydrolases (urease, β-glucosidase) in a contaminated soil treated with wet olive pomace similar to nonamended soil in the first week after its addition. Nevertheless, medium- to long-term stimulatory effects of raw wastes have been reported to be associated with the gradual degradation of toxic compounds (Clemente et al., 2007a; Burgos et al., 2010). De la Fuente et al. (2008 and 2011) showed that the concentrations of water soluble phenolic compounds in soils treated with wet olive pomace or its water extracts significantly decreased 2 months after its addition to a contaminated soil. The reduction of OMWs toxicity with time was reported by Clemente et al. (2007a), who found a positive evolution of the concentrations of B_C and microbial biomass N (B_N) and simultaneously a reduction of qCO_2 in a soil indirectly affected by mining activities treated with wet olive pomace. These results are in agreement with those obtained by Burgos et al. (2010), who also observed significant increases of B_C concentrations and dehydrogenase and arylsulfatase activities in two soils amended with fresh wet olive pomace throughout 40 weeks of incubation.

Treated OMWs have been found to provoke remarkable positive effects on soil microbial properties when applied to TEs contaminated soils, especially when they are used as compost (Fornes et al., 2009; Clemente et al., 2012; Pardo et al., 2014b,c,d; Hmid et al., 2015). Elevated B_C and B_N values, dehydrogenase and hydrolase activities and respiration rates, and the reduction of stress indicators in soils treated with composted OMWs have been described at laboratory (Romero et al., 2005; Alburquerque et al., 2011; Pardo et al., 2011), greenhouse (de la Fuente et al., 2011; Curaqueo et al., 2014; Pardo et al., 2014b), and field scale (Clemente et al., 2012; Pardo et al. 2014c). Fornes et al. (2009) found increases in the populations of heterotrophic bacteria, actinomycetes and fungi, and in their activity (FDA activity) after olive pomace-compost application to two TEs contaminated soils. Curaqueo et al. (2014) observed that olive pomace-compost addition in combination with arbuscular mycorrhizal fungi (AMF) inoculation to a mine tailing improved dehydrogenase, alkaline phosphatase, β-glucosidase, urease, and protease-BAA activities with respect to inoculation alone. The long duration of the beneficial influence of compost on soil microbial communities was proven by Pardo et al. (2014c) in a field experiment, as 2.5 years after treatment of a mine soil, significantly higher concentration (B_C and B_N values), and activity [hydrolase activities and average well color development value from Biolog EcoPlates (AWCD)], and better status of soil microbial communities (dehydrogenase activity, basal respiration, and qCO_2) than in the nonamended soil were found. In this case, the long term supply of nutrients was essential for the prolonged stimulation of soil microorganisms.

The effect on soil microorganisms of biochar obtained from OMWs, despite the few studies done, is apparently quite promising. The application of olive kernel-biochar improved the richness, diversity (H index), and overall functional activity of the heterotrophic cultivable portion of the soil bacterial communities (AWCD) of a metal contaminated soil, especially at the higher doses of biochar and equilibration period used as a result of higher metal immobilization (Hmid et al., 2015).

Moreover, the application of raw and composted OMWs can alter the structure of soil microbial communities as a consequence of the modification of soil characteristics (pH, metals and nutrients availability), but also the composition of the amendments may play a role on this regard (Fornes et al., 2009; Alburquerque et al., 2011). Several species of fungi show resistance to potentially toxic phenolic compounds and it has been observed that both Gram-positive and Gram-negative bacteria are negatively affected by raw OMWs, modifying the fungi/bacteria ratio in the soil (Moreno et al., 2013). Higher ratios of B_C/B_N in a contaminated soil treated with wet olive pomace with respect to compost treatment were observed by de la Fuente et al. (2011), showing a proliferation of fungi populations. The increase of soil fungi/bacterial ratio has been also observed in soils amended with OMWs compost, as a consequence of fungi being able to degrade lignin compounds, abundant in this type of composts, more easily than bacteria (Alburquerque et al., 2011).

9.3.4 PLANT ESTABLISHMENT AND TRACE ELEMENT UPTAKE

Vegetation is frequently scarce or null in TEs polluted soils and they are, consequently, heavily exposed to erosion and leaching processes, which can lead to the dispersion of the contaminants. In this sense, one of the ultimate objectives of TEs contaminated sites restoration is to facilitate the development of the plant community and to restore the ability of soil to function as a self-sustaining ecosystem (Vangronsveld et al., 2009). Several studies have reported that the application of OMWs to TEs contaminated soils favors the establishment of a vegetation cover (increasing ground cover, biomass production and species richness) and improves plant nutritional status (Clemente et al., 2012; Martínez-Fernández and Walker, 2012; Curaqueo et al., 2014; Pardo et al., 2014a,c,d).

In addition, OMWs are able to reduce TEs uptake and accumulation in plant tissues (Clemente et al., 2007b; Pardo et al., 2014a). Raw materials are normally less effective than compost (de la Fuente et al., 2011), although it must be taken into account that the plant uptake mechanisms may vary according to the plant species and age, the concentration of TEs, the concentration of soluble organic chelating agents, or the stability of the metal-organic matter complexes in soil (Baldock and Nelson, 2000).

Beneficial effects on plant yield after fresh OMWs application have been reported in crops growing in noncontaminated soils (López-Piñeiro et al., 2008; Nasini et al., 2013). However, in contaminated soils, the use of OMWW and wet olive pomace generally affects plant growth negatively as a consequence of the high concentration of phenolic compounds (de la Fuente et al., 2011). For example, a drastic inhibition of the germination of *Dittrichia viscosa* and high mortality of the few plants that germinated were reported by Nogales and Benítez (2006) after the application of wet olive pomace to an artificially contaminated soil. However, these authors found a significant stimulation of plant growth when using compost and vermicompost of wet olive pomace, reaching with the latter an aerial biomass production 45% higher than that in plants from control soils. Similarly, Clemente et al. (2007b) and de la Fuente et al. (2011) observed in field and pot experiments, inadequate growth of *Beta vulgaris* and *Beta maritima*, low biomass production, and higher leaf concentrations of Fe, Mn, and Zn (phytotoxic levels) in plants from olive pomace-amended soils with respect to plants growing in untreated soils, likely as a result of the oxidation of phenolic compounds of wet olive pomace. Contrastingly, Rodríguez et al. (2016) described an improvement of total plant biomass of *Lupinus albus* and a decrease of Zn and Pb shoot and root concentrations with the application of wet olive pomace to two mine soils. In this case, the formation of metal complexes with the organic compounds of wet olive pomace reduced metal uptake. Murillo et al. (2005) observed that the application of fulvic acids from OMWW to two soils affected by a mine spill did not significantly modify the availability of TEs, only slightly increasing As mobility. However, As mobilization did not result in an enrichment of As in leaves of *Olea europaea*, while the fulvic acids added improved plant growth, chlorophyll content, and photosynthesis rate.

Regarding composts from OMW:s, the liming effect, the reduction of TEs mobility and the long term supply of nutrients that their addition generally provoke in contaminated soils allow plant establishment, and improve plant growth and nutritional status (Clemente et al., 2005; Nogales and Benítez, 2006; de la Fuente et al., 2011; Pardo et al., 2014a, 2016). For example, the spontaneous growth of native species and the percentage of ground cover were increased in an olive pomace-compost treated contaminated soil after 2 years of field application compared to control untreated soil (Pardo et al., 2014d). Also, lower TEs leaf concentrations were found by Pardo et al. (2016) in plants of *Atriplex halimus* growing in two tailings amended with a red mud derivate in combination with olive pomace-compost than in plants receiving only the inorganic material. These authors described that the main factors governing TEs accumulation in that plant species were pH, extractable and pore water TEs concentrations, and the concentration of organic-C (total and dissolved in pore water). Curaqueo et al. (2014) found higher biomass production of *Tetraclinis articulata* in a mine tailing treated with wet olive pomace-compost in combination with AMF inoculation, and also a decrease of TEs solubility and accumulation in plant roots and shoots.

Clemente et al. (2005 and 2012) found significant positive correlations between the concentrations of As and Cu in the leaves of *Brassica juncea* and *Atriplex halimus* and soil TOC, 2 and 6 months after olive pomace-compost addition to contaminated soils. Contrastingly, Pardo et al. (2016) found that although the addition of exogenous soluble organic matter with olive pomace-compost to two mine tailings increased As solubility, dissolved organic C favored the reduction of As accumulation in plants according to a regression model.

Moreover, besides the direct influence of organic matter on TEs availability, some studies have proven that the TEs accumulation depends on the nutritional status of plants, as healthier plants are able to overtake the soil toxicity. In agreement with this, a reduction in TEs accumulation concomitant with high shoot nutrients concentrations have been found in plants growing in olive pomace-compost treated soils (Clemente et al., 2012; Pardo et al., 2014a, 2016). For example, Pardo et al. (2014a,b) observed that leaf concentrations of As, Cd, Cu, and Fe in *Lolium perenne* growing in olive pomace-compost amended soil was influenced by K content in its tissues. Concerning the use of biochar, Hmid et al. (2015) described significant reductions of TEs concentration in plant roots and shoots with increasing olive kernel-biochar rates and time, which resulted in lower activities of the antioxidant enzymes (an indication of stress amelioration) in plant tissues.

9.3.5 ECOTOXICITY

Ecotoxicological assays have been considered useful tools to evaluate the environmental impact of TEs contaminated soils in the ecosystem, and thus complete the study of the efficiency or suitability of a remediation technique (Pardo et al., 2014c). Direct and indirect ecotoxicological tests can reflect soil toxicity by showing the effects of the interactions between the contaminants, the soil matrix, and their living organisms (Maisto et al., 2011).

The toxic properties of raw OMWs have been described by several studies, which have been directly associated to their phenolic substances content (Sassi et al., 2006; Karaouzas et al., 2011). The high acute toxicity of OMWW has been reported for the aquatic macroinvertebrates *Gammarus pulex* and *Hydropsyche peristerica* (LC50 values within the range 2.6–3.4% and 3.6–3.8%, respectively), the microinvertebrates *Daphnia magna* and *Thamnocephalus platyurus* (LC50: 1.1–12.4% and 0.7–12.5%), and for the luminescent bacteria *Vibrio fischeri* (EC50: 0.2–1.2%) (Paixao et al., 1999; Rouvalis et al., 2004; Karaouzas et al., 2011). Once in the soil, OMWs may produce negative effects on germination and plant growth, which are gradually reduced with time (Mekki et al., 2007; Saadi et al., 2007). In agreement with this, Buchmann et al. (2015) reported a steady decrease of the phytotoxicity on *Lepidium sativum* germination of an OMWW-treated soil as result of phenolic compounds degradation processes.

The toxicity of fresh OMWs in TEs contaminated soils has been reported in different experiments, provoking the decrease of soil microbiological activity and/or plant growth (Romero et al., 2005; Nogales and Benítez, 2006). However, the application of ecotoxicological tests to assess the efficacy of the phytoremediation process in OMWs-treated contaminated soils is scarce, and most of them have aimed to evaluate the effect of processed OMWs (Pardo et al., 2011, 2014b,c; Beesley et al., 2014; Hmid et al., 2015). In general, the use of OMWs compost improves the soil habitat function and allows reducing the potential risks associated with the different exposure routes of TEs contaminated soils, minimizing their environmental impact in surrounding ecosystems (Pardo et al., 2011, 2014b,c). Lower direct and indirect phytotoxicity of contaminated soils for *Lactuca sativa, Lolium perenne, Zea mays*, and *L. sativum* after olive pomace-compost application have been described in laboratory (Pardo et al., 2011), greenhouse (Pardo et al., 2014b), and field experiments (Pardo et al., 2014c), as higher LC50 values for emergence and germination, and EC50 values for growth and root elongation with respect to control soil were found. In all these studies, this stimulation was related to the release of available nutrients, the increase of soil pH, and the decrease of TEs solubility. For instance, toxicity for *T. platyurus* and *V. fischeri* significantly diminished after olive pomace-compost addition to two highly contaminated mine soils, both in the short- (4.5 months) and long- (2.5 years) term (Pardo et al., 2014b,c), this proving the reduction of the risk of TEs leaching and contamination of surface and groundwater. Also, Pardo et al. (2014b) found that EC50 values of *V. fischeri* luminesce inhibition after 15 and 30 min. of exposition negatively correlated with Cu, Fe, and Pb concentrations dissolved in pore water, and positively with soil pH. The reduction of the potential risks of TEs lixiviation was also showed by Beesley et al. (2014), who described a significant decrease in the toxicity of a contaminated soil toward *L. sativum* seed germination and *V. fischeri* luminescence as a consequence of the decrease in TEs availability after the application of olive pomace-compost, alone or in combination with biochar.

Short-term negative effects on survival, growth, and reproduction and avoidance behavior have been observed in the earthworm *Eisenia foetida* in soils after the application of biochars (Liesch et al. 2010; Li et al. 2011). These toxicity symptoms have been attributed to the quick soil pH modification, but also to the ingestion of biochar by earthworms and to possible dermic damages. Nevertheless, such symptoms disappear with time. Accordingly, Hmid et al. (2015) did not find any toxic effect of olive kernel-biochar after 120 days of equilibration in soil, and even a reduction in the mortality of *E. foetida* in a TEs contaminated soil (100% survival rate was reached versus 12.3% of nontreated soil) was reported, together with a significantly increase in worms weight and the number of cocoons produced. It is also remarkable that the increase of olive kernel-biochar rate from 5% to 10% and 15% did not improve these aspects but neither decreased them, showing therefore that the biochar created better habitat conditions in the contaminated soil.

9.4 FINAL CONSIDERATIONS

Significant environmental problems are associated to olive oil production industry due to the large amount of solid and liquid wastes generated throughout the oil extraction process, their production in a short period of time within the year (campaigns normally cover from November to February), their physicochemical characteristics (that depend on the oil extraction method), and the lack of optimized solutions implanted to manage this type of wastes (Azbar et al., 2004; Alburquerque et al., 2004). In this sense, the use of OMWs, fresh or processed, in TEs contaminated soils remediation strategies has been shown as an environment friendly and sustainable approach for their recycling, this being an interesting alternative to conventional landfill or incineration treatments. Their application as amendments may modify TEs solubility and bioavailability in the soil, provide essential nutrients and organic matter, change soil structure, and therefore, improve the habitat function in soil (Table 9.2). This would stimulate the development of soil microbial communities and promote the establishment of a healthy and self-sustaining vegetation cover, which would reactivate the nutrients cycles (usually inhibited in contaminated soils) and also protect the soil preventing the dispersion of the contaminants in the

Table 9.2 Summarized Effects of OMWs Use in the Restoration of TEs Contaminated Soils

Type of Material		Solubility of TEs	OM and Nutrients	Soil Microorganisms	Plant Growth and TEs Uptake	Ecotoxicology	References
Fresh	OMWW		Inhibition of nitrification	Short term toxicity; modification of communities' structure	Stimulation of plan growth; inhibition of germination and plant growth	High toxicity to microorganisms and aquatic invertebrates; short-term phytotoxicity	Paixao et al. (1999); Murillo et al. (2005); Sassi et al. (2006); Mekki et al. (2007); Saadi et al. (2007); Di Serio et al. (2008); Karaouzas et al. (2011); Moreno et al. (2013); Rouvalis et al. (2004); Buchmann et al. (2015)
	Wet olive pomace	Increased availability and leaching	Increased soluble OM and K	Short-term toxicity: decrease of richness and activity, increased stress; medium- to long-term stimulation; modification of communities' structure	Inhibition of germination and plant growth; long-term stimulation of plan growth; increase of TEs accumulation in leaves		Romero et al. (2005); Nogales and Benítez (2006); Sassi et al. (2006); Clemente et al. (2007a, 2007b); de la Fuente et al. (2008, 2011); Walker and Bernal (2008); Burgos et al. (2010); Di Bene et al. (2013); Rodríguez et al. (2016)

Table 9.2 Summarized Effects of OMWs Use in the Restoration of TEs Contaminated Soils (*cont.*)

Type of Material		Solubility of TEs	OM and Nutrients	Soil Microorganisms	Plant Growth and TEs Uptake	Ecotoxicology	References
Processed	Olive pomace compost	Immobilization; Cu mobilization; As solubilization	Long-term release of C, N, K and P; K and N leaching	Stimulation of richness, diversity, and activity; modification of communities' structure	Improved plan growth and nutritional status; reduction of TEs accumulation; increased Cu leaf concentration	Decreased direct and indirect soil phytotoxicity; decrease of soil indirect toxicity to microorganisms and invertebrates	Clemente et al. (2005, 2007a, 2012); Romero et al. (2005); Nogales and Benítez (2006); Fornes et al. (2009); Burgos et al. (2010); Alburquerque et al. (2011); Pardo et al. (2011, 2014a,b,c,d, 2016); Martínez-Fernández and Walker (2012); Moreno-Jiménez et al. (2013); Beesley et al. (2014); Curaqueo et al. (2014)
	Olive kernel biochar	Immobilization		Stimulation of richness, diversity, and activity	Reduction of TEs accumulation and plant stress	Decreased toxicity to earthworms	Hmid et al. (2015)

ecosystem. Therefore, the OMWs application would positively influence the overall ecosystem functioning and would help to minimize the potential environmental impact that the presence of TEs in the soil may imply.

However, some details may have to be thoroughly evaluated in order to elude any possible negative effects in the environment (Table 9.2). Specifically, the application form (if they are applied raw or transformed) and application rates need to be carefully considered in order to avoid any adverse effect related mainly to their concentration of phenolic substances, which may cause modifications of soil pH and redox conditions that could provoke mobilization of certain elements (like As or Mn) and cause toxicity, and to minimize the risk of nutrient (like N or K) leaching.

REFERENCES

Adriano, D.C., 2001. Trace Elements in Terrestrial Environments. Biogeochemistry, Bioavailability, and Risk of Metals, second ed. Springer-Verlag, New York.

Adriano, D.C., Wenzel, W.W., Vangronsveld, J., Bolan, N.S., 2004. Role of assisted natural remediation in environmental cleanup. Geoderma 122, 121–142.

Aguilar-Carrillo, J., Garrido, F., Barrios, L., García-González, M.T., 2009. Induced reduction of the potential leachability of As, Cd and Tl in an element-spiked acid soil by the application of industrial by-products. Geoderma 149, 367–372.

Alburquerque, J.A., Gonzálvez, J., García, D., Cegarra, J., 2004. Agrochemical characterisation of "alperujo", a solid by-product of the two-phase centrifugation method for olive oil extraction. Bioresour. Technol. 91, 195–200.

Alburquerque, J.A., Gonzálvez, J., García, D., Cegarra, J., 2006. Composting of a solid olive-mill by-product and the potential of the resulting compost for cultivating pepper under commercial conditions. Waste Manag. 6, 620–626.

Alburquerque, J.A., Calero, J.M., Barrón, V., Torrent, J., del Campillo, M.C., Gallardo, A., Villar, R., 2014. Effects of biochars produced from different feedstocks on soil properties and sunflower growth. J. Plant Nutr. Soil Sci. 177, 16–25.

Alburquerque, J.A., de la Fuente, C., Bernal, M.P., 2011. Improvement of soil quality after "alperujo" compost application to two contaminated soils characterised by differing heavy metal solubility. J. Environ. Manag. 92, 733–741.

Alburquerque, J.A., Salazar, P., Barrón, V., Torrent, J., del Campillo, M.C., Gallardo, A., Villar, R., 2013. Enhanced wheat yield by biochar addition under different mineral fertilization levels. Agron. Sustain. Dev. 33, 475–484.

Alloway, B.J., 2013. Introduction. In: Alloway, B.J. (Ed.), Heavy Metals in Soils. third ed. Springer, Dordrecht, pp. 3–10.

Andrados, A., De Marco, I., Lopez-Urionabarrenechea, A., Solar, J., Caballero, B., 2015. Avoiding tar formation in biocoke production from waste biomass. Biomass Bioenerg. 74, 172–179.

Azbar, N., Bayram, A., Filibeli, A., Muezzinoglu, A., Sengul, F., Ozer, A., 2004. A review of waste management options in olive oil production. Crit. Rev. Environ. Sci. Technol. 34, 209–247.

Belaqziz, M., El-Abbassi, A., Lakhal, E.K., Agrafioti, E., Galanakis, C.M., 2016. Agronomic application of olive mill wastewaters: effects on maize production and soil properties. J. Environ. Manag. 171, 158–165.

Baldock, J.A., Nelson, P.N., 2000. Soil organic matter. In: Sumner, M.E. (Ed.), Handbook of Soil Science. CRC Press, Boca Raton (FL), pp. B25–B84.

Beesley, L., Inneh, O., Norton, G.J., Moreno-Jiménez, E., Pardo, T., Clemente, R., Dawson, J.J.C., 2014. Assessing the influence of compost and biochar amendments on the mobility and toxicity of metals and arsenic in a naturally contaminated mine soil. Environ. Pollut. 186, 195–202.

Benavente, V., Calabuig, E., Fullana, A., 2015. Upgrading of moist agro-industrial wastes by hydrothermal carbonization. J. Anal. Appl. Pyrol. 113, 89–98.

Bernal, M.P., Clemente, R., Walker, D.J., 2007. The role of organic amendments in the bioremediation of heavy metal-polluted soils. In: Gore, R.W. (Ed.), Environmental Research at the Leading Edge. Nova Science Publishers Inc, New York, pp. 1–57.

Bhargava, A., Carmona, F.F., Bhargava, M., Srivastava, S., 2012. Approaches for enhanced phytoextraction of heavy metals. J. Environ. Manag. 105, 103–120.

Bolan, S.N., Park, J.H., Robinson, B., Naidu, R., Huh, K.Y., 2011. Phytostabilization: a Green approach to contaminant containment. Adv. Agron. 112, 145–204.

Brunetti, G., Plaza, C., Senesi, N., 2005. Olive pomace amendment in Mediterranean conditions: effect on soil and humic acid properties and wheat (*Triticum turgidum* L.) yield. J. Agric. Food Chem. 53, 6730–6737.

Buchmann, C., Felten, A., Peikert, B., Muñoz, K., Bandow, N., Dag, A., Schaumann, G.E., 2015. Development of phytotoxicity and composition of a soil treated with olive mill wastewater (OMW): an incubation study. Plant Soil 386, 99–112.

Burgos, P., Madejón, P., Cabrera, F., Madejón, E., 2010. By-products as amendment to improve biochemical properties of trace element contaminated soils: effects in time. Int. Biodeterior. Biodegradation 64, 481–488.

Cayuela, M.L., Bernal, M.P., Roig, A., 2004. Composting olive mill waste and sheep manure for orchard use. Compost. Sci. Util. 12, 130–136.

Chowdhury, A.K.Md.M.B., Akratos, C.S., Vayenas, D.V., Pavlou, S., 2013. Olive mill waste composting: a review. Int. Biodeterior. Biodegradation 85, 108–119.

Clemente, R., Bernal, M.P., 2006. Fractionation of heavy metals and distribution of organic carbon in two contaminated soils amended with humic acids. Chemosphere 64, 1264–1273.

Clemente, R., de la Fuente, C., Moral, R., Bernal, M.P., 2007a. Changes in microbial biomass parameters of a heavy metal-contaminated calcareous soil during a field remediation experiment. J. Environ. Qual. 36, 1137–1144.

Clemente, R., Escolar, A., Bernal, M.P., 2006. Heavy metals fractionation and organic matter mineralisation in contaminated calcareous soil amended with organic materials. Bioresour. Technol. 97, 1894–1901.

Clemente, R., Pardo, T., Madejón, P., Madejón, E., Bernal, M.P., 2015. Food byproducts as amendments in trace elements contaminated soils. Food Res. Int. 73, 176–189.

Clemente, R., Paredes, C., Bernal, M.P., 2007b. A field experiment investigating the effects of olive husk and cow manure on heavy metal availability in a contaminated calcareous soil. Agric. Ecosyst. Environ. 118, 319–326.

Clemente, R., Walker, D.J., Bernal, M.P., 2005. Uptake of heavy metals and As by Brassica juncea grown in a contaminated soil in Aznalcóllar (Spain): the effect of soil amendments. Environ. Pollut. 138, 46–58.

Clemente, R., Walker, D.J., Pardo, T., Martínez-Fernández, D., Bernal, M.P., 2012. The use of a halophytic plant species and organic amendments for the remediation of a trace elements-contaminated soil under semi-arid conditions. J. Hazard. Mater. 223-224, 63–71.

Conesa, H.M., Evangelou, M.W.H., Robinson, B.H., Schulin, R., 2012. A critical view of current state of phytotechnologies to remediate soils: still a promising tool? Sci. World J. 2012, 173829.

Curaqueo, G., Schoebitz, M., Borie, F., Caravaca, F., Roldán, A., 2014. Inoculation with arbuscular mycorrhizal fungi and addition of composted olive-mill waste enhance plant establishment and soil properties in the regeneration of a heavy metal-polluted environment. Environ. Sci. Pollut. Res. 21, 7403–7412.

de la Fuente, C., Clemente, R., Bernal, M.P., 2008. Changes in metal speciation and pH in olive processing waste and sulphur-treated contaminated soil. Ecotoxicol. Environ. Saf. 70, 207–215.

de la Fuente, C., Clemente, R., Martínez-Alcalá, I., Tortosa, G., Bernal, M.P., 2011. Impact of fresh and composted solid olive husk and their water-soluble fractions on soil heavy metal fractionation, microbial biomass and plant uptake. J. Hazard. Mater. 186, 1283–1289.

Di Bene, C., Pellegrino, E., Debolini, M., Silvestri, N., Bonari, E., 2013. Short- and long-term effects of olive mill wastewater land spreading on soil chemical and biological properties. Soil Biol. Biochem. 56, 21–30.

Di Serio, M.G., Lanza, B., Mucciarella, M.R., Russi, F., Iannucci, E., Marfisi, P., Madeo, A., 2008. Effects of olive mill wastewater spreading on the physico-chemical and microbiological characteristics of soil. Int. Biodeterior. Biodegradation 62, 403–407.

Dickinson, N.M., Baker, A.J.M., Doronila, A., Laidlaw, S., Reeves, R.D., 2009. Phytoremediation of inorganics: realism and synergies. Int. J. Phytoremediation 11, 97–114.

Evanko, C.R., Dzombak, D.A., 1997. Remediation of Metal-Contaminated Soils and Groundwater. Technology Repost TE-97-01. Ground-water remediation Technologies Analysis Center (GWRTAC), Pittsburgh, PA, USA.

FAOSTAT 2013, Food and Agriculture Organisation database. Available from: http://faostat.fao.org

Farrell, M., Jones, D.L., 2010. Use of composts in the remediation of heavy metal contaminated soil. J. Hazard. Mater. 175, 575–582.

Fitz, W., Wenzel, W.W., 2002. Arsenic transformations in the soil-rhizosphere-plant system: fundamentals and potential application to phytoremediation. J. Biotechnol. 99, 259–278.

Fornes, F., García de la Fuente, R., Belda, R.M., Abad, M., 2009. Alperujo" compost amendment of contaminated calcareous and acidic soils: effects on growth and element uptake by five *Brassica species*. Bioresour. Technol. 100, 3982–3990.

García-Gomez, A., Bernal, M.P., Roig, A., 2002. Growth of ornamental plants in two composts prepared from agroindustrial wastes. Bioresour. Technol. 83, 81–87.

García-Gomez, A., Roig, A., Bernal, M.P., 2003. Composting of the solid fraction of olive mill wastewater with olive leaves: organic matter degradation and biological activity. Bioresour. Technol. 86, 59–64.

Gaskin, J.W., Steiner, C., Harris, K., Das, K.C., Bibens, B., 2008. Effect of low temperature pyrolysis conditions on biochar for agricultural use. Trans. ASABE 51, 2061–2069.

Hachicha, S., Cegarra, J., Sellamia, F., Hachicha, R., Drirac, N., Medhioua, K., Ammar, E., 2009. Elimination of polyphenols toxicity from olive mill wastewater sludge by its co-composting with sesame bark. J. Hazard. Mater. 161, 1131–1139.

Hmid, A., Al Chami, Z., Sillen, W., De Vocht, A., Vangronsveld, J., 2015. Olive mill waste biochar: a promising soil amendment for metal immobilization in contaminated soils. Environ. Sci. Pollut. Res. 22, 1444–1456.

Hmid, A., Montelli, D., Fiore, S., Fanizzi, F.P., Al Chami, Z., Dumontet, S., 2014. Production and characterization of biochar from three-phase olive mil waste through slow pyrolysis. Biomass Bioenerg. 71, 330–339.

Huang, P.M., 1990. Role of soil minerals in transformations of natural organics and xenobiotics in soil. Bollag, J.M., Strotzky, G. (Eds.), Soil Biochemistry, vol. 6, Dekker, New York, pp. 29–115.

Kabata-Pendias, A., Adriano, C., 1995. Trace metals. In: Rechcigl, J.E. (Ed.), Soil Amendments and Environmental Quality. CRC Lewis Publications, Boca Raton, pp. 139–167.

Kabata-Pendias, A., Pendias, H., 2001. Trace Elements in Soils and Plants, third ed. CRC Press, Boca Raton.

Kahn, A.G., Kuek, C., Chaudhry, T.M., Khoo, C.S., Hayes, W.J., 2000. Role of plants, mycorrhizae and phytochelators in heavy metal contaminated land remediation. Chemosphere 41, 197–207.

Karaka, A., 2004. Effect of organic wastes on the extractability of cadmium, copper, nickel, and zinc in soil. Geoderma 122, 297–303.

Karaouzas, I., Cotou, E., Albanis, T.A., Kamarianos, A., Skoulikidis, N.T., Giannakou, U., 2011. Bioassays and biochemical biomarkers for assessing olive mill and citrus processing wastewater toxicity. Environ. Toxicol. 26, 669–676.

Kidd, P., Barceló, J., Bernal, M.P., Navari-Izzo, F., Poschenrieder, C., Shilev, S., Clemente, R., Monterroso, C., 2009. Trace element behaviour at the root–soil interface: Implications in phytoremediation. Environ. Exp. Bot. 67, 243–259.

Lee, S.S., Lim, J.E., El-Azeem, S.A.M.A., Choi, B., Oh, S.-E., Ok, Y.S., 2013. Heavy metal immobilization in soil near abandoned mines using eggshell waste and rapeseed residue. Environ. Sci. Pollut. Res. 20, 1719–1726.

Lehman, J., Joseph, S., 2009. Biochar for environmental management: science and technology. Earthscan, London, pp 1–12.

Li, D., Hockaday, W.C., Masiello, C.A., Alvarez, P.J.J., 2011. Earthworm avoidance of biochar can be mitigated by wetting. Soil Biol. Biochem. 43, 1732–1737.

Liesch, A.M., Weyers, S.L., Gaskin, J.W., Das, K.C., 2010. Impact of two different biochars on earthworm growth and survival. Ann. Environ. Sci. 4, 1–9.

López-Piñeiro, A., Albarrán, A., Rato Nunes, J.M., Barreto, C., 2008. Short and medium-term effects of two-phase olive mill waste application on olive grove production and soil properties under semiarid Mediterranean conditions. Bioresour. Technol. 99, 7982–7987.

Madrid, L., Díaz-Barrientos, E., 1998. Release of metals from homogeneous soil columns by wastewater from an agricultural industry. Environ. Pollut. 101, 43–48.

Mahmoud, M., Janssen, M., Haboub, N., Nassour, A., Lennartz, B., 2010. The impact of olive mill wastewater application on flow and transport properties in soils. Soil Tillage Res. 107, 36–41.

Mahmoud, M., Janssen, M., Peth, S., Horn, R., Lennartz, B., 2012. Long-term impact of irrigation with olive mill wastewater on aggregate properties in the top soil. Soil Tillage Res. 124, 24–31.

Maisto, G., Manzo, S., De Nicola, F., Carotenuto, R., Rocco, A., Alfani, A., 2011. Assessment of the effects of Cr, Cu, Ni and Pb soil contamination by ecotoxicological tests. J. Environ. Monitor. 13, 3049–3056.

Martínez-Fernández, D., Walker, D.J., 2012. The effects of soil amendments on the growth of Atriplex halimus and Bituminaria bituminosa in heavy metal-contaminated soils. Water Air Soil Pollut. 223, 63–72.

Mekki, A., Dhouib, A., Sayadi, S., 2007. Polyphenols dynamics and phytotoxicity in a soil amended by olive mill wastewaters. J. Environ. Manag. 84, 134–140.

Mench, M., Lepp, N., Bert, V., Schwitzguébel, J.P., Gawronski, S.W., Schröder, P., Vangronsveld, J., 2010. Successes and limitations of phytotechnologies at field scale: outcomes, assessment and outlook from COST Action 859. J. Soils Sediments 10, 1039–1070.

Mendez, M.O., Maier, R., 2008. Phytostabilization of mine tailings in arid and semiarid environments—an emerging remediation technology. Environ. Health Persp. 116, 278–283.

Moreno, J.L., Bastida, F., Sánchez-Monedero, M.A., Hernández, T., García, C., 2013. Response of soil microbial community to a high dose of fresh olive mill wastewater. Pedosphere 23, 281–289.

Moreno-Jiménez, E., Clemente, R., Mestrot, A., Meharg, A.A., 2013. Arsenic and selenium mobilisation from organic matter treated mine spoil with and without inorganic fertilisation. Environ. Pollut. 173, 238–244.

Morillo, J.A., Antizar-Ladislao, B., Monteoliva-Sánchez, M., Ramos-Cormenzana, A., Russell, N.J., 2009. Bioremediation and biovalorisation of olive-mill wastes. Appl. Microbiol. Biotechnol. 82, 25–39.

Murillo, J.M., Madejón, E., Madejón, P., Cabrera, F., 2005. The response of wild olive to the addition of a fulvic acid-rich amendment to soils polluted by trace elements (SW Spain). J. Arid Environ. 63, 284–303.

Naidu, R., Krishnamurti, G.S.R., Bolan, N.S., Wenzel, W., Megharaj, M., 2001. Heavy metal interactions in soils and implications for soil microbial biodiversity. In: Prasad, M.N.V. (Ed.), Metals in the Environment. Marcel Dekker Inc, New York, pp. 401–431.

Nasini, L., Gigliotti, G., Balduccini, M.A., Federici, E., Cenci, G., Proietti, P., 2013. Effect of solid olive-mill waste amendment on soil fertility and olive (*Olea europaea* L.) tree activity. Agric. Ecosyst. Environ. 164, 292–297.

Nogales, R., Benítez, E., 2006. Absorption of zinc and lead by *Dittrichia viscosa* grown in a contaminated soil amended with olive-derived wastes. Bull. Environ. Contamin. Toxicol. 76, 538–544.

Novak, J.M., Lima, I., Baoshan, X., Gaskin, J.W., Steiner, C., Das, K.C., Watts, D.W., Busscher, W.J., Schomberg, H., 2009. Characterization of designer biochar produced at different temperatures and their effects on a loamy sand. Ann. Environ. Sci. 3, 195–206.

Olmo, M., Alburquerque, J.A., Barrón, V., del Campillo, M.C., Gallardo, A., Fuentes, M., Villar, R., 2014. Wheat growth and yield responses to biochar addition under Mediterranean climate conditions. Biol. Fert. Soils 50, 1177–1187.

Paixao, S.M., Mendonça, E., Picado, A., Anselmo, A.M., 1999. Acute toxicity evaluation of olive mill wastewaters: a comparative study of three aquatic organisms. Environ. Toxicol. 14, 263–269.

Pardo, T., Bernal, M.P., Clemente, R., 2014a. Efficiency of soil organic and inorganic amendments on the remediation of a contaminated mine soil: I. Effects on trace elements and nutrients solubility and leaching risk. Chemosphere 107, 121–128.

Pardo, T., Bes, C., Bernal, M.P., Clemente, R., 2016. Alleviation of environmental risks associated to severely contaminated mine tailings using amendments: modelling of trace elements speciation, solubility and plant accumulation. Environ. Toxicol. Chem., in press, doi:10.1002/etc.3434

Pardo, T., Clemente, R., Alvarenga, P., Bernal, M.P., 2014b. Efficiency of soil organic and inorganic amendments on the remediation of a contaminated mine soil: II. Biological and ecotoxicological evaluation. Chemosphere 107, 101–108.

Pardo, T., Clemente, R., Bernal, M.P., 2011. Effects of compost, pig slurry and lime on trace element solubility and toxicity in two soils differently affected by mining activities. Chemosphere 84, 642–650.

Pardo, T., Clemente, R., Epelde, L., Garbisu, C., Bernal, M.P., 2014c. Evaluation of the phytostabilisation efficiency in a trace elements contaminated soil using soil health indicators. J. Hazard. Mater. 268, 68–76.

Pardo, T., Martínez-Fernández, D., Clemente, R., Bernal, M.P., Walker, D.J., 2014d. The use of olive-mill waste compost to promote the plant vegetation cover in a trace element-contaminated soil. Environ. Sci. Pollut. Res. 21, 1029–1038.

Paredes, C., Bernal, M.P., Cegarra, J., Roig, A., 2002. Biodegradation of olive mill wastewater sludge by its co-composting with agricultural wastes. Bioresour. Technol. 85, 1–8.

Paredes, C., Cegarra, J., Roig, A., Sánchez-Monedero, M.A., Bernal, M.P., 1999. Characterization of olive mill wastewater (alpechín) and its sludge for agricultural purpose. Bioresour. Technol. 67, 111–116.

Pierantozzi, P., Zampini, C., Torres, M., Isla, M.I., Verdenelli, R.A., Meriles, J.M., Maestri, D., 2012. Physico-chemical and toxicological assessment of liquid wastes from olive processing-related industries. J. Sci. Food Agric. 92, 216–223.

Piotrowska, A., Iamarino, G., Antonietta, M., Gianfreda, L., 2006. Short-term effects of olive mill wastewater (OMW) on chemical and biochemical properties of a semiarid Mediterranean soil. Soil Biol. Biochem. 38, 600–610.

Piotrowska, A., Rao, M.A., Scotti, R., Gianfreda, L., 2011. Changes in soil chemical and biochemical properties following amendment with crude and dephenolized olive mill waste water (OMW). Geoderma 161, 8–17.

Qian, J., Li, D., Zhan, G., Zhang, L., Su, W., Gao, P., 2012. Simultaneous biodegradation of Ni–citrate complexes and removal of nickel from solutions by *Pseudomonas alcaliphila*. Bioresour. Technol. 116, 66–73.

Ramos-Cormenzana, A., Monteoliva-Sánchez, M., López, M.L., 1995. Bioremediation of alpechín. Int. Biodeterior. Biodegradation 35, 249–268.

Renella, G., Landi, L., Ascher, J., Ceccherini, M.T., Pietramellara, G., Mench, M., 2008. Long-term effects of aided phytostabilisation of trace elements on microbial biomass and activity, enzyme activities, and composition of microbial community in the Jales contaminated mine spoils. Environ. Pollut. 152, 702–712.

Robinson, Brett, H., Bañuelos, G., Conesa, H.M., Evangelou, M.W.H., Schulin, R., 2009. The phytomanagement of trace elements in soil. Crit. Rev. Plant Sci. 28, 240–266.

Rodis, P.S., Karathanos, V.T., Mantzavinou, A., 2002. Partitioning of olive oil antioxidants between oil and water phases. J. Agric. Food Chem. 50, 596–601.

Rodríguez, L., Gómez, R., Sánchez, V., Alonso-Azcárate, J., 2016. Chemical and plant tests to assess the viability of amendments to reduce metal availability in mine soils and tailings. Environ. Sci. Pollut. Res. 23, 6046–6054.

Romero, E., Benítez, E., Nogales, R., 2005. Suitability of wastes from olive-oil industry for initial reclamation of a Pb/Zn mine tailing. Water Air Soil Pollut. 165, 153–165.

Rouvalis, A., Iliopoulou-Georgudaki, J., Lyberatos, G., 2004. Application of two microbiotests for acute toxicity evaluation of olive mill wastewaters. Fresenius Environ. Bull. 13, 458–464.

Saadi, I., Laor, Y., Raviv, M., Medina, S., 2007. Land spreading of olive mill wastewater: effects on soil microbial activity and potential phytotoxicity. Chemosphere 66, 75–83.

Sassi, A. Ben, Boularbah, A., Jaouad, A., Walker, G., Boussaid, A., 2006. A comparison of olive oil Mill Wastewaters (OMW) from three different processes in Morocco. Process Biochem. 41, 74–78.

Singh, B., Singh, B.P., Cowie, A.L., 2010. Characterisation and evaluation of biochars for their application as a soil amendment. Aust. J. Soil Res. 48, 516–525.

Srivastava, P.K., Vaish, A., Dwivedi, S., Chakrabarty, D., Singh, N., Tripathi, R.D., 2011. Biological removal of arsenic pollution by soil fungi. Sci. Total Environ. 409, 2430–2442.

Vangronsveld, J., Herzig, R., Weyens, N., Boulet, J., Adriaensen, K., Ruttens, A., Thewys, T., Vassilev, A., Meers, E., Nehnevajova, E., Van der Lelie, D., Mench, M., 2009. Phytoremediation of contaminated soils and groundwater: lessons from the field. Environ. Sci. Pollut. Res. 16, 765–794.

Walker, D.J., Bernal, M.P., 2008. The effects of olive mill waste compost and poultry manure on the availability and plant uptake of nutrients in a highly saline soil. Bioresour. Technol. 99, 396–403.

Wenzel, W.W., Salt, D., Smith, R., Adriano, D.C., 1999. Phytoremediation: a plant microbe based remediation system. In: Adriano, D.C., Bollag, J.M., Frankenberger, Jr., W.T., Sims, R.C. (Eds.), Bioremediation of Contaminated Soils. Soil Science Society of America, Madison, pp. 456–508.

Zabaniotou, A., Rovas, D., Libutti, A., Monteleone, M., 2015. Boosting circular economy and closing the loop in agriculture: case study of a small-scale pyrolysis–biochar based system integrated in an olive farm in symbiosis with an olive mill. Environ. Dev. 14, 22–36.

CHAPTER

RECOVERY OF BIOACTIVE COMPOUNDS FROM OLIVE MILL WASTE

10

Charis M. Galanakis*, Kali Kotsiou**

**Department of Research & Innovation, Galanakis Laboratories, Chania, Greece; **Department of Chemistry, Section of Industrial and Food Chemistry, University of Ioannina, Ioannina, Greece*

10.1 INTRODUCTION

Olive oil is obtained from olive fruit by mechanical procedures, whereas its production involves one of the following extraction processes: (1) discontinuous (press) extraction, (2) three-phase centrifugal extraction, or (3) two-phase centrifugal extraction. Each of these processes generates in different forms and compositions (Rahmanian et al., 2014). The traditional olive pressing and the three phases continuous systems produce three streams: olive oil, olive cake (or kernel), and olive mill wastewater (OMWW). The annual world OMWW production is estimated between 10 and 30 million m^3 (El-Abbassi et al., 2012). The discontinuous process (not used often anymore) produces less but more concentrated wastewater (0.5–1 m^3 per 1000 kg) than the centrifugation process (1–1.5 m^3 per 1000 kg) (Paraskeva and Diamadopoulos, 2006). The two-phase centrifugal system was introduced during the 1990s in which the olive paste is separated into phases of olive oil and wet pomace (sludge by-product) that enables reduction of the volume of OMWW (Dermeche et al., 2013). Wet olive pomace is usually further extracted with *n*-hexane yielding olive cake oil, although it has no significant value because of the required energy for the drying process.

OMWW is a dark-colored, acidic ($3<$ pH value <5.9) suspension of three phases: water, oil, and solids (smashed particles of olive paste and kernel). It has a characteristic unpleasant odor and high organic content, whereas it is claimed to be one of the most polluting waste produced by the agro-food industries (Davies et al., 2004). Typically OMWW consists of: 83–94% water, 0.4–2.5% mineral salts, 0.03–1.1% lipids, and 4–16% organic compounds, such as carbohydrates (2–8 g/100 g), pectin, mucilage, lignin, and tannins (Obied et al., 2005; Niaounakis and Halvadakis, 2006; Galanakis et al., 2010d; Rahmanian et al., 2014). These phytochemicals are distributed between the aforementioned phases. Phenols distribution is governed by their generally low oil per water partition coefficients (ranging from 6×10^{-4} to 1.5) and thus they favor concentrating into the wastewater instead of olive oil during processing. Nevertheless, some of the phenols are bound to DF or proteins and they could be "trapped" inside the matrix of the smashed particles (Galanakis et al., 2010e). Due to its composition, OMWWs are reported to exhibit antimicrobial, ecotoxic, and phytotoxic properties due to their high content of phenols and long-chain fatty acids (DellaGreca et al., 2001; Greco et al., 2006), thus their treatment and disposal raise serious environmental concerns, especially to the Mediterranean region where 97% of the worldwide olive oil is produced (Paraskeva and Diamadopoulos, 2006; Belaqziz et al., 2016).

Olive Mill Waste. http://dx.doi.org/10.1016/B978-0-12-805314-0.00010-8

The characteristics of OMWW are variable, depending on the extraction method, origin, type, and maturity of olives, and climatic conditions (Paraskeva and Diamadopoulos, 2006). The waste streams of the discontinuous process and two-phase centrifugal system have similar characteristics with OMWW in spite of their content, differ only on the water content and the concentration degree of the solutes.

The difficulties in OMWW management arise from the fact that olive oil production is seasonal (during olive harvest), so the treatment process should operate in a noncontinuous mode. The large number of small olive mills, spread across the Mediterranean region, make individual on-site treatment options unaffordable (Paraskeva and Diamadopoulos, 2006). Besides, the high phenolic nature of OMWW and its organic contents make it highly resistant to biodegradation (Zirehpour et al., 2014). Like all food processing wastes, OMWW and other olive oil processing by-products have long been considered as a matter of treatment, minimization, and prevention due to the environmental effects induced by their disposal. On the other hand, similarly to all fruit processing by-products, they account as a cheap source of valuable components that could be recovered and utilized as natural food additives (Galanakis, 2012; Prokopov et al., 2015). For instance, phenols are characterized by antioxidant properties with potential health-benefits (Obied et al., 2005). Thus, from the early 2000s, an increasing interest not only in safely disposing of OMWWs but also in recovering valuable nutrients for nutraceutical applications has been emerged (Bouzid et al., 2005; Galanakis et al., 2010b,e; Patsioura et al., 2011; Galanakis, 2012; Ramos et al., 2013). Perspectives originate from the enormous amounts that are discharged in the Mediterranean and the existing technologies, which promise the recovery, recycling, and sustainability of high-added value ingredients inside food chain (Rahmanian et al., 2014).

10.2 TARGET BIOACTIVE COMPOUNDS OF OMWW

The composition of OMWW is very complex and variations are observed between examined samples (Bianco et al., 2003; Obied et al., 2005). Fruit cultivar, maturity, irrigation, storage time, and extraction techniques are factors that strongly affect the presence and the concentration of specific phenolic and other compounds in OMWWs (Lesage-Meessen et al., 2001; Mulinacci et al., 2001; De Marco et al., 2007; Obied et al., 2008; Aviani et al., 2012).

10.2.1 BIOACTIVE PHENOLS

Broadly distributed in the plant kingdom and abundant in our diet, phenols are today among the most discussed categories of natural antioxidants (Boskou, 2006). They include one or more hydroxyl groups (polar part) attached directly to an aromatic ring (nonpolar part) and are often found in plants as esters or glycosides, rather than as free molecules. This stereochemistry distinguishes them according to their polarity variance (Galanakis et al., 2013a). Recent epidemiological studies have shown the inverse association between risk of cancer, cardiovascular diseases, diabetes, several age related chronic diseases, and intake of diet rich in phenols and antioxidants (Yang et al., 2008; Karadag et al., 2009; Ilyasoglu et al., 2011).

Phenols can exist inherently in the olive fruit or be generated during olive oil production (Obied et al., 2005). OMWW contains typically from 0.5 to 24 g phenols/L, which is about 98% of the phenols present in olive fruit (Rodis et al., 2002), since only 2% of them pass in the oil phase during extraction process (Rahmanian et al., 2014). Olive fruit contains phenolic acids and alcohols, secoiridoids and flavonoids, while more than 50 and 40 relevant compounds have been identified in OMWW and olive oil, respectively (Obied et al., 2007b; Boskou, 2008). Phenolic acids include *o*- and *p*-coumaric, cinnamic, caffeic, ferulic, gallic, sinapic, chlorogenic, protocatechuic, syringic, vanillic, and elenolic acids (Balice and Cera, 1984; Mulinacci et al., 2001; Fiorentino et al., 2003; D'Alessandro et al., 2005). The most typical phenolic alcohols are tyrosol and hydroxytyrosol (Knupp et al., 1996; DellaGreca et al., 2004). The qualitative and quantitative HPLC analysis of raw OMWW is similar to olive oil, whereas hydroxytyrosol and tyrosol are the most abundant phenolic compounds similar with olive oil (Boukhoubza et al., 2009; Bayram et al., 2013). One liter of crude OMWW provides 4 g of dry extract and 1 g of pure hydroxytyrosol. OMWW also contain oleuropein, dimethyl-oleuropein, verbascoside, catechol, 4-methyl catechol, p-cresol, resorcinol, and high amounts of secoiridoid derivatives like hydroxytyrosyl acyclodihydroelenolate (HT-ACDE), comselogoside, and dialdehyde 3,4-dihydroxyphenyl-ethanol-elenolic ether linked to hydroxytyrosol. This ether is generated during olive fruit malaxation by the hydrolysis of oleuropein and dimethyl oleuropein (Capasso et al., 1992; Lo Scalzo and Scarpati, 1993; Vinciguerra et al., 1997; Servili et al., 1999; Japon-Lujan and Luque de Castro, 2007; Obied et al., 2007b, 2009). Finally, OMWW contains numerous flavonoids such as apigenin, hesperidin, cyanidin flavone, anthocyanin, and quercetin (Visioli et al., 1999; Obied et al., 2005, 2007a).

OMWW phenols are well known for their unique antioxidant properties for human health that promise their reutilization as additives in foodstuff and cosmetics (Galanakis, 2012). Indeed, one of the most established activities of OMWW phenols is their ability to capture free radicals (both in vitro and in vivo). This ability has been studied using several radical-generator compounds such as $DPPH^{\bullet}$ (1,1-diphenyl-2-picrylhydrazyl) (Lesage-Meessen et al., 2001), $ABTS^{+}$ (2,2'-asinobis (3-ethylbenz-thiazoline-6-sulfonic) diammonium salt) (De Marco et al., 2007) or hyperoxide anion (Visioli et al., 1999). OMWW phenols have also shown a scavenging ability against hypochlorous acid (HClO) (Visioli et al., 2002) as well as ferric reducing ability (McDonald et al., 2001). Besides, several studies have reported the antimicrobial (i.e., against *Staphylococcus aureus*, *Bacillus subtilis*, *Bacillus cinerea*, *Escherichia coli,* and *Pseudomonas aeruginosa*) of olive polyphenols (Boskou, 2006; Obied et al., 2007b; Yangui et al., 2010).

10.2.2 HYDROXYTYROSOL

Hydroxytyrosol is the most known olive phenol in spite of its beneficial properties, for example, it exerts advanced antiradical properties compared to vitamins E and C (Fernández-Bolaños et al., 2006). The antioxidant ability of hydroxytyrosol has been proven in the plasma and liver of rats (Visioli et al., 2001), while its cardio-protective effect has been successfully assayed on human cells (Leger et al., 2000). Together with other *o*-phenolics (e.g., oleuropein and caffeic acid) are known to exert protective effect against low-density lipoprotein (LDL) oxidation in vitro and appeared that it can be most effective at low concentrations to protect human erythrocytes and DNA against oxidative damages (Bouzid et al., 2005; Covas et al., 2006; De Marco et al., 2007). The beneficial effect of hydroxytyrosol as an effective hypoglycemic and antioxidant agent in alleviating oxidative stress and free radicals as well as in enhancing enzymatic defences in diabetic rats has also been demonstrated (Allouche et al., 2004). Hydroxytyrosol-rich olive oils have been approved by the European Food Safety Authority for their ability to maintain healthy LDL cholesterol level and lipid antioxidation (EFSA, 2011). The natural hydroxytyrosol concentrates obtained from OMWW and other olive by-products (using clean technologies) may also exert antioxidant, antimicrobial, antiinflammatory, and anticarcinogenic properties superior to those observed for pure hydroxytyrosol, in equivalent concentration (De Magalhaes Nunes Da Ponte et al., 2008).

10.2.3 DIETARY FIBER

DF includes different components of the cell wall (e.g., cellulose, hemicellulose, lignin, and pectin) and nonstructural constituents (e.g., gums and mucilages). In its latest definition, DF also includes industrial additives, such as modified cellulose or pectin, commercial gums, and algae polysaccharides (Galanakis, 2011). DF is known to play a vital role in many physiological processes and in the prevention of several diseases, whereas its consumption has been related to a decreased incidence of several type of cancer (Rodríguez et al., 2006). Its water soluble fraction has been linked with blood cholesterol reduction and the diminution of glucose intestinal absorption. On the other hand, its insoluble part has been correlated to both water absorption and intestinal regulation (Grigelmo-Miguel et al., 1999). Olives contain an appreciable amount of DF (e.g., 5–6 g/100 g fresh olive), while their content and composition varies among the different varieties, harvest season, and cultivars (Jiménez et al., 2000). Olive DF components in whole fruit include highly (>80%) esterified pectin (polyuronides and arabinans), hemicelluloses (rich in xylans, xyloglucans, glucuronoxylans, and mannans), cellulose, and lignin (Coimbra et al., 1994, 1999).

Insoluble fiber components (e.g., lignin, cellulose, and hemicelluloses) exist mainly in olive endocarp and therefore can be found in higher amounts in olive kernel or in lower amounts in wet olive pomace. Water insoluble DF has been targeted as a source of fermentable sugars, saccharides, microcrystalline, or powdered cellulose. For instance, Valiente et al. (1995) saccharified DF from dried olive pomace and used it as an additive in bakery products, showing ability to improve cooking properties and texture. Pectin is mainly found in olive pulp and thus it is subsequently transferred to OMWW and wet olive pomace during olive oil processing. Its neutral fraction is constituted of arabinans, whereas its acidic one is composed of homogalacturonans and rhamnogalacturonans (Jiménez et al., 1994). Pectin material, recovered from OMWW and wet olive pomace, showed promising rheological and functional properties (e.g., water holding, gelling ability, and cation exchange capacity) and thus could be utilized in confectionary as stabilizing agent or fat replacement in low fat products (Vierhuis et al., 2003; Galanakis et al., 2010a,d,c). Despite these promising applications, OMWW and wet olive pomace do not contain such a high amount of water soluble DF compared to other sources (e.g., apples, oranges). Thus, the design of a methodology only for their recovery is principally not economically feasible. However, their recovery could be conducted as a side process of a methodology that primary targets to recapture phenols from OMWW.

10.3 AN INTEGRATED APPROACH FOR THE RECOVERY OF BIOACTIVE COMPOUNDS FROM OLIVE OIL PROCESSING BY-PRODUCTS

Several techniques have been individually or in combination suggested to recover phenols, DF, and other valuable compounds from olive oil processing by-products. Typical systems include extraction with solvents, adsorption onto resins and membrane systems, membrane systems, supercritical fluid extraction (SFE), ultrasound-assisted extraction, microwave-assisted extraction, subcritical water extraction, high voltage electrical discharges, pulsed electric field, and others (Serpil and Alper, 2009; Rahmanian et al., 2014; Roselló-Soto et al., 2015a,b). In another point of view, the "5-Stage Universal Recovery Process" (Galanakis, 2015a) provides an integral strategy in order to combine these techniques and purify phenols prior to their final formation to a product. Following this approach, the recovery of any target compound from food by-products can be accomplished in five distinct stages that follow the principles of analytical chemistry. The "5-Stages Universal Recovery Process" includes five distinct stages: (1) macroscopic pretreatment, (2) separation of high-molecular from low-molecular compounds, (3) extraction, (4) purification/isolation, and (5) encapsulation or product formation (Galanakis, 2012; Galanakis et al., 2015b). This procedure is selected if two (at least) different components are recovered or one of them is a small molecule, for example, an antioxidant. On the other hand, when the target compound concerns a longer molecule (e.g., protein or pectin), the second stage may be omitted.

10.3.1 PRETREATMENT OF OLIVE OIL PROCESSING BY-PRODUCTS

At first, OMWW, wet olive pomace or other processing by-products should be pretreated with an aim of allowing easier handling and faster diffusion of extractants over the next stages. Pretreatment is also able to modify the endogenous enzymes of the substrate accordingly to the needs of the recovery procedure (e.g., to activate or deactivate them). For instance, in the case of OMWW (a suspension of water, fats, oils, and other solids), pretreatment includes a skimming step to remove fats and oils and concentration (vacuum or thermal) to reduce water content (Galanakis et al., 2010a,d). The latest particles can restrict mechanical processing such as substrate flow, mixing, and homogenization, or cause deterioration of the substrate (Galanakis et al., 2010a). Microfiltration (MF) can also be applied so as to remove remaining oil and suspended solids. Thermal concentration at high temperatures (e.g., 80°C) could deactivate key enzymes such polyphenol oxidases and pectin methyl esterase (Galanakis, 2012). On the other hand, wet olive pomace needs malaxation (e.g., using pectinolytic enzymes) with or without steam explosion. The latest process increases its susceptibility to enzymatic hydrolysis by converting hemicelluloses into soluble carbohydrates (Rodriguez et al., 2007; Galanakis, 2011). In some occasions, the partial removal of water from wet olive pomace is needed, too. Solid by-products such as olive kernel should be delignified, either by chloride treatment or steam explosion (Fernández-Bolaños et al., 2001; Galanakis, 2011). Saccharification of lignicellulosic material from olive cake has alternatively been conducted by hydrolysis in the presence of diluted sulfuric acid and heating (Asli and Qatibi, 2009). Hydrothermal treatment (e.g., at 160°C for 60 min or using H_2SO_4 at 200–220°C for 5 min) has also been used to pretreat wet olive pomace and increase the yield of recovered hydroxytyrosol in the final product (Fernández-Bolaños et al., 2002b; Rubio-Senent et al., 2013).

10.3.2 SEPARATION OF SMALLER AND LARGER COMPOUNDS

Alcohol precipitation is the most popular method for the separation of micromolecules (i.e., antioxidants, acids, or ions) from macromolecules (i.e., pectin, DFs, or hydrocolloids), which are collected in the so-called alcohol insoluble residue. This method is cheap, nontoxic, and easy to use. However, it is not selective, thus not being able to separate the complexes and links between smaller and larger components (Galanakis, 2012). Ultrafiltration (UF) comprises a key physicochemical and nondestructive technique to concentrate macromolecules and release smaller molecules in the permeate stream, respectively (Galanakis, 2015b). Indeed, many studies suggest to recover phenols from OMWW using different membrane technologies in a sequential design (Section 10.5). For instance, phenols pass through UF membranes due to their low molecular weights (up to 540.5 g/mol, which is the MW of oleuropein). The permeate obtained containing phenols, is then subjected to nanofiltration (NF) and reverse osmosis (RO) for their clarification from coextracted compounds (fourth stage of the 5-Stages Universal Recovery Process).

10.3.3 EXTRACTION

The extraction of phenols from natural substrates is typically conducted using organic solvents that provide a physical carrier for their transportation between different phases (i.e., solid, liquid, and vapor) (Galanakis, 2012). Liquid–liquid solvent extraction is an easy to operate technique and it can be applied even in small, family-owned olive oil mills. It is affected by the diffusion coefficient and the dissolution rate of compounds until they reach the equilibrium concentration inside the solvent, whereas the most important parameters include the type of the solvent, the extraction temperature, and time. Phenols are polar compounds that match perfectly with highly-polar solvents. For example, ethyl acetate is a promising solvent for the liquid–liquid extraction of flavonoid aglycons (Lesage-Meessen et al., 2001; De Leonardis et al., 2007; Sannino et al., 2013). Phenolic acids are easily solubilized in polar protic mediums like hydroalcoholic mixtures, while respective fractions can be obtained on the basis of polarity by varying alcohol concentration (Tsakona et al., 2010; Galanakis et al., 2013a). In general, methanol–water mixtures have been referred to extract phenols with the highest yield and widest array. On the other hand, the recovery of certain phenolic terpenes has been referred to occur preferably with nonpolar solvents (i.e., hexane, petroleum ether). One of the main disadvantage of this process is that most of the organic solvents are undesirable for industrial exploitation as they are toxic, nonedible, and raise environmental, health, and safety concerns (Wang and Weller, 2006; Serpil and Alper, 2009). Ethanol possesses a lot of advantages: it is cheap, reusable as well as nontoxic and the corresponding extracts could be utilized directly in the beverage industry (Galanakis et al., 2010e). In addition, it inhibits enzyme activity at low pH conditions (pH 2–3) and precipitates larger compounds, such as DF.

The recovered residue is typically used for the extraction of water soluble and insoluble DF. For instance, Coimbra et al. (1995) delignified olive seed hull in order to recover hemicelluloses and cellulose with sequential solvent and alkali extraction. Thereafter, glucuronoxylan and acidic xylan rich fractions were isolated by further alkali treatment and graded ethanol precipitation. Ultimately, fermentable sugars and oligosaccharides were generated after acid hydrolysis of the acidic xylan fraction. Heredia et al. (1996) proposed an improvement by replacing hot with cold ethanol, showing better results in spite of purifying hemicelluloses A and cellulose purification from the delignified cell wall material. Cardoso et al. (2003a,b) extracted a low methoxy pectin from the alcohol insoluble residue of the wet olive pomace using nitric acid as an extractant. The recovered material was further purified with chelating agents and recovered with graded ethanol precipitation.

The emergent need for more efficient technologies has given rise to a deeper interest in new nonthermal extraction techniques that promise to reduce solvent consumption, shorten the processing time, increase recovery yield, control the Maillard reactions, improve the product quality, and enhance functionality of extracts. These include ultrasonic-assisted, microwave-assisted, pressurized liquid, and SFE as well as electrotechnologies [e.g., high voltage electrical discharges, ultrasound (US), or pulsed electric field] (Galanakis, 2013; Deng et al., 2014; Barba et al., 2015; Galanakis et al., 2015a; Wong et al., 2015; Zinoviadou et al., 2015). In particular, ultrasonic-assisted extraction is based on the disruption of plant cell membranes as a result of the formation of cavitation bubbles generated via ultrasound high frequency (20 kHz) waves (O'Sullivan et al., 2016). The optimum operating conditions (solid/solvent ratio = 500 mg/10 mL, 50% ethanol, and extraction time of 60 min) for ultrasound-assisted extraction of phenols from OMWW have been reported by Sahin and Samli (2013). High voltage electric discharge (HVED) is an alternative extraction technique that requires low energy input (60–80 kJ/kg) to disrupt cell tissues and subsequently enhance phenols extraction from plant matrices due to direct energy release into the medium (Boussetta and Vorobiev, 2014). Similarly, pulsed electric field (PEF) accelerates mass transfer via tissue breakdown, which is induced by the application of a critical electrical potential to cell membranes (Vorobiev and Lebovka, 2010). Recently, Roselló-Soto et al. (2015a) compared HVED, PEF, and US for the recovery of phenols and proteins from olive kernels. According to their results, HVED technology was demonstrated to be more effective than US and PEF technologies in spite of phenols extraction (at equivalent energy inputs).

On the other hand, Japón-Luján et al. (2006) compared microwave assisted extraction of oleuropein, verbacoside, apigenin-7-glucoside, and luteolin-7-glucoside with conventional stirring and heating extraction from olive leaves. At this case, microwaves demonstrated phenols extraction within 8 min, under optimal working conditions of 200 W microwave power and 80% ethanol. Other investigators suggested the optimal conditions for the advanced recovery of phenols from olive leaves using microwaves were: 6 min at 80°C using methanol:water (80:20, v/v) as a solvent (Taamalli et al., 2012). Pressurized liquid extraction is another advanced extraction that has been widely used for the extraction of phenols from OMWW. For instance, Herrero et al. (2011) separated phenols from olive leaves using this technique and found that the extracts with the highest antioxidant capacity were obtained using hot pressurized water at 200°C and liquid ethanol at 150°C. Supercritical fluid techniques uses a liquid–gas solvent (typically CO_2) above its critical point in order to extract compounds with high efficiency. Therefore, Stavroulias and Panayiotou (2005) used a supercritical fluid technique to extract squalene from wet olive pomace. The highest squalene recovery was achieved using 17.5 MPa pressure, 33°C temperature, and 5.4 g/min. CO_2 flow rate as optimum condition for squalene extraction. Although they show promising techniques, emerging technologies have not yet been fully developed to industrial applications for the recovery of valuable compounds from food processing by-products. This is happening due to the fact that in many cases they require high capital cost or their approach is too complicated to be used in small scale olive oil industries.

10.3.4 CLARIFICATION OR ISOLATION

This stage targets either the purification or the isolation of the target compounds from the coextracted impurities with membranes as noted earlier. Adsorption, supported by activated carbons, resins, or polysaccharides is another attractive process that enables the isolation of selected low molecular weight phenols from dilute solutions with high capacity and insensitivity to toxic substances (Galanakis, 2012). For instance, Fernández-Bolaños et al. (2002a) recovered hydroxytyrosol (by 99.5% per weight) olive pomace using chromatographic columns filled with two resins: a nonactivated ionic and an XAD-type nonionic. Removal of phenols from OMWW has also been conducted with a high yield using (95%) sand filtration and subsequent treatment with powdered activated carbon in a batch system (Sabbah et al., 2004). Achak et al. (2009) used another food processing by-product (banana peel) as a low-cost biosorbent to remove phenols from OMW. The adsorption rate of phenols was increased from 60% to

88% as a function of increasing banana peel dosage from 10 to 30 g/L, respectively. More recently, four resins (XAD7, XAD16, IRA96, and Isolute ENV+) have been tested as adsorbents of phenols from OMW. Among the assayed materials, ENV+ showed the highest recovery of phenols (after elution with acidified ethanol) and hydroxytyrosol recovery of 77% (after elution with nonacidified ethanol) (Bertin et al., 2011).

10.3.5 ENCAPSULATION

The last formation stage includes the encapsulation of bioactive compounds inside a coating material, preserves their stability, masks undesirable organoleptic characteristics, and protects them against environmental stresses (Galanakis, 2012). In the case of phenols from olive oil processing by-products, the most commonly used coating materials include polysaccharides like cyclodextran and maltodextrin. Whey protein has also been investigated for this purpose. Spray drying is the most used encapsulation technique for the industrial formation of phenols from OMWW as it is an economic, flexible, and continuous operation. The same technique could be applied to dry DF recovered from OMWW, whereas at this case, coating material is not needed. Other drying technologies include melt extrusion and freeze-drying, while not-drying methods like liposome and emulsion entrapment are typically used when phenols are destined to come in contact the lipid phase of a food product and protect from oxidation (Fang and Bhandari, 2010).

10.4 A CASE SCENARIO FOR THE RECOVERY OF BIOACTIVE COMPOUNDS FROM OMWW USING THE 5-STAGES UNIVERSAL RECOVERY PROCESS

Fig. 10.1 shows an example for the integral recovery of valuable compounds from OMWW, as adapted to the *5-Stages Universal Recovery Process* (Galanakis et al., 2010a,b). In particular, pretreatment includes three processes: centrifugation, skimming, and vacuum concentration. Thereafter, the treatment of the concentrated and defatted substrate with acids and ethanol produces two materials. The first one is an alcohol insoluble residue rich in DF and the second one an ethanolic extract rich in phenols. Isolation of the latter compounds can be performed using resin adsorption or chromatography, whereas UF (25 kDa) treatment has also been suggested in order to remove the higher and autooxidized classes of phenolic compounds (Galanakis et al., 2010b, 2015b). On the other hand, purification of the water soluble fraction of the residue (mainly pectin) from high cation concentrations (mainly potassium) can be conducted using again a 25 kDa-membrane (Galanakis, 2015b). This process allows to improve the functional properties of the pectin and improve the taste of the final product (Galanakis et al., 2010b). The recovered pectin material has been shown to restrict oil uptake of low fat meatballs during deep fat frying and thereby giving rise to meatballs with sustained reduced fat content (Galanakis et al., 2010a,c).

10.5 MEMBRANE APPLICATIONS FOR THE RECOVERY OF PHENOLS

Membrane techniques have been developed for the treatment of liquid wastes from food industries (Mudimu et al., 2012). They aim not only at purifying the wastewater, but also at recovering valuable compounds. Their advantages include small area-requirement, low energy consumption, water-reuse,

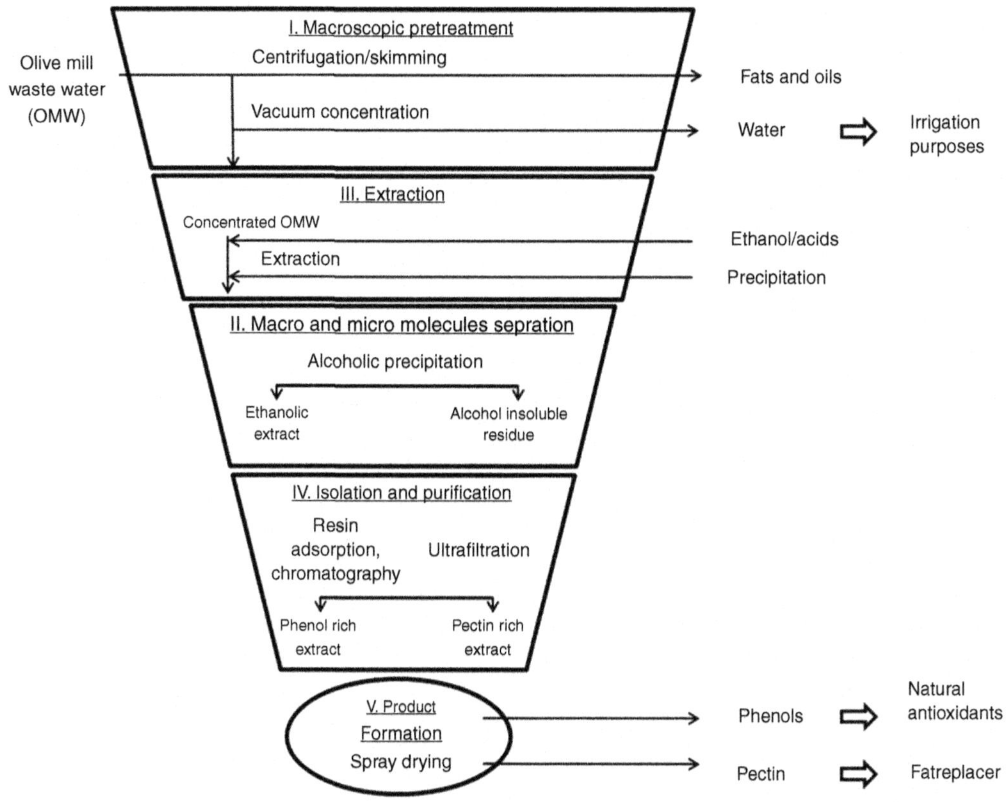

FIGURE 10.1 Recovery of Bioactive Compounds from Olive Mill Wastewater and Reutilization in Different Products

Galanakis, C.M., Martinez-Saez, N., del Castillo, M.D., Barba, F.J., Mitropoulou, V.S., 2015b. Patented and commercialized applications. In: Galanakis, C.M., (Ed.), Food Waste Recovery. Academic Press, San Diego, 337–360 (Chapter 15).

by-products recovery, stabilization of effluent, and absence of organic solvents. Moreover, membrane processes allow to adjust the treatment with regard to the particular wastewater. However, up today, membrane technologies have not been proved to be efficient in spite of OMWW treatment. This is happening due to their high operational cost (due to the continuous discharge or cleaning of membranes) and other complicated problems, such as concentration polarization and fouling. The latest phenomena reduce the permeate fluxes and cause repeated shutdown for cleaning and washing the membranes. Concentration polarization and fouling depend on several factors, most importantly membrane characteristics and feed solution properties (Galanakis et al., 2010b, 2012). A common problem is that OMWW is loaded almost directly (after the removal of solid particles) to the UF units. However, OMWW contains an increased amount of high molecular weight solutes (e.g., pectin, protein) that interact with the membrane surface together with the polar micromolecules (e.g., phenols, sugars), thus developing the earlier phenomena. Although not able to solve the environmental problem of OMWW disposal, membrane technologies comprise a valuable tool in terms of food and ingredients' separations (Patsioura et al., 2011; Galanakis et al., 2013b, 2014, 2015b), and thus provide important solutions when implemented this way. In order to overcome the mentioned problems, the utilization of membrane technologies should be carefully designed within the frame of the "5-Stages Universal Recovery Process," as described earlier (Section 10.4).

Table 10.1 shows relevant processes that have been proposed for the treatment of OMWW. Usual membrane separations include MF, UF, NF, and/or RO. As a first step, OMWW is pretreated (e.g., by centrifugation or common filtering) to remove suspended solids, reducing its chemical oxygen demand (COD) and making it easier to handle. MF and UF can be applied as pretreatment stages, too. The permeates of the latter processes are treated with NF and/or RO to generate concentrated solutions of phenolic compounds, suitable for food, pharmaceutical, and cosmetic applications. The final effluents of NF and RO processes can be safely discharged in aquatic systems or used for irrigation purposes.

Turano et al. (2002) presented an approach for OMWW treatment, including a preliminary centrifugation step, in which the suspended solids are removed, and in a selective separation step of the centrifuge supernatant, carried out by UF. The centrifuge operated at a velocity of 4000 rpm for 10 min, whereas UF was performed using a Polysulfone flat-sheet membrane, with a molecular weight cut-off (MWCO) of 17 kDa, at a temperature of 25°C and a pressure of 1.5 bar. At the end of the process, 90% reduction of COD was achieved. UF permeate contained phenols ranging from 0.2 to 0.3 g/L depending on the feed concentration.

Pizzichini and Russo (2005) suggested a sequential fractionation of OMWW constituents using a combination of centrifugation, MF, UF, NF, and RO treatment (WO 2005/123603). The process starts with pH adjustment to a value between 3 and 4.5, which aims at preventing the oxidation of the phenols by deactivating the polyphenol-oxidase enzyme, promoting the transformation of oleuropein into hydroxytyrosol and creating the optimal conditions for subsequent enzymatic treatment. Moreover, organic compounds are degraded through enzymatic hydrolysis so that their degradation products can be easily separated via the upcoming centrifugation. After the centrifugation step, OMWW is subjected to MF where ceramic membranes (separation limits ranging from 0.1 to 1.4 μm and surface area of 0.35 m^2 per ceramic block) were used. The obtained MF permeate is subjected to UF, in spiral-wound polymeric membranes (cut-offs ranging between 1 and 20 kDa) made of one of the following materials: polysulfone, polyethersulfone polyamide, or regenerated cellulose acetate. The permeate is further treated by means of NF, using spiral-shaped polymeric membranes made of composite polyamide or nylon (cut-offs ranging from 150 to 250 Da). Finally, the obtained NF permeate is subjected to RO, using spiral-shaped polymeric membranes made of composite polyamide. This final operation yields concentrate liquid rich in phenols (30 g/L) and a permeate of purified water. Through this operation setup, 12 L of concentrate is obtained from 200 L of OMWW.

In the patent WO 2005/003037, Castanas et al. (2005) described a system composed of sub stratums selected from turf, sand, and sawdust. Optionally one or more resins (e.g., cationic, anionic, mixed type, or polyvinylpolypyrrolidone) could be added. OMWW passes through different combinations of filters to obtain an effluent that can be used for agricultural irrigation, while the phenols are retained on resins. After separating resins from the other filters and washing, a recovery of 40.8% in phenols could be achieved.

Villanova et al. (2006) focused on the recovery of tyrosol and hydroxytyrosol from OMWW (US 2006/0070953). The process included: rough filtration (RF), MF, UF, NF, and RO, whereas the described pilot plant consisted of 3 units. The first one consisted of seven RF modules ranging from 4.76 to 0.033 mm. The second one consisted of two modules, which were both equipped with a 120 kDa MF membrane. The final unit consisted of four different modules in the following sequence: two UF membranes with cut-offs of 120–20 kDa and 20–1 kDa, an NF membrane (1000–350 kDa cut-off), and

Table 10.1 Summary of Membrane Techniques for OMW Treatment and Phenolic Compounds Recovery

Technique Description	Membrane Characteristics	Effect on COD	Phenolic Compounds Recovery	References
Centrifugation, ultrafiltration	Polysulphone flat-sheet, MWCO of 17 kDa	90%	0.2–0.3 g/L in the UF permeate	Turano et al. (2002)
Combination of centrifugation, microfiltration, and ultrafiltration for COD followed by nanofiltration and reverse osmosis for phenols recovery.	MF: ceramic membrane, separation limits 0.1–1.4 μm; UF: Spiral-wound polymeric (polysulfone, polyethersulfone polyamide or regenerated cellulose acetate) membrane, MWCO 1–20 kDa; NF: Spiral-shaped polymeric composite (polyamide or nylon) membrane, MWCO 150–250 Da; RO: Spiral-shaped polymeric (composite polyamide) membrane	After RO COD lower than 100 ppm of O_2	30 g/L in the RO concentrate	Patent WO 2005/123603 (Pizzichini and Russo, 2005)
Sub stratums and filters of resins	Substratums: turf, sand, and sawdust; Filters of resins: cationic, anionic, mixed type or polyvinylpolypyrrolidone	Not reported	40.8% recovery	Patent WO 2005/003037 (Castanas et al., 2005)
Combination of rough filtration, microfiltration, ultrafiltration, nanofiltration, and reverse osmosis	7 RF modules 4.76–0.033 mm; 2 MF modules, MWCO 120 kDa; 2 UF membranes, MWCO 120–20 kDa and 20–1 kDa; NF: MWCO 1000–350 kDa; RO: MWCO 350 kDa	Lower than 100 mg O_2/L	>1 g/L of hydroxytyrosol and >0.6 g/L of tyrosol in the RO extract	Patent US 2006/0070953 (Villanova et al., 2006)
Microfiltraion, ultrafiltration, and reverse osmosis	MF: ceramic membrane, 0.45 μm, 0.8 μm, 500 Kd UF: Ceramic 1 kDa and polymeric 6, 20, and 80 kDa membranes; RO: Polymeric hydronautics (composite polyamide) membrane	Not reported	464.870 ppm of low MW phenolic in the RO concentrate	Russo (2007)
Ultrafiltration, nanofiltration, and reverse osmosis	UF: Zirconium oxide membrane, 100 nm; NF: Spiral-wound polymeric membrane, MWCO 200 Da; RO: Spiral-wound polymeric membrane, MWCO 100 Da	21% and 97% reduction after UF and NF, respectively	9.962 g/L in the NF retentate (feed concentration 0.725 g/L); 6.782 g/L in the RO concentrate (feed concentration 1.018 g/L)	Paraskeva et al. (2007)

(*Continued*)

Table 10.1 Summary of Membrane Techniques for OMW Treatment and Phenolic Compounds Recovery *(cont.)*

Technique Description	Membrane Characteristics	Effect on COD	Phenolic Compounds Recovery	References
Ultrafiltration and nanofiltration	UF: GE Osmonics GM4040F membrane (composite thin film Desal-5); NF: GE Osmonics DK4040F membrane (composite thin film Desal-5)	50% and 77% reduction after UF and NF, respectively	1.64 mg/L in the UF permeate (initial feedstock concentration 3.1 mg/L) 16.0 mg/L in the NF residue	Stoller (2008)
Supercritical fluid extraction, nanofiltration, and reverse osmosis individually or in integrated mode	NF: Desal DK type membrane (supplied by GE Osmonics), MWCO 250 Da; RO: Dow Filmtec SW 30 membrane	Not reported	Recovery of hydroxytyrosol of about 70% at the NF step.	Patent US 2008/0179246 (De Magalhaes Nunes Da Ponte et al., 2008)
Membrane distillation	Polytetrafluoroethylene (PTFE) membrane with mean pore size of 198.96 nm or polyvinylidene fluoride (PVDF) membrane with mean pore size: 283.15 nm.	Not reported	Concentration factor of 1.72 and 1.4 using PTFE and PVDF membranes respectively	El-Abbassi et al. (2009)
MF and NF as pretreatments, osmotic distillation (OD) or vacuum membrane distillation (VMD) for polyphenol recovery.	MF: tubular Al_2O_3 membrane (pore sizes 200 nm); NF: Nadir N30F spiral-wound membrane made of polyethersulfone, MWCO 578 Da OD: Polyethylene membrane, containing microporous polypropylene hollow-fibers (mean pore size about 30 nm)	Not reported	78% of the initial phenols were recovered in the permeate stream of the MF 160.32 ppm in the NF permeate 0.5 g/L after OD	Garcia-Castello et al. (2010)
Microfiltration, ultrafiltration, and reverse osmosis	MF: Tubular polypropylene membrane, MWCO between 0.1 and 0.3 μm; UF: Spiral membranes of polyamide and traces of polysulfone, MWCO of 7 kDa RO: Spiral thin film membrane (TFM) composed of Durasan and polysulfone, MWCO ~ 100 Da	Final reduction of 98%	19.3 g/L total phenols in the RO concentrate (initial OMW concentration 4.9 g/L)	Servili et al. (2011)
A sequence of two ultrafiltration processes followed by an nanofiltration process	UF: A hollow fiber membrane with nominal pore size of 0.02 μm and a flat-sheet composite fluoropolymer membrane with MWCO of 1000 Da; NF: Spiral-wound membrane module, MWCO of 90 Da	Not reported	960 mg/L in the NF retantate (feed concentration: 625 mg/L; initial concentration:1409 mg/L)	Cassano et al. (2013)

Table 10.1 Summary of Membrane Techniques for OMW Treatment and Phenolic Compounds Recovery *(cont.)*

Technique Description	Membrane Characteristics	Effect on COD	Phenolic Compounds Recovery	References
Sequential sieving, UF, NF, RO, resin adsorption/desorption	Sieving: Stainless steel filters with final pore size of 0.125 mm UF: Ceramic (zirconia) with a pore diameter 100 nm NF: Spiral wound, polymeric RO: Spiral wound, polymeric Resins: nonionic resins (XAD4, XAD16, and XAD7HP)	6.47 g/L COD level in the final effluent	165 g in the UF concentrate, 74 g in the NF concentrate, and 25 g of free simple phenols in the RO concentrate (initial 265 g of phenols) 28 g/L after rotary evaporation of RO concentrate or 378 g/L when the RO concentrate was treated with resins and vacuum evaporation	Zagklis and Paraskeva (2014) and Zagklis et al. (2015)
MF	MF: Ceramic membrane (MWCO~100,000 Da, membrane porosity 0.1 μm) UF: A polyethersulfone membrane (MWCO of 5000 Da) NF: Polyamide membrane (MWCO of 200 Da)	Not reported	2456–5284 μg/mL in MF fraction; 1404–3065 μg/mL in UF fraction (55–65% recovery); 373–1583 μg/mL in NF fraction (15–33% recovery)	D'Antuono et al. (2014)
MF, UF, distillation	MF: Ceramic membrane, pore size 0.2 μm UF: Organic membrane, MWCO 8KD	48.44 g/L COD level in the UF filtrate	7.2 g/L hydroxytyrosol in the UF permeate (initial concentration 0.23 g/L)	Hamza and Sayadi (2015)
MF	MF: Polyvinyl-fluoride membrane, pore size 200 nm UF: Ceramic membrane, MWCO 20 kD NF: DMS DK membrane (composite), MWCO 150–300 Da	34.8% reduction after MF; 44.9% reduction after UF; In the NF step, COD was reduced from 33 to 6.9 g/L	2.3 g/L in the MF retentate; 2.0 g/L in the UF retentate; 2.4 g/L in the UF retentate	Abdel-Shafy et al. (2015)

Adapted from Galanakis, C.M., Martinez-Saez, N., del Castillo, M.D., Barba, F.J., Mitropoulou, V.S., 2015b. Patented and commercialized applications. In: Galanakis, C.M. (Ed.), Food Waste Recovery. Academic Press, San Diego, pp. 337–360 (Chapter 15).

an RO membrane (350 kDa cut-off). According to the described process, the osmosis extract contained more than 1 g/L of hydroxytyrosol and more than 0.6 g/L of tyrosol, which could be later isolated with purity higher than 98% by chromatography in inverted phase on a preparatory column.

Russo (2007), based on patent WO 2005/123603, applied several MF and UF membranes and noted that among the tested ones, ceramic 1 Kd and polymeric 6 Kd showed no differences on the selectivity of hydroxytyrosol. MF and UF permeates were concentrated using RO consisting of a polymeric hydronautics membrane (composite polyamide) with an area of 7 m^2. The RO membrane produced a retentate containing more than 370 mg hydroxytyrosol/L (concentration factor of 2.5).

Paraskeva et al. (2007) investigated the possibility of complete fractionation of OMWW using combination of membrane processes, too. Their pilot plant was operated for a complete olive harvesting period, in a typical Greek olive mill, using the three-phase decanter technology. At first, wastes were filtered with a polypropylene screen (80 μm) to remove suspended solids. UF membrane (made of zirconium oxide with pore sizes of 100 nm and a surface area of 0.24 m^2) was applied for the reduction of organic substances using a pressure of 1.75 bar. After UF process reductions of 97% in suspended solids, 17% in total organic carbon (TOC), 21% in COD, and 34% in phenols were achieved. Better performance of UF unit was obtained at higher temperatures (50°C). Further operation steps included NF or RO, which both used spiral-wound polymeric membranes (area of 2.5 m^2) with a MWCO of 200 and 100 Da, respectively. The NF retentate contained ~10 g/L of phenols (feed concentration 0.725 g/L). The NF step showed reduction efficiencies of 99% in total suspended solids (TSS), 97% in TOC, 97% in COD, and 98% in phenols.

Stoller (2008) used a combination of UF and NF processes for the treatment of OMWW. After an aerobic biological operation carried out in continuous mode, the process stream was sent to an UF batch process where OMWW was filtered using the GE Osmonics GM4040F membrane (composite thin film Desal-5). The UF permeate was then subjected to NF using the GE Osmonics DK4040F membrane (composite thin film Desal-5). The UF step, showed a rejection value of 50% for COD, while significantly reduced the amount of contained phenols (only 53% pass through the membrane), resulting in permeate containing 1.64 mg/L (initial feedstock concentration 3.1 mg/L). Due to this reduction the concentration factor for further treatments was limited.

In the patent US 2008/0179246, a process for obtaining a hydroxytyrosol-rich concentrate from olive tree residues and subproducts was described (De Magalhaes Nunes Da Ponte et al., 2008). The technique used SFE, NF, and RO, individually or in an integrated mode. The recovery of hydroxytyrosol in the concentrate stream ranges between 15% and 98% (mass fraction), whereas other bioactive compounds such as luteolin, caffeic, and *p*-coumaric acids are identified. On the other hand, El-Abbassi et al. (2009) investigated a direct contact membrane distillation unit. Two types of commercial flat-sheet hydrophobic membranes were used. The first one was made of polytetrafluoroethylene (PTFE) with mean pore size of 198.96 nm and the second one was made of polyvinylidene fluoride (PVDF) with mean pore size of 283.15 nm. After 9 h of DCMD operating time, the PTFE membrane showed a better polyphenols separation coefficient ($\alpha = 99\%$) than the PVDF membrane ($\alpha = 89\%$). The DCMD processing of OMWW allowed to reach concentration factors of 1.72 for the PTFE membrane and 1.4 for the PVDF membrane (initial concentration of the feed 4 g/L).

Garcia-Castello et al. (2010), proposed an integrated membrane system for OMWW treatment. The process included MF and NF as pretreatments for reduction of the organic pollution. The NF permeate contained 160.32 ppm of phenols. A more concentrated solution was obtained after treating the NF permeate by osmotic distillation (OD), in a compact plant with a polyethylene membrane, containing microporous polypropylene hollow-fibers (mean pore size about 30 nm). The obtained solution was enriched in phenols (about 0.5 g/L) and particularly hydroxytyrosol (0.28 g/L).

Thereafter, Servili et al. (2011), used a three-phase membrane system for the reduction of OMWW pollution and the recovery of hydrophilic phenols from fresh olive vegetation waters in an industrial plant. The process included a consecutive filtration with MF and UF in combination with a prior enzymatic treatment. The treatment was carried out at controlled temperature (20°C), under N_2 atmosphere for reducing O in the headspace of the containers for the wastewaters collection and storage during and after the filtration. For the MF, a tubular polypropylene membrane was used (cut-off ranging from 0.1

to 0.3 μm and a total surface area of 8 m^2). UF was carried out using two spiral membranes made of polyamide and traces of polysulfone (cut-off of 7 kDa and a total area of 16 m^2). For the RO process, a spiral thin film membrane (TFM) composed of Durasan and polysulfone, was applied. The cut-off was approximately 100 Da (the total surface area was 9 m^2). The entire process reduced the initial COD by 98%. The RO concentrate contained 19.3 g/L total phenols (initial OMWW concentration 4.9 g/L), namely 3, 4-DHPEA (0.03 g/L), *p*-HPEA (0.01 g/L), 3, 4DHPEA-EDA (16.9 g/L), and verbascoside (2.4 g/L).

Cassano et al. (2013) applied a process including a sequence of two UF steps followed by a NF step. OMWWs were pretreated in a laboratory pilot UF unit, equipped with a hollow fiber membrane (nominal pore size 0.02 μm), in order to remove suspended solids and reduce fouling phenomena in the subsequent membrane operations. The UF permeate was then submitted to a second UF process, using a flat-sheet composite fluoropolymer membrane with a molecular weight cut-off of 1000 Da. UF membranes revealed rejections of about 26% and 31% in phenols. The permeate stream obtained from this operation was submitted to a NF performed in a bench plant equipped with a spiral-wound membrane module, with MWCO of 90 Da. The NF membrane showed a 93% rejection toward total phenols and 100% rejection toward low MW polyphenols. The concentration of total phenols in the retentate obtained, through NF process, was 960 mg/L, while the feed concentration was 625 mg/L and the initial OMWW concentration was 1409 mg/L.

A cost effective system for their maximum exploitation of agroindustrial wastes or by-products, such as OMWW, using a combined process of membrane filtration and other physicochemical processes, was suggested by Zagklis and Paraskeva (2014). As a preliminary step, sequential sieving was performed, through stainless steel filters with final pore size of 0.125 mm. This process was followed by UF, NF, and RO in pilot scale units (cross-flow mode in batch operation). The UF module was ceramic (zirconia) with a pore diameter 100 nm. The NF and RO modules were spiral wound, polymeric. From the initial 265 g of phenols contained in the waste, 165 g (62%) remained in the UF concentrate, 74 g (28%) remained in the NF concentrate, and 25 g of free simple phenols (9%) remained in the RO concentrate. The NF unit was proven to be capable of removing compound with molecular weight larger than 468 (including complex phenols), by 90%. The disadvantage of this process was the relatively high COD (6.47 g/L) of the final effluent, as the main aim of the procedure was the concentration of the phenols and not the optimum characteristics of the final effluent. In a subsequent work, the RO concentrate obtained after NF, which contained the low-molecular-weight compounds, was further treated with resin adsorption/desorption (Zagklis et al., 2015). Nonionic resins (XAD4, XAD16, and XAD7HP) were used, for the recovery of phenols and their separation from carbohydrates. The phenols, recovered through this process, were concentrated through vacuum evaporation reaching a final concentration of 378 g/L (gallic acid equivalents) and 84.8 g/L hydroxytyrosol.

More recently, Hamza and Sayadi (2015), recovered simple phenols with high biological value, using a combination of enzymatic hydrolysis (by a β-glucosidase) and membrane filtration technology. The freshly collected OMWW samples were treated with β-glucosidase (at a concentration of 5 IU mL) in a slowly steered batch reactor at 50°C for 120 min. Subsequent, MF and UF were performed in a ceramic membrane with a cut-off of 0.2 μm (area 0.2 m^2) and in an organic membrane with a cut-off of 8 kD (area 0.0825 m^2), respectively. Finally, the UF permeate was concentrated by distillation, in a pilot-scale evaporation unit (45°C, 100 mbar/2 h). At the end of the process, the UF filtrate exhibited a COD level of 48.44 g/L, while the concentration of the UF permeate in hydroxytyrosol reached 7.2 g/L (initial concentration of hydroxytyrosol of the OMWW 0.23 g/L), after evaporation. Finally,

Abdel-Shafy et al. (2015), used integrated membrane systems for the recovery and concentration of phenols from OMWW followed enzymatic (using β-glucosidase) and acid hydrolysis in order to release higher concentration of pure hydroxytyrosol. OMWW which contained 2.5 g/L of phenols, was directly subjected to the MF unit without any preliminary centrifugation. The MF membrane was made of polyvinyl-fluoride (with mean pore size of 200 nm) and resulted in 77.6% and 34.8% reductions of TSS and COD, respectively. The concentration of phenols in the retentate was 2.3 g/L. The UF system was equipped with a ceramic membrane (cut-off of 20 kD) and resulted in a removal of 44.9% and 100% for the COD and TSS, respectively, whereas phenols concentration of the retentate was 2.0 g/L. Finally, in the NF unit step, COD was reduced from 33 to 6.9 g/L, TSS was removed by 100% TSS and retentate concentration in phenols reached 2.4 g/L. Finally, by employing β-glucosidase acid hydrolysis on NF retentate, an amount of 1.3 g/L of pure HT could be released.

10.6 THE CURRENT STATE OF THE ART IN THE MARKET OF PHENOLS RECOVERED FROM OLIVE PROCESSING BY-PRODUCTS

Nowadays, numerous patents (WO/2002/0218310, US/2002/0198415, US 2002/0058078, WO2004/005228, US 6414808, EP-A 1 582 512) have been obtained for the extraction of target phenolic compounds (e.g., oleuropein, hydroxytyrosol) from olive oil processing by-products (Galanakis et al., 2015b). Table 10.2 shows a collection of them that lead to commercial applications. In most cases, a slow acidic hydrolysis (for 2–12 months) of the initial substrate is proposed with a final aim of converting more than 90% of the present oleuropein to hydroxytyrosol (Liu et al., 2008). Phenolea Complex is a natural hydrophilic extract (without using any kind of organic solvent) obtained directly from OMWW. After the milling process, OMWW is collected, pretreated and sequentially subjected to tangential MF (using ceramic membranes) and vacuum evaporation of the permeate (FDA, 2012). The commercial recovery of hydroxytyrosol from wet olive pomace has been proposed using chromatographic columns filled with two resins: a nonactivated ionic and an XAD-type nonionic (Fernández-Bolaños et al., 2002a). Hydroxytyrosol has shown improved advanced antiradical properties compared to vitamin E and C, and thus could be used as functional supplement, food preservative in bakery products, or life prolonging agents. Another well-known methodology is the one proposed by (Crea, 2002). More specifically, OMWW is pretreated with acid and then an incubation process follows aiming at the conversion of oleuropein to hydroxytyrosol. Further extraction of hydroxytyrosol could be performed using SFE and a column operating in the counter-current mode, where a nonselective porous membrane is the barrier interface between the hydroxytyrosol containing fluid and the dense gas. Final stage (product formation) includes freeze or spray drying. The commercially available product from CreAgri is named Hidrox. Other commercially available products include:

1. Olnactiv (Glanbia, Milan, Italy), Oleaselect, and Opextan (Indena, Milan, Italy);
2. Olive Braun Standard 500 (containing 1.0–2.2 g of hydroxytyrosol and 0.2–0.7 g of tyrosol/kg) from Naturex;
3. Olive phenols from Albert isliker (containing 22–23 g hydroxytyrosol and 6.5–8.0 g tyrosol/kg), Prolivols (containing 35% polyphenols, 2% hydroxytyrosol, and 3% tyrosol) from Seppic Inc; and
4. Olive polyphenols NLT from Lalilab Inc. (containing 2.0–6% hydroxytyrosol and 0.7–1.1% tyrosol).

Table 10.2 Patented Methodologies Leading to Commercial Applications of Olive Oil Processing By-Products (Galanakis et al., 2015b)

Food Waste Source	Patents Application Number	Applicant/ Company	Title/Treatment Steps	Products/ Brand Names	Potential/ Commercialized Applications	Inventors/ References
Olive mill waste	PCT/ US2001/ 027132	CreAgri, Inc (Hayard, USA)	Method of obtaining a hydroxytyrosol-rich composition from vegetation water	Hydroxytyrosol/Hidrox	Food supplements and cosmetics	Crea (2002)
	PCT/ ES2002/ 000058	Consejo Superior de Investigaciones Cientificas (Madrid, Spain)/ Genosa I + D S.A. (Malaga, Spain)	Method for obtaining purified hydroxytyrosol from products and by-products derived from olive tree	Hydroxytyrosol (99.5%)/ Hytolive	Conserving foods, functional ingredient in bread	Fernández-Bolaños et al. (2002a)
	GR2010/ 1006660	Polyhealth (Larissa, Greece)	Ultrafiltration, ion exchange resin adsorption, solvent elution, spray drying	Medoliva (hydroxytyrosol, tyrosol, caffeic acid, and *p*-coumaric acid)	Food supplements and antioxidants, cosmetics, personal care products	Petrotos et al. (2010)
	PCT/ SE2007/ 001177	Phenoliv AB (Lund, Sweden)	Olive waste recovery	Olive phenols and dietary fibers containing powders	Natural antioxidants in foodstuff and fat replacement in meatballs, respectively	Tornberg and Galanakis (2008)
	US patent 6361803 B1	Usana, Inc. (USA)	Wastewater from olive oil production	Antioxidant compounds	Food supplements and antioxidants, cosmetics, personal care products	Cuomo and Rabovskiy (2004)
Olive leaves extracts	EP 1582512 A1	Cognis IP Management GmbH, (Düsseldorf, Germany)	Olive waste recovery	Olive hydroxytyrosol	Natural antioxidants in foodstuff	Beverungen (2005)
	US Patent/2005/ 0103711 A1		Olive waste recovery	Oleuropein aglycon	Natural antioxidant	Emmons and Guttersen (2005)

In general, olive pulp extracts have been approved by FDA which GRAS status (GRN No. 459) to be used as antioxidants in numerous foods (e.g., baked goods, beverages, and dressings) up to a level of 3 g/Kg in the final food (Galanakis et al., 2015b).

10.7 FUTURE PERSPECTIVES

Solvent extraction has long been applied to recover oil from olive kernels, which are considered as an established commodity similarly to olive fruit. On the other hand, OMWW and wet olive pomace considered for many decades only as materials of treatment, minimization, and prevention. Nowadays, researchers and olive oil producers focus on the recovery of polyphenols from OMWW and wet olive pomace. Indeed, their commercial implementation is already a reality since at least five companies around the world produce phenol rich extracts from these sources and trade them as natural preservatives or bioactive additives in food products and cosmetics (Galanakis and Schieber, 2014). The prospects of these applications are discussed thoroughly in Chapters 11 and 12, respectively. It is important to state that the earlier approach cannot solve alone the environmental problems induced by OMWW disposal. Indeed, more integral policies are needed as noted in Chapters 2–9. However, it could provide profit for the involved olive oil industries and certain solutions within the biorefinery concept (Chapter 3).

The recovery of phenols from OMWW and wet olive pomace could be conducted with several technologies, most of them being already patented. The main idea is to pretreat the initial material and convert oleuropein to hydroxytyrosol, prior extracting phenols with solvent and/or other technologies. Purification and product formation allow obtaining purer extracts with advanced antioxidant properties. The beneficial health claims of these extracts are still under debate in Europe, but in other regions (e.g., USA, Asia) have been accepted. Potential health claim proof of these products will open the door in new opportunities and bring revolution in the olive oil industry. Besides, experience has shown that a project focused on the recovery technologies without investigating and establishing definite food-targeted applications of the final product, is doomed to fail (Galanakis, 2012). Therefore, the development of certain applications for the recovered phenols is as much important as the development of the recovery process itself.

REFERENCES

Abdel-Shafy, H.I., Schories, G., Mohamed-Mansour, M.S., Bordei, V., 2015. Integrated membranes for the recovery and concentration of antioxidant from olive mill wastewater. Desalin. Water Treat. 56, 305–314.

Achak, M., Hafidi, A., Ouazzani, N., Sayadi, S., Mandi, L., 2009. Low cost biosorbent "banana peel" for the removal of phenolic compounds from olive mill wastewater: Kinetic and equilibrium studies. J. Hazard. Mater. 166, 117–125.

Allouche, N., Fki, I., Sayadi, S., 2004. Toward a high yield recovery of antioxidants and purified hydroxytyrosol from olive mill wastewaters. J. Agric. Food Chem. 52, 267–273.

Asli, A., Qatibi, A.-I., 2009. Ethanol production from olive cake biomass substrate. Biotechnol. Bioprocess Eng. 14, 118–122.

Aviani, I., Raviv, M., Hadar, Y., Saadi, I., Dag, A., Ben-Gal, A., Yermiyahu, U., Zipori, I., Laor, Y., 2012. Effects of harvest date, irrigation level, cultivar type and fruit water content on olive mill wastewater generated by a laboratory scale 'Abencor' milling system. Bioresour. Technol. 107, 87–96.

Balice, V., Cera, O., 1984. Acidic phenolic fraction of the olive vegetation water determined by a gas-chromatographic method. Grasas Aceites 35, 178–180.

Barba, F.J., Galanakis, C.M., Esteve, M.J., Frigola, A., Vorobiev, E., 2015. Potential use of pulsed electric technologies and ultrasounds to improve the recovery of high-added value compounds from blackberries. J. Food Eng. 167 (Part A), 38–44.

Bayram, B., Ozcelik, B., Schultheiss, G., Frank, J., Rimbach, G., 2013. A validated method for the determination of selected phenolics in olive oil using high-performance liquid chromatography with coulometric electrochemical detection and a fused-core column. Food Chem. 138, 1663–1669.

Belaqziz, M., El-Abbassi, A., Lakhal, E.K., Agrafioti, E., Galanakis, C.M., 2016. Agronomic application of olive mill wastewater: Effects on maize production and soil properties. J. Environ. Manag. 171, 158–165.

Bertin, L., Ferri, F., Scoma, A., Marchetti, L., Fava, F., 2011. Recovery of high added value natural polyphenols from actual olive mill wastewater through solid phase extraction. Chem. Eng. J. 171, 1287–1293.

Beverungen, C., 2005. Process for obtaining hydroxytyrosol from olive leaves extracts. EP Patent EP 1582512 A1.

Bianco, A., Buiarelli, F., Cartoni, G., Coccioli, F., Jasionowska, R., Margherita, P., 2003. Analysis by liquid chromatography-tandem mass spectrometry of biophenolic compounds in olives and vegetation waters, part I. J. Sep. Sci. 26, 409–416.

Boskou, D., 2006. Sources of natural phenolic antioxidants. Trends Food Sci. Technol. 17, 505–512.

Boskou, D., 2008. Phenolic compounds in olives and olive oil. In: Boskou, D. (Ed.), Olive Oil Minor Constituents and Health. CRC Press, Boca Raton, pp. 11–44.

Boukhoubza, F., Jail, A., Korchi, F., Idrissi, L.L., Hannache, H., Duarte, J.C., Hassani, L., Nejmeddine, A., 2009. Application of lime and calcium hypochlorite in the dephenolisation and discolouration of olive mill wastewater. J. Environ. Manag. 91, 124–132.

Boussetta, N., Vorobiev, E., 2014. Extraction of valuable biocompounds assisted by high voltage electrical discharges: a review. Comptes Rendus Chimie 17, 197–203.

Bouzid, O., Navarro, D., Roche, M., Asther, M., Haon, M., Delattre, M., Lorquin, J., Labat, M., Asther, M., Lesage-Meessen, L., 2005. Fungal enzymes as a powerful tool to release simple phenolic compounds from olive oil by-product. Process Biochem. 40, 1855–1862.

Capasso, R., Evidente, A., Scognamiglio, F., 1992. A simple thin layer chromatographic method to detect the main polyphenols occurring in olive oil vegetation waters. Phytochem. Anal. 3, 270–275.

Cardoso, S.M., Coimbra, M.A., Lopes da Silva, J.A., 2003a. Calcium-mediated gelation of an olive pomace pectic extract. Carbohydr. Polym. 52, 125–133.

Cardoso, S.M., Coimbra, M.A., Lopes da Silva, J.A., 2003b. Temperature dependence of the formation and melting of pectin–Ca2+ networks: a rheological study. Food Hydrocolloid. 17, 801–807.

Cassano, A., Conidi, C., Giorno, L., Drioli, E., 2013. Fractionation of olive mill wastewaters by membrane separation techniques. J. Hazard. Mater. 248-249, 185–193.

Castanas, E., Andricopoulos, N., Mposkou, G., Vercauteren, J., 2005. A method for the treatment of olive mill waste waters. World Intellectual Property Organization, WO/2005/003037.

Coimbra, M.A., Waldron, K.W., Selvendran, R.R., 1994. Isolation and characterisation of cell wall polymers from olive pulp (*Olea europaea* L.). Carbohydr. Res. 252, 245–262.

Coimbra, M.A., Waldron, K.W., Selvendran, R.R., 1995. Isolation and characterisation of cell wall polymers from the heavily lignified tissues of olive (*Olea europaea*) seed hull. Carbohydr. Polym. 27, 285–294.

Coimbra, M.A., Barros, A., Rutledge, D.N., Delgadillo, I., 1999. FTIR spectroscopy as a tool for the analysis of olive pulp cell-wall polysaccharide extracts. Carbohydr. Res. 317, 145–154.

Covas, M.-I., de la Torre, K., Farré-Albaladejo, M., Kaikkonen, J., Fitó, M., López-Sabater, C., Pujadas-Bastardes, M.A., Joglar, J., Weinbrenner, T., Lamuela-Raventós, R.M., de la Torre, R., 2006. Postprandial LDL phenolic content and LDL oxidation are modulated by olive oil phenolic compounds in humans. Free Radic. Biol. Med. 40, 608–616.

Crea, R., 2002. Method of obtaining a hydroxytyrosol-rich composition from vegetation water. World Intellectual Property Organization, WO/2002/0218310.

Cuomo, J., Rabovskiy, A.B., 2004. Antioxidant compositions extracted from a wastewater from olive oil production. US Patent 6361803 B1.

D'Alessandro, F., Marucchini, C., Minuti, L., Zadra, C., Tatocchi, A., 2005. GC/MS-SIM analysis of phenolic compounds in olive oil waste waters. Ital. J. Food Sci. 17, 83–88.

D'Antuono, I., Kontogianni, V.G., Kotsiou, K., Linsalata, V., Logrieco, A.F., Tasioula-Margari, M., Cardinali, A., 2014. Polyphenolic characterization of olive mill wastewaters, coming from Italian and Greek olive cultivars, after membrane technology. Food Res. Int. 65 (Part C), 301–310.

Davies, L.C., Vilhena, A.M., Novais, J.M., Martins-Dias, S., 2004. Olive mill wastewater characteristics: modelling and statistical analysis. Grasas Aceites 55, 233–241.

De Leonardis, A., Macciola, V., Lembo, G., Aretini, A., Nag, A., 2007. Studies on oxidative stabilisation of lard by natural antioxidants recovered from olive-oil mill wastewater. Food Chem. 100, 998–1004.

De Magalhaes Nunes Da Ponte, M.L., Dos Santos, J.L.C., Matias, A.A.F., Nunes, A.V.M.M., Duarte, C.M.M., Crespo, J.P.S.G., 2008. Method of obtaining a natural hydroxytyrosol-rich concentrate from olive tree residues and subproducts using clean technologies. Patent US 2008/0179246.

De Marco, E., Savarese, M., Paduano, A., Sacchi, R., 2007. Characterization and fractionation of phenolic compounds extracted from olive oil mill wastewaters. Food Chem. 104, 858–867.

DellaGreca, M., Monaco, P., Pinto, G., Pollio, A., Previtera, L., Temussi, F., 2001. Phytotoxicity of low-molecular-weight phenols from olive mill waste waters. Bull. Environ. Contamin. Toxicol. 67, 0352–0359.

DellaGreca, M., Previtera, L., Temussi, F., Zarrelli, A., 2004. Low-molecular-weight components of olive oil mill waste-waters. Phytochem. Anal. 15, 184–188.

Deng, Q., Zinoviadou, K.G., Galanakis, C.M., Orlien, V., Grimi, N., Vorobiev, E., Lebovka, N., Barba, F.J., 2014. The effects of conventional and non-conventional processing on glucosinolates and its derived forms, isothiocyanates: extraction, degradation, and applications. Food Eng. Rev. 7, 357–381.

Dermeche, S., Nadour, M., Larroche, C., Moulti-Mati, F., Michaud, P., 2013. Olive mill wastes: biochemical characterizations and valorization strategies. Process Biochem. 48, 1532–1552.

EFSA, 2011. Polyphenols in olive related health claims, 9(4), 2033. European Food Safety Authority (EFSA), Parma, Italy.

El-Abbassi, A., Hafidi, A., García-Payo, M.C., Khayet, M., 2009. Concentration of olive mill wastewater by membrane distillation for polyphenols recovery. Desalination 245, 670–674.

El-Abbassi, A., Kiai, H., Hafidi, A., García-Payo, M.C., Khayet, M., 2012. Treatment of olive mill wastewater by membrane distillation using polytetrafluoroethylene membranes. Sep. Purif. Technol. 98, 55–61.

Emmons, W., Guttersen, C., 2005. Isolation of oleuropein aglycon from olive vegetation water. US Patent/2005/0103711 A1.

Fang, Z., Bhandari, B., 2010. Encapsulation of polyphenols – a review. Trends Food Sci. Technol. 21, 510–523.

FDA, 2012. Amendments to notification of GRAS determination for olive pulp extract (OPE)Phenolea® complex for use in foods as antioxidant. GRAS exemption claim. Available from: http://www.fda.gov/ucm/groups/fdagov-public/@fdagov-foods-gen/documents/document/ucm346897.pdf

Fernández-Bolaños, J., Felizón, B., Heredia, A., Rodrnguez, R., Guillén, R., Jiménez, A., 2001. Steam-explosion of olive stones: hemicellulose solubilization and enhancement of enzymatic hydrolysis of cellulose. Bioresour. Technol. 79, 53–61.

Fernández-Bolaños, J., Heredia, A., Rodríguez, G., Rodríguez, R., Guillén, R., Jiménez, A., 2002a. Method for obtaining purified hydroxytyrosol from products and by-products derived from the olive tree. World Intellectual Property Organization, WO/2002/064537.

Fernández-Bolaños, J., Rodríguez, G., Rodríguez, R., Heredia, A., Guillén, R., Jiménez, A., 2002b. Production in large quantities of highly purified hydroxytyrosol from liquid-solid waste of two-phase olive oil processing or "Alperujo". J. Agric. Food Chem. 50, 6804–6811.

Fernández-Bolaños, J., Rodríguez, G., Rodríguez, R., Guillén, R., Jiménez, A., 2006. Extraction of interesting organic compounds from olive oil waste. Grasas Aceites 57, 95–106.

Fiorentino, A., Gentili, A., Isidori, M., Monaco, P., Nardelli, A., Parrella, A., Temussi, F., 2003. Environmental effects caused by olive mill wastewaters: toxicity comparison of low-molecular-weight phenol components. J. Agric. Food Chem. 51, 1005–1009.

Galanakis, C.M., 2011. Olive fruit dietary fiber: components, recovery and applications. Trends Food Sci. Technol. 22, 175–184.

Galanakis, C.M., 2012. Recovery of high added-value components from food wastes: conventional, emerging technologies and commercialized applications. Trends Food Sci. Technol. 26, 68–87.

Galanakis, C.M., 2013. Emerging technologies for the production of nutraceuticals from agricultural by-products: a viewpoint of opportunities and challenges. Food Bioprod. Process. 91, 575–579.

Galanakis, C.M., 2015a. The Universal Recovery Strategy, Food Waste Recovery. Academic Press, San Diego, pp. 59–81 (Chapter 3).

Galanakis, C.M., 2015b. Separation of functional macromolecules and micromolecules: from ultrafiltration to the border of nanofiltration. Trends Food Sci. Technol. 42, 44–63.

Galanakis, C.M., Schieber, A., 2014. Editorial. Special issue on recovery and utilization of valuable compounds from food processing by-products. Food Res. Int. 65 (Part C), 299–300.

Galanakis, C.M., Tornberg, E., Gekas, V., 2010a. A study of the recovery of the dietary fibres from olive mill wastewater and the gelling ability of the soluble fibre fraction. LWT Food Sci. Technol. 43, 1009–1017.

Galanakis, C.M., Tornberg, E., Gekas, V., 2010b. Clarification of high-added value products from olive mill wastewater. J. Food Eng. 99, 190–197.

Galanakis, C.M., Tornberg, E., Gekas, V., 2010c. Dietary fiber suspensions from olive mill wastewater as potential fat replacements in meatballs. LWT Food Sci. Technol. 43, 1018–1025.

Galanakis, C.M., Tornberg, E., Gekas, V., 2010d. The effect of heat processing on the functional properties of pectin contained in olive mill wastewater. LWT Food Sci. Technol. 43, 1001–1008.

Galanakis, C.M., Tornberg, E., Gekas, V., 2010e. Recovery and preservation of phenols from olive waste in ethanolic extracts. J Chem. Technol. Biotechnol. 85, 1148–1155.

Galanakis, C.M., Fountoulis, G., Gekas, V., 2012. Nanofiltration of brackish groundwater by using a polypiperazine membrane. Desalination 286, 277–284.

Galanakis, C.M., Goulas, V., Tsakona, S., Manganaris, G.A., Gekas, V., 2013a. A knowledge base for the recovery of natural phenols with different solvents. Int. J. Food Prop. 16, 382–396.

Galanakis, C.M., Markouli, E., Gekas, V., 2013b. Recovery and fractionation of different phenolic classes from winery sludge using ultrafiltration. Sep. Purif. Technol. 107, 245–251.

Galanakis, C.M., Chasiotis, S., Botsaris, G., Gekas, V., 2014. Separation and recovery of proteins and sugars from Halloumi cheese whey. Food Res. Int. 65 (Part C), 477–483.

Galanakis, C.M., Barba, F.J., Prasad, K.N., 2015a. Cost and safety issues of emerging technologies against conventional techniques. In: Galanakis, C.M. (Ed.), Food Waste Recovery: Processing Technologies and Industrial Techniques, Academic Press, San Diego.

Galanakis, C.M., Martinez-Saez, N., del Castillo, M.D., Barba, F.J., Mitropoulou, V.S., 2015b. Patented and commercialized applications. In: Galanakis, C.M. (Ed.), Food Waste Recovery, Academic Press, San Diego, 337–360 (Chapter 15).

Garcia-Castello, E., Cassano, A., Criscuoli, A., Conidi, C., Drioli, E., 2010. Recovery and concentration of polyphenols from olive mill wastewaters by integrated membrane system. Water Res. 44, 3883–3892.

Greco, Jr., G., Colarieti, M.L., Toscano, G., Iamarino, G., Rao, M.A., Gianfreda, L., 2006. Mitigation of olive mill wastewater toxicity. J. Agric. Food Chem. 54, 6776–6782.

Grigelmo-Miguel, N., Abadías-Serós, M.a.I., Martín-Belloso, O., 1999. Characterisation of low-fat high-dietary fibre frankfurters. Meat Sci. 52, 247–256.

Hamza, M., Sayadi, S., 2015. Valorisation of olive mill wastewater by enhancement of natural hydroxytyrosol recovery. Int. J. Food Sci. Technol. 50, 826–833.

Heredia, A., Guillén, R., Sánchez, C., Jiménez, A., Fernández-Bolaños, J., 1996. Changes in pectic polysaccharides during elaboration of table olives. In: Visser, J., Voragen, A.G.J. (Eds.), Progress in Biotechnology. Elsevier, The Netherlands, pp. 569–576.

Herrero, M., Temirzoda, T.N., Segura-Carretero, A., Quirantes, R., Plaza, M., Ibanez, E., 2011. New possibilities for the valorization of olive oil by-products. J. Chromatogr. A 1218, 7511–7520.

Ilyasoglu, H., Ozcelik, B., Van Hoed, V., Verhe, R., 2011. Cultivar characterization of Aegean olive oils with respect to their volatile compounds. Sci. Horticult. 129, 279–282.

Japon-Lujan, R., Luque de Castro, M.D., 2007. Static-dynamic superheated liquid extraction of hydroxytyrosol and other biophenols from alperujo (a semisolid residue of the olive oil industry). J. Agric. Food Chem. 55, 3629–3634.

Japón-Luján, R., Luque-Rodríguez, J.M., Luque de Castro, M.D., 2006. Multivariate optimisation of the microwave-assisted extraction of oleuropein and related biophenols from olive leaves. Anal. Bioanal. Chem. 385, 753–759.

Jiménez, A.N.A., Guillén, R., Fernández-Bolaños, J., Heredia, A., 1994. Cell wall composition of olives. J. Food Sci. 59, 1192–1196.

Jiménez, A., Rodríguez, R., Felizón, B., Fernández-Caro, I., Guillén, R., Fernández-Bolaños, J., Heredia, A., 2000. Cell wall polysaccharides implied in green olive behaviour during the pitting process. Eur. Food Res. Technol. 211, 181–184.

Karadag, A., Ozcelik, B., Saner, S., 2009. Review of methods to determine antioxidant capacities. Food Anal. Methods 2, 41–60.

Knupp, G., Rücker, G., Ramos-Cormenzana, A., Garrido Hoyos, S., Neugebauer, M., Ossenkop, T., 1996. Problems of identifying phenolic compounds during the microbial degradation of olive mill wastewater. Int. Biodeterior. Biodegradation 38, 277–282.

Leger, C.L., Kadiri-Hassani, N., Descomps, B., 2000. Decreased superoxide anion production in cultured human promonocyte cells (THP-1) due to polyphenol mixtures from olive oil processing wastewaters. J. Agric. Food Chem. 48, 5061–5067.

Lesage-Meessen, L., Navarro, D., Maunier, S., Sigoillot, J.C., Lorquin, J., Delattre, M., Simon, J.L., Asther, M., Labat, M., 2001. Simple phenolic content in olive oil residues as a function of extraction systems. Food Chem. 75, 501–507.

Liu, J., Schalch, W., Wang-Schmidt, Y., Wertz, K., 2008. Novel use of hydroxytyrosol and olive extracts/concentrates containing it. WO Patent App. WO/2008/128552.

Lo Scalzo, R., Scarpati, M.L., 1993. A new secoiridoid from olive wastewaters. J. Nat. Prod. 56, 621–623.

McDonald, S., Prenzler, P.D., Antolovich, M., Robards, K., 2001. Phenolic content and antioxidant activity of olive extracts. Food Chem. 73, 73–84.

Mudimu, O.A., Peters, M., Brauner, F., Braun, G., 2012. Overview of membrane processes for the recovery of polyphenols from olive mill wastewater. Am. J. Environ. Sci. 8, 195–201.

Mulinacci, N., Romani, A., Galardi, C., Pinelli, P., Giaccherini, C., Vincieri, F.F., 2001. Polyphenolic content in olive oil waste waters and related olive samples. J. Agric. Food Chem. 49, 3509–3514.

Niaounakis, M., Halvadakis, C.P., 2006. Olive Processing Waste Management: Literature Review and Patent Survey. Elsevier.

Obied, H.K., Allen, M.S., Bedgood, D.R., Prenzler, P.D., Robards, K., Stockmann, R., 2005. Bioactivity and analysis of biophenols recovered from olive mill waste. J. Agric. Food Chem. 53, 823–837.

Obied, H.K., Bedgood, Jr., D.R., Prenzler, P.D., Robards, K., 2007a. Chemical screening of olive biophenol extracts by hyphenated liquid chromatography. Anal. Chim. Acta. 603, 176–189.

Obied, H.K., Bedgood, Jr., D.R., Prenzler, P.D., Robards, K., 2007b. Bioscreening of Australian olive mill waste extracts: biophenol content, antioxidant, antimicrobial and molluscicidal activities. Food Chem. Toxicol. 45, 1238–1248.

Obied, H.K., Bedgood, D., Mailer, R., Prenzler, P.D., Robards, K., 2008. Impact of cultivar, harvesting time, and seasonal variation on the content of biophenols in olive mill waste. J. Agric. Food Chem. 56, 8851–8858.

Obied, H.K., Prenzler, P.D., Konczak, I., Rehman, A.U., Robards, K., 2009. Chemistry and bioactivity of olive biophenols in some antioxidant and antiproliferative in vitro bioassays. Chem. Res. Toxicol. 22, 227–234.

O'Sullivan, J., Murray, B., Flynn, C., Norton, I., 2016. The effect of ultrasound treatment on the structural, physical and emulsifying properties of animal and vegetable proteins. Food Hydrocolloid. 53, 141–154.

Paraskeva, P., Diamadopoulos, E., 2006. Technologies for olive mill wastewater (OMW) treatment: a review. J. Chem. Technol. Biotechnol. 81, 1475–1485.

Paraskeva, C.A., Papadakis, V.G., Tsarouchi, E., Kanellopoulou, D.G., Koutsoukos, P.G., 2007. Membrane processing for olive mill wastewater fractionation. Desalination 213, 218–229.

Patsioura, A., Galanakis, C.M., Gekas, V., 2011. Ultrafiltration optimization for the recovery of β-glucan from oat mill waste. J. Membr. Sci. 373, 53–63.

Petrotos, K.B., Goutsidis, P.E., Christodouloulis, K., 2010. Method of total discharge of olive mill vegetation waters with co-production of polyphenol powder and fertilizer. Greek Patent 1006660.

Pizzichini, M., Russo, C., 2005. Process for recovering the components of olive mill wastewater with membrane technologies. World Intellectual Property Organization, WO/2005/123603.

Prokopov, T., Goranova, Z., Baeva, M., Slavov, A., Galanakis Charis, M., 2015. Effects of powder from white cabbage outer leaves on sponge cake quality. Int. Agrophys. 29, 493–500.

Rahmanian, N., Jafari, S.M., Galanakis, C.M., 2014. Recovery and removal of phenolic compounds from olive mill wastewater. J. Am. Oil Chem. Soc. 91, 1–18.

Ramos, P., Santos, S.A.O., Guerra, Â.R., Guerreiro, O., Felício, L., Jerónimo, E., Silvestre, A.J.D., Neto, C.P., Duarte, M., 2013. Valorization of olive mill residues: Antioxidant and breast cancer antiproliferative activities of hydroxytyrosol-rich extracts derived from olive oil by-products. Ind. Crop. Prod. 46, 359–368.

Rodis, P.S., Karathanos, V.T., Mantzavinou, A., 2002. Partitioning of olive oil antioxidants between oil and water phases. J. Agric. Food Chem. 50, 596–601.

Rodriguez, G., Rodriguez, R., Jimenez, A., Guillen, R., Fernandez-Bolanos, J., 2007. Effect of steam treatment of alperujo on the composition, enzymatic saccharification, and in vitro digestibility of alperujo. J. Agric. Food Chem. 55, 136–142.

Rodríguez, R., Jiménez, A., Fernández-Bolaños, J., Guillén, R., Heredia, A., 2006. Dietary fibre from vegetable products as source of functional ingredients. Trends Food Sci. Technol. 17, 3–15.

Roselló-Soto, E., Barba, F.J., Parniakov, O., Galanakis, C.M., Lebovka, N., Grimi, N., Vorobiev, E., 2015a. High voltage electrical discharges, pulsed electric field, and ultrasound assisted extraction of protein and phenolic compounds from olive kernel. Food Bioprocess Technol. 8, 885–894.

Roselló-Soto, E., Galanakis, C.M., Brnčić, M., Orlien, V., Trujillo, F.J., Mawson, R., Knoerzer, K., Tiwari, B.K., Barba, F.J., 2015b. Clean recovery of antioxidant compounds from plant foods, by-products and algae assisted by ultrasounds processing. Modeling approaches to optimize processing conditions. Trends Food Sci. Technol. 42, 134–149.

Rubio-Senent, F., Rodríguez-Gutiérrez, G., Lama-Muñoz, A., Fernández-Bolaños, J., 2013. Phenolic extract obtained from steam-treated olive oil waste: characterization and antioxidant activity. LWT Food Sci. Technol. 54, 114–124.

Russo, C., 2007. A new membrane process for the selective fractionation and total recovery of polyphenols, water and organic substances from vegetation waters (VW). J. Membr. Sci. 288, 239–246.

Sabbah, I., Marsook, T., Basheer, S., 2004. The effect of pretreatment on anaerobic activity of olive mill wastewater using batch and continuous systems. Process Biochem. 39, 1947–1951.

Sahin, S., Samli, R., 2013. Optimization of olive leaf extract obtained by ultrasound-assisted extraction with response surface methodology. Ultrason. Sonochem. 20, 595–602.

Sannino, F., De Martino, A., Capasso, R., El Hadrami, I., 2013. Valorisation of organic matter in olive mill wastewaters: recovery of highly pure hydroxytyrosol. J. Geochem. Explor. 129, 34–39.

Serpil, T., Alper, K., 2009. Recovery of phenolic antioxidants from olive mill wastewater. Recent Patents on Chemical Engineering 2, 230–237.

Servili, M., Baldioli, M., Selvaggini, R., Macchioni, A., Montedoro, G., 1999. Phenolic compounds of olive fruit: one- and two-dimensional nuclear magnetic resonance characterization of Nuzhenide and its distribution in the constitutive parts of fruit. J. Agric. Food Chem. 47, 12–18.

Servili, M., Esposto, S., Veneziani, G., Urbani, S., Taticchi, A., Di Maio, I., Selvaggini, R., Sordini, B., Montedoro, G., 2011. Improvement of bioactive phenol content in virgin olive oil with an olive-vegetation water concentrate produced by membrane treatment. Food Chem. 124, 1308–1315.

Stavroulias, S., Panayiotou, C., 2005. Determination of optimum conditions for the extraction of squalene from olive pomace with supercritical CO_2. Chem. Biochem. Eng. Quart. 19, 373–381.

Stoller, M., 2008. Technical optimization of a dual ultrafiltration and nanofiltration pilot plant in batch operation by means of the critical flux theory: a case study. Chem. Eng. Process. 47, 1165–1170.

Taamalli, A., Arraez-Roman, D., Barrajon-Catalan, E., Ruiz-Torres, V., Perez-Sanchez, A., Herrero, M., Ibanez, E., Micol, V., Zarrouk, M., Segura-Carretero, A., Fernandez-Gutierrez, A., 2012. Use of advanced techniques for the extraction of phenolic compounds from Tunisian olive leaves: phenolic composition and cytotoxicity against human breast cancer cells. Food Chem. Toxicol. 50, 1817–1825.

Tornberg, E., Galanakis, C.M., 2008. Olive Waste Recovery. World Intellectual Property Organization, WO/2008/082343.

Tsakona, S., Galanakis, C.M., Gekas, V., 2010. Hydro-ethanolic mixtures for the recovery of phenols from Mediterranean plant materials. Food Bioprocess Technol. 5, 1384–1393.

Turano, E., Curcio, S., De Paola, M.G., Calabrò, V., Iorio, G., 2002. An integrated centrifugation–ultrafiltration system in the treatment of olive mill wastewater. J. Membr. Sci. 209, 519–531.

Valiente, C., Arrigoni, E., Esteban, R.M., Amadò, R., 1995. Chemical composition of olive by-product and modifications through enzymatic treatments. J. Sci. Food Agric. 69, 27–32.

Vierhuis, E., Korver, M., Schols, H.A., Voragen, A.G.J., 2003. Structural characteristics of pectic polysaccharides from olive fruit (*Olea europaea* cv moraiolo) in relation to processing for oil extraction. Carbohydr. Polym. 51, 135–148.

Villanova, L., Villanova, L., Fasiello, G., Merendino, A., 2006. Process for the recovery of tyrosol and hydroxytyrosol from oil mill wastewaters and catalytic oxidation method in order to convert tyrosol in hydroxytyrosol. Patent US 2006/0070953.

Vinciguerra, V., D'Annibale, A., Gàcs-Baitz, E., Monache, G.D., 1997. Biotransformation of tyrosol by whole-cell and cell-free preparation of Lentinus edodes. J. Mol. Catal. B Enzym. 3, 213–220.

Visioli, F., Romani, A., Mulinacci, N., Zarini, S., Conte, D., Vincieri, F.F., Galli, C., 1999. Antioxidant and other biological activities of olive mill waste waters. J. Agric. Food Chem. 47, 3397–3401.

Visioli, F., Caruso, D., Plasmati, E., Patelli, R., Mulinacci, N., Romani, A., Galli, G., Galli, C., 2001. Hydroxytyrosol, as a component of olive mill waste water, is dose- dependently absorbed and increases the antioxidant capacity of rat plasma. Free Radic. Res. 34, 301–305.

Visioli, F., Galli, C., Galli, G., Caruso, D., 2002. Biological activities and metabolic fate of olive oil phenols. Eur. J. Lipid Sci. Technol. 104, 677–684.

Vorobiev, E., Lebovka, N., 2010. Enhanced extraction from solid foods and biosuspensions by pulsed electrical energy. Food Eng. Rev. 2, 95–108.

Wang, L., Weller, C.L., 2006. Recent advances in extraction of nutraceuticals from plants. Trends Food Sci. Technol. 17, 300–312.

Wong, W.H., Lee, W.X., Ramanan, R.N., Tee, L.H., Kong, K.W., Galanakis, C.M., Sun, J., Prasad, K.N., 2015. Two level half factorial design for the extraction of phenolics, flavonoids and antioxidants recovery from palm kernel by-product. Ind. Crop. Prod. 63, 238–248.

Yang, C.S., Sang, S., Lambert, J.D., Lee, M.J., 2008. Bioavailability issues in studying the health effects of plant polyphenolic compounds. Mol. Nutr. Food Res. 52 (Suppl. 1), S139–S151.

Yangui, T., Sayadi, S., Rhouma, A., Dhouib, A., 2010. Potential use of hydroxytyrosol-rich extract from olive mill wastewater as a biological fungicide against *Botrytis cinerea* in tomato. J. Pest Sci. 83, 437–445.

Zagklis, D.P., Paraskeva, C.A., 2014. Membrane filtration of agro-industrial wastewaters and isolation of organic compounds with high added values. Water Sci. Technol. 69, 202–207.

Zagklis, D.P., Vavouraki, A.I., Kornaros, M.E., Paraskeva, C.A., 2015. Purification of olive mill wastewater phenols through membrane filtration and resin adsorption/desorption. J. Hazard. Mater. 285, 69–76.

Zinoviadou, K.G., Galanakis, C.M., Brnčić, M., Grimi, N., Boussetta, N., Mota, M.J., Saraiva, J.A., Patras, A., Tiwari, B., Barba, F.J., 2015. Fruit juice sonication: implications on food safety and physicochemical and nutritional properties. Food Res. Int. 77 (Part 4), 743–752.

Zirehpour, A., Rahimpour, A., Jahanshahi, M., Peyravi, M., 2014. Mixed matrix membrane application for olive oil wastewater treatment: process optimization based on Taguchi design method. J. Environ. Manag. 132, 113–120.

CHAPTER

APPLICATIONS OF RECOVERED BIOACTIVE COMPOUNDS IN FOOD PRODUCTS

11

Gianluca Veneziani*, Enrico Novelli, Sonia Esposto*, Agnese Taticchi*, Maurizio Servili***

**Department of Agricultural, Food and Environmental Sciences, University of Perugia, Perugia, Italy;*
***Department of Comparative Biomedicine and Food Science, University of Padova, Legnaro, Italy*

11.1 INTRODUCTION

The entire plant world is extremely rich in an extraordinary quantity and variety of molecules characterized by important biological activities (Manach et al., 2004; Shahidi, 1997). Every single part of a plant (roots, flowers, leaves, fruits, seeds, etc.) could represent an unexpected source of compounds, which can play an important role in the different sectors of human life. As a result, the majority of agrifood processing by-products contains substances featuring important biological properties (antimicrobial, antiviral, antioxidant, antiradical activity, antiinflammatory, antiatherosclerotic, chemo-preventive, anticancer, etc.). Subsequently, these substances should be recovered and used in different applications (Khan et al., 2014; Lou-Bonafonte et al., 2012; Fuccelli et al., 2014; Marimoutou et al., 2015; Pietta et al., 2003; Servili et al., 2014; Tomas-Menor et al., 2015; Wang et al., 2009). This viewpoint meets a double goal: to stimulate and promote the use of natural, bioactive compounds as well as valorize by-products in a way that guarantees a more sustainable agrifood production.

Phenols, the secondary metabolites of plants, include a broad range of very heterogeneous substances. However, they are all clearly identifiable by the presence of at least one aromatic ring with one or more hydroxyl substituent, and at the same time are characterized by different biological properties and potential health benefits. The abundance of these molecules in vegetables, fruits, olive oil, wine, and other food products, are one of the strong points of the Mediterranean diet. The latter is based on the large consumption of plant-food polyphenols, which is attributed to a reduced risk of chronic diseases, longevity, and human health (Bahadoran et al., 2013; Casaburi et al., 2013; Chirumbolo, 2014; Giacosa et al., 2013; Isemura and Timmermann, 2013; Iriti and Varoni, 2014; Laudadio et al., 2015; Medina-Remon et al., 2015; Pandey and Rizvi, 2009; Servili et al., 2014; Vasto et al., 2014).

One of the main food products of the Mediterranean diet is extra virgin olive oil (EVOO), which is the major source of fat. Its recognized, health-protective effects are ascribed not only to the high, polyunsaturated, fatty acid (PUFAs) content, but also to the minor, bioactive components, which give a significant, quantitative, and qualitative contribution to the daily intake of polyphenols in the traditional Mediterranean diet (Cases et al., 2015; Vasto et al., 2014).

Olive Mill Waste. http://dx.doi.org/10.1016/B978-0-12-805314-0.00011-X

The olive drupe contains 1–3% phenolic compounds of the fresh pulp weight (Garrido Fernández et al., 1997), but their concentration is strictly influenced by the cultivar and the related growing area. Other variables like pedoclimatic factors, fruit ripening, cultivation techniques, water resources, fertilization, and soil management have demonstrated an important impact, too (Inglese et al., 2011; Servili et al., 2004). The phenolic fraction consists of different groups: phenolic acids, phenolic alcohols, flavonoids, and secoiridoids (Rovellini et al., 1997; Servili et al., 2004, 2009; Tsimidou et al., 1996). The last group of chemical compounds is not only a feature of *Olearaceae* that includes *Olea europaea* (L.), but is also the most abundant, and is characterized by substances with high biological properties. Oleuropein, demethyloleuropein, ligstroside, and nüzhenide are the most abundant secoiridoid glucosides present in the olive fruit, which also contains the strictly cultivar dependent verbascoside, a derivative of the hydroxycinnamic acid. Other quantitatively and qualitatively relevant phenolic compounds are several aglycon derivatives of secoiridoids and the main phenyl alcohols hydroxytyrosol, (3,4-dihydroxyphenyl) ethanol (3,4 DHPEA) and tyrosol, *p*-(hydroxyphenyl) ethanol (*p*-HPEA). During the mechanical oil extraction process, the main secoiridoid glucosides are converted into aglycon derivatives, due to the hydrolytic reaction catalyzed by endogenous enzymes of the olive fruit (β-Glucosidases). The most abundant secoiridoid aglycons present in EVOO are the dialdehydic form of decarboxymethyl elenolic acid linked to 3,4-DHPEA or *p*-HPEA (3,4-DHPEA-EDA or *p*-HPEA-EDA), an isomer of oleuropein aglycon (3,4-DHPEA-EA) and ligstroside aglycon (*p*-HPEA-EA). The olive oil concentrations of the phenolic fraction are very variable with a range of 0.5–2%, highly dependent on genetic, agronomic, and technological aspects. In addition to secoiridoids, the other quantitatively significant group is represented by lignans, which include (+)-1-acetoxypinoresinol and (+)-1-pinoresinol (Brenes et al., 2000). However, the latter are characterized by a lower variability compared to the other phenolic compounds (Selvaggini et al., 2014; Veneziani et al., 2015). As highlighted earlier, only a small amount of phenolic compounds are transferred into the oily phase, whereas the largest fraction is present in the relevant by-products of the mechanical olive oil extraction process (Servili et al., 2011a), such as olive mill wastewater (OMWW). The OMWW is very rich in phenolic compounds (Araújo et al., 2015; Cardoso et al., 2011; Comandini et al., 2015; De Marco et al., 2007; Obied et al., 2007; Servili et al., 1999; Vougogiannopoulou et al., 2015), which are the main molecules present in the EVOO. They are characterized by high hydrophilicity and have significant biological properties (Alesci et al., 2014; Angelino et al., 2011; Araújo et al., 2015; Bitler et al., 2007; Cardinali et al., 2011; Lampronti et al., 2013; Obied et al., 2005; Pantano et al., 2016; Schaffer et al., 2010). The phenolic concentration of OMWWs, in the range of 0.5–24 g/L (Niaounakis and Halvadakis, 2004), is strictly related to several factors. The impact of these variables, assessed in some articles, also included the different olive oil extraction systems (Bianco et al., 2003; D'Antuono et al., 2014; Mulinacci et al., 2001; Obied et al., 2008a). Table 11.1 shows the main phenolic compounds found in OMWW, including the most abundant tyrosol, hydroxytyrosol, 3,4-DHPEA-EDA, and verbascoside, marked by a high antioxidant activity.

The technologies adopted in the olive oil mechanical extraction process are mainly referable to the three-phase and two-phase centrifugal decanters. The first system (still diffusely used in Italy and Greece) results in the production of olive oil, olive kernel, and a liquid by-product, commonly known as OMWW. The two-phase decanter (widely used in Spain) results in the production of olive oil and a thick, viscous by-product, the so-called wet olive pomace (or alperujo). Due to the large volume of water added for the malaxation of the olive paste, the three-phase unit is considered as a less competitive system as it generates huge amounts of OMWW that should be appropriately managed by the mill company (Uceda et al., 2006).

Table 11.1 The Main Phenolic Compounds Found in Olive Mill Wastewaters

Phenyl alcohols	*Secoiridoids*
Hydroxytyrosol (3,4-DHPEA)[a,b,c,d,e,f,g,h]	Demethyloleuropein[f,h]
Tyrosol (*p*-HPEA)[a,b,c,d,e,f,g,h]	Ligstroside[a,d]
	Oleuropein[a,c,e,f,g,h]
Phenolic acids/aldehydes	3,4-DHPEA-EDA[a,c,d,f,h]
Caffeic acid[a,b,c,d,e,f,g,h]	3,4-DHPEA-EA[a,b,c]
Cinnamic acid[a,b,f]	*p*-HPEA-EDA[a,c]
p-Cumaric acid[a,b,c,e,f,g]	
Elenolic acid[b]	*Phenylethanoid glycoside*
Ferulic acid[a,f,g]	Verbascoside[a,c,d,f,h]
Gallic acid[a,g]	
p-Hydroxybenzoic acid[b,f,g,h]	*Flavonoids*
p-Hydroxyphenylacetic acid[f,g]	Apigenin[a,f]
Protocatechuic acid[a,f]	Apigenin-7-*O*-glucoside[a,d,f]
Siringic acid[f,g]	Luteolin[a,d,f]
Vanillic acid[a,b,d,e,f,g,h]	Luteolin-7-*O*-glucoside[a,f,h]
Vanillin[a,b,g]	Quercetin[f]
	Rutin[a,c,e,f,h]
p-Cumaric acid derivative	
Comselogoside[a,c,e]	

3,4-DHPEA, (3,4-dihydroxyphenyl) ethanol; 3,4-DHPEA-EDA, dialdehydic form of decarboxymethyl elenolic acid linked to (3,4-dihydroxyphenyl) ethanol; p-*HPEA-EDA, dialdehydic form of decarboxymethyl elenolic acid linked to* (p-*hydroxypheny1) ethanol;* p-*HPEA,* (p-*hydroxyphenyl) ethanol; 3,4-DHPEA-EA, isomer of the oleuropein aglycon.*
[a]Araújo et al. (2015); [b]Comandini et al. (2015); [c]D'Antuono et al. (2014); [d]De Marco et al. (2007); [e]Obied et al. (2005, 2007); [f]Niaounakis and Halvadakis (2004); [g]Bianco et al. (2003); [h]Servili et al. (1999).

Both types of wastes have a dark color, with an acidic pH and a high concentration of salts and phenols. OMWW is an emulsion consisting of oil, mucilage, and pectin, with pH ranges from 4.5 to 6 (Hanifi and El-Hadrami, 2008; Saadi et al., 2007; Karpouzas et al., 2010). It contains 3–16% of organic compounds and is characterized by high values of biological oxygen demand (BOD) and chemical oxygen demand (COD), which range between 35 and 110 g/L and from 40 to 195 g/L, respectively (Servili et al., 2011a). These characteristics make OMWW a by-product with a high environmental impact. For example, it can inhibit seed germination, alter plant growth (Saez et al., 1992), affect the microbial diversity of the soil (Karpouzas et al., 2010) and modify the physical and chemical characteristics of the soil (Ben Sassi et al., 2006). The OMWW toxicity mechanism is still not completely clear. In vegetable cells they appear to exert a noncovalent membrane interaction (Greco et al., 2006). By means of structural changes induced in the inner mitochondrial membrane in the eukaryote cells, OMWW causes a decrease in the phosphorylation efficiency of mitochondria (Peixoto et al., 2008). The low pH and the osmotic stress caused by the high ionic strength may play a role in acute OMWW toxicity (Hanifi and El Hadrami, 2009). Yangui et al. (2010) showed the fungicidal action of hydroxytyrosol-rich OMWW extract not only on the soil-borne plant pathogen, *Verticillium dahliae,* but also against bacteria such as *P. syringae* and *X. campestris*. For each of the aforementioned microorganisms, the minimal inhibitory concentration ranged between 7.18 and 14.36 mg/L.

11.2 VALORIZATION OF OMWW FOR RECOVERY PURPOSES

OMWW represents a disposal cost for the oil industry, which should at least be limited to guarantee an improvement in the sector. In addition to the introduction of innovative technologies (Abenoza et al., 2013; Clodoveo et al., 2013; Jiménez et al., 2007; Leone et al., 2014; Puértolas and Martínez de Marañón, 2015; Veneziani et al., 2015) to further improve VOO quality, the new challenge of the olive oil industry is to exploit processing by-products and resolve the serious problem of OMWW management (Barbera et al., 2013; Fiorentino et al., 2003). Indeed, following the new trend "green food processing," the scientific world is cooperating with the olive oil sector, to develop protocols and technological innovations to turn by-products into high added-value products, rather than dealing with disposal cost. To this line, the high concentration of phenolic compounds contained in these by-products is a good starting point for the development of food ingredients with important biological activities, prior utilizing them in agri-food, nutrition and feeding sectors (Araújo et al., 2015; Carofiglio et al., 2015; De Leonardis et al., 2007; Dal Bosco et al., 2012; Esposto et al., 2015; Estaun et al., 2014; Fki et al., 2005; Gravador et al., 2015; Luciano et al., 2013; Mele et al., 2014; Novelli et al., 2014; Ragni et al., 2003; Servili et al., 2011a, 2011b; Troise et al., 2014; Tufarelli et al., 2013; Yangui et al., 2015).

11.3 RECOVERY AND FORMULATION OF HIGH ADDED-VALUE COMPOUNDS FROM OMWW

The liquid by-products of the mechanical olive oil extraction process can be used as source of different compounds, which can be further exploited to produce new, biologically active foods and/or ingredients. Some of which are characterized by antioxidant and antimicrobial properties that may have various applications in several sectors of food industries, replacing food additives of synthetic origin. For instance, some approaches suggest the combination of solvent extraction with membrane technologies to recover simultaneously two different phenol- and pectin-rich extracts (Galanakis et al., 2010a,b,c). Other phenol-rich concentrates could be obtained directly by membrane filtration system composed by a final treatment step with reverse osmosis (Servili et al., 2011a). The recovered and partially purified phenolic compounds can be stabilized by means of lyophilization and/or a spray-dried process to produce powder phenolic formulations. Fig. 11.1 explains the chromatograms of different phenolic formulation recovered by fresh vegetation water. The target compounds, mainly represented by 3,4-DHPEA-EDA, verbascoside, 3,4-DHPEA, and *p*-HPEA are identified by HPLC technique at a wavelength of 278 nm.

All forms of new ingredients obtained (liquid concentrates, extracts, lyophilized, and spray-dried forms) are characterized by a different purification grade accompanied with different times and conditions of storage, in terms of stability. The great variability that quantitatively and qualitatively affects

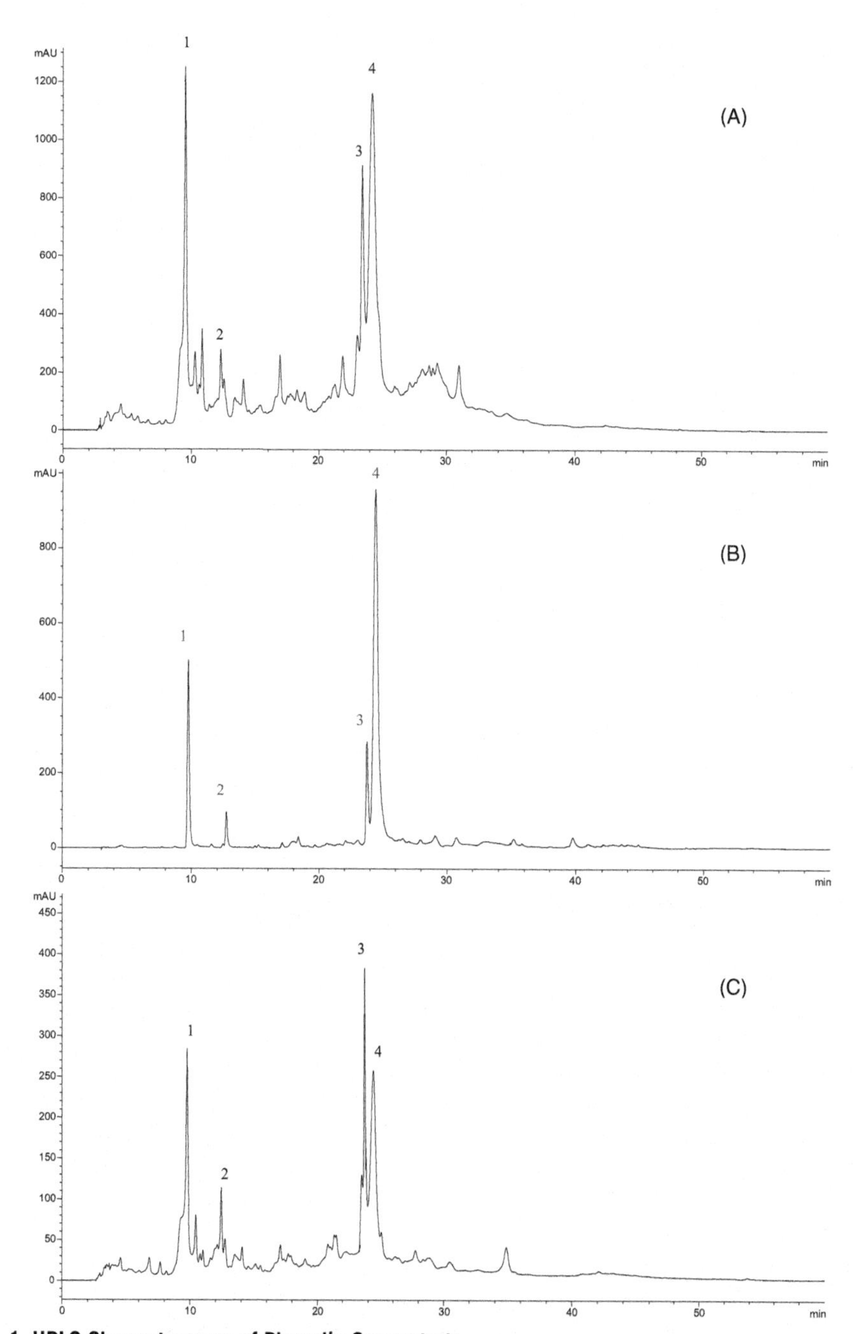

FIGURE 11.1 HPLC Chromatograms of Phenolic Concentrate

(A) Purified phenolic extract, (B) phenolic powder, and (C) recorded with DAD at 278 nm. *Peak numbers: 1*, Hydroxytyrosol (3,4-DHPEA); *2*, Tyrosol (*p*-HPEA); *3*, Verbascoside; *4*, Dialdehydic form of decarboxymethyl elenolic acid linked to 3,4-DHPEA (3,4-DHPEA-EDA).

the phenolic concentration leads to standardization problems for the new formulations. The cultivar, pedoclimatic conditions, fruit ripening, processing conditions, and hydrolysis degree of OMWW (linked to its freshness) are all variables affecting the different concentrations of the main phenolic compounds present in the final product. Subsequently, the new generated formulations may have different levels of biological activities as a result of the different quantities of specific phenolic compounds (Obied et al., 2005).

11.4 ANTIOXIDANT AND FUNCTIONAL APPLICATIONS IN FOODS

In recent years, the number of studies dealing with the reutilization of compounds (recovered from OMWW) in foodstuff has been increased (Galanakis, 2012; Rahmanian et al., 2014; Galanakis, 2015). These attempts try to utilize the bioactivities of phenolic compounds and the technological properties of other ones (e.g., gelling properties of pectin). For example, hydroxytyrosol, which is the most powerful antioxidant compound of OMWW (Angelino et al., 2011; Baldioli et al., 1996; Carrasco-Pancorbo et al., 2005), has shown advanced antiradical properties compared to vitamin E and C, and thus has been used to prevent oxidation of lipids in fish products (Fernández-Bolaños et al., 2006). Thereby, it could be used as functional supplement, food preservative in bakery products or life prolonging agents (Liu et al., 2008). In USA, olive pulp extracts have been approved by FDA with GRAS status (GRN No. 459) for being used as antioxidants in baked goods, beverages, cereals, sauces and dressings, seasonings, snacks, and functional foods at a level up to 3,000 mg/kg in the final food (FDA, 2014; Galanakis et al., 2015). The commercial implementation of OMWW to recover polyphenols is nowadays a reality (Galanakis, 2012; Rahmanian et al., 2014). Indeed, at least five companies around the world recover polyphenols from OMWW (Galanakis and Schieber, 2014) and sell them as natural preservatives or bioactive additives in food products.

11.4.1 ANTIOXIDANT PROPERTIES OF OMWW-PHENOLS

As widely described in Table 11.1, the OMWW is composed by several phenolic compounds characterized by different biological properties and a different level of antioxidant activity. The most abundant phenolic compounds: secoiridoids aglycon, verbascoside, and *phenyl alcohols* (tyrosol and hydroxytyrosol) are also the main molecules responsible of the high antioxidant properties of olive oil by-product. These compounds, the same present in the virgin olive oil, were extensively studied by many authors (Artajo et al., 2006; Baldioli et al., 1996; Carrasco-Pancorbo et al., 2005; Hassanzadeh et al., 2014; Pellegrini et al., 2001; Trujillo et al., 2006) that, evaluating the antioxidant activity and the related oxidative stability of olive oil with several tests (e.g., rancimat, DPPH (1,1,-diphenyl-2-picrylhydrazyl), OSI (oxidative stability instrument), flow injection analysis (FIA)—amperometry), highlighted the importance of *o*-diphenols pointing out the strongest antioxidant property of 3,4-DHPEA and oleuropein derivatives (3,4-DHPEA-EDA, and 3,4-DHPEA-EA).

Visioli et al. (1999) analyzed three phenolic extracts obtained from OMWW that showed a high antioxidant capacity compared to the most common antioxidants employed in foodstuff with a diverse degree of activity due to the different phenolic composition and degrees of purity. In another study, verbascoside and 3,4-dihydroxyphenylethyl alcohol–deacetoxyelenolic acid dialdehyde (3,4-DHPEA-DEDA) were indicated as the most effective antioxidant biophenols in OMWW Correggiola cultivars analyzing the values of radical-scavenging activity of DPPH reagent (Obied et al., 2008b). Also Angelino et al. (2011) showed the high antioxidant activity of a purified phytocomplex from OMWW obtained by olive oil mechanical extraction process of Leccino cultivar harvested at two different ripening stage. 3,4-DHPEA exhibited a greater antioxidant capacity, evaluated by oxygen radical absorbance capacity method, than 3,4-DHPEA-EDA, justified by a steric hindrance of the latter molecule during the reaction with the peroxyl radical. More recently a review article by Ciriminna et al. (2016) also focused the attention on the high antioxidant activity of phenolic compounds of OMWW with a particular attention to their impact on health benefit connected to food industry, pharmaceutical, and cosmetic sector.

11.4.2 FORTIFICATION OF OILS AND PRESERVATION OF FATS WITH POLYPHENOLS

One of the most significant bioactive properties of phenols is represented by their antioxidant effect, which has been exploited to preserve the stability of some vegetable and animal fats. The introduction of phenolic compounds in foodstuffs could represent an environmentally sustainable alternative to the use of synthetic antioxidants (Fki et al., 2005; Lafka et al., 2011), such as butylated hydroxyanisole (BHA, E-320), butylated hydroxytoluene (BHT, E-321), tert-butylhydroquinone (TBHQ, E-319), and propyl gallate (PG, E-310). The latest compounds are subjected to a maximum limit of concentrations in foodstuffs, due to their potential negative effect on human health (Araújo et al., 2015; EFSA, 2011, 2012, 2014, 2016).

Esposto et al. (2015) evaluated the effectiveness of phenolic extract (obtained by the purification of phenolic concentrate recovered from fresh OMWW) in controlling the oil stability during a simulated frying process (until 12 h at 180°C). The phenolic extract was added at different concentrations to a refined olive oil (100, 200, 400, and 1200 mg/kg of polyphenols) and its impact was also compared to a refined olive oil containing BHT and to an EVOO with a high content of polyphenols. This extract recovered from OMWW, mainly consisting of tyrosol, hydroxytyrosol, 3,4-DHPEA-EDA, and verbascoside, demonstrated a better capacity to preserve α-tocopherol content than BHT, reducing their oxidation if used at a concentration of at least 400 mg polyphenols/kg. The other significant effect concerned the reduction of negative volatile compounds in refined olive oil during the frying process, accompanied by a consequent improvement of the organoleptic quality of fried foodstuff (Fig. 11.2). These results demonstrated the possibility of using this phenolic formulation as a bioactive ingredient in oils before frying or cooking processes.

The efficacy of polyphenols (recovered from OMWW in a powder containing 10% polyphenols and 5% hydroxytyrosol) for the prevention of oil oxidation has also been investigated against other natural antioxidants (e.g., ascorbic acid, tocopherols mixture, and α-tocopherol). In particular, all antioxidants were mixed or emulsified at different concentrations (500–3000 mg/L) with an extra virgin and a refined kernel olive oil, prior heated in the oven at 100°C (30 min) and 160°C (120 min). The oxidation of treated oils was monitored by determining peroxide value, π-anisidine values, total polar components, "totox" index and extinction coefficient (K_{270}). The obtained results indicated that olive polyphenols in the lower and higher concentrations range (500 and 3000 mg/L, respectively) were able to reduce the oxidation of both heated oils. Ascorbic acid at higher concentrations (2000 and 3000 mg/L) was more efficient than the olive polyphenols, especially in the case of olive kernel oil. On the other hand, tocopherols formulation showed low effectiveness against most oxidation indexes, probably due to their instability at high cooking temperatures.

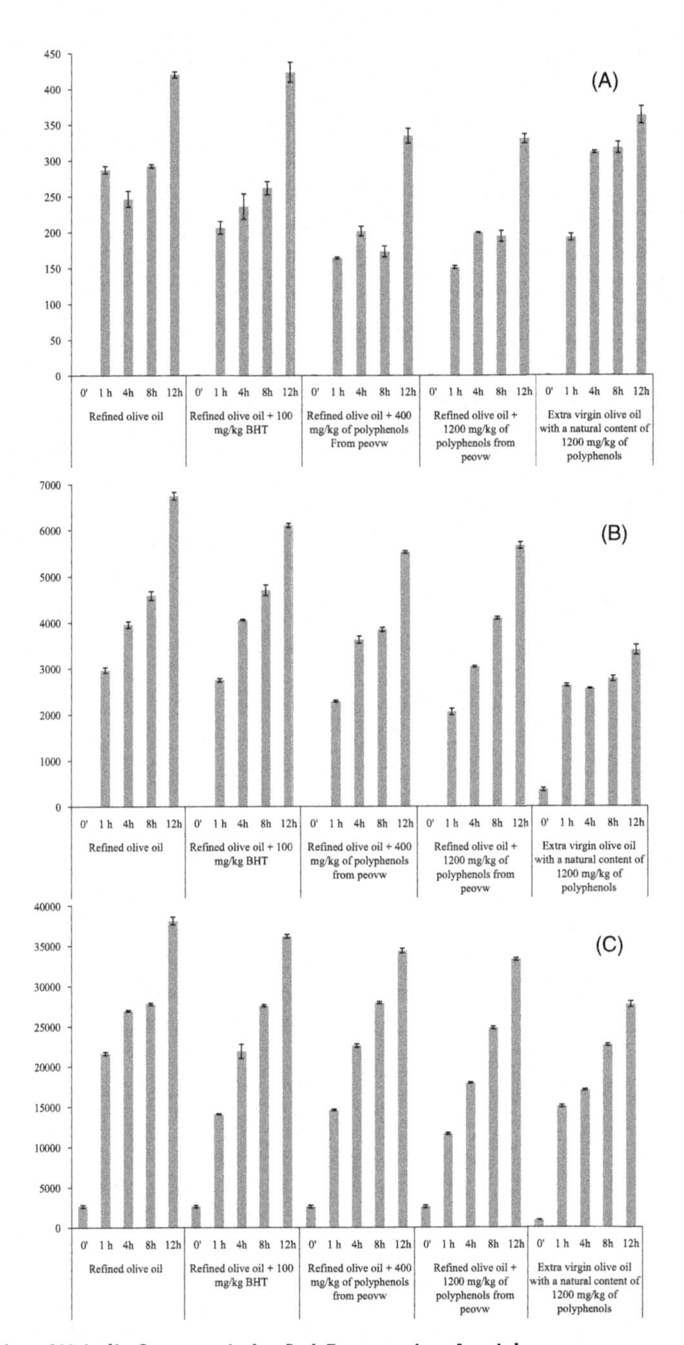

FIGURE 11.2 Evolution of Volatile Compounds (μg/kg) Expressed as Acrolein

(A) Saturated and unsaturated C_4—C_6 aldehydes, (B) saturated and unsaturated C_7—C_{11} aldehydes, and (C) in refined olive oil with the addition of 100 mg/kg of Butil Hidroxy toluene (BHT), 400 and 1200 mg/kg of polyphenols from OVWPE (phenolic extract recovered by olive vegetation water) and in 1200 mg/kg of naturally contained EVOO polyphenols during a simulated frying. Data are the mean of two independent experiments analyzed twice (Esposto et al., 2015).

In another study, natural antioxidants extracted from olive leaves were dissolved in ethanol/water and included in samples of sunflower oil, soybean oil, and their blend (1:1, w/w) to evaluate the improvement of oil stability during deep-fat frying (Zahran et al., 2015). The result showed a reduction in the changes in the physicochemical properties of all oil samples and explained how the phenolic extract at 1000 ppm can effectively stabilize unsaturated vegetable oils, compared to the BHT legal limit. The growing interest in natural antioxidant versus synthetic compounds was also confirmed by Yangui et al. (2015), who evaluated the great antioxidant activity of phenolic compound recovered from OMWW and used to enrich olive oil. In fact, the study highlighted its high free radical scavenging activity, which enables the formation of toxic complexes in oils and relative foodstuffs to be prevented. The potential capacity of oxidative stabilization due to phenolic compounds extracted by olive oil vegetation water was also tested on lard. The latter is an edible fat derived from pigs appreciated as an important ingredient for cooking (De Leonardis et al., 2007). The phenolic extract, mainly consisting of hydroxytyrosol and low amounts of tyrosol, caffeic acid, and ferulic acid, showed a positive impact on the peroxide value and oxidative resistances during lard storage, without any cytotoxic effect on cell growth at the applied doses (100–200 ppm). Another scientific work analyzed the antioxidant power of a phenolic extract obtained by olive pomace and other processing waste (peanut skins and pomegranate peels). This extract was added to ghee at a different concentration and compared with the application of a synthetic antioxidant (BHA). The ghee, a class of clarified butter, showed good oxidative stability as a result of phenolic treatment, whereas its major quality parameters usually alter during storage.

Some applications exploit not only the antioxidant properties of OMWW-phenols, but also other biological activities that are able to improve the health benefit and quality of foods. For instance, Servili et al. (2011a) used a crude phenolic concentrate (CPC) to increase the bioactive phenol content in virgin olive oil. CPC obtained by a filtration plant, equipped with three consecutive membrane-filtration steps with decreasing molecular cut-off values (microfiltration, 0.1–0.3 μm; ultrafiltration, 7 kDa; reverse osmosis). The phenolic fraction of reverse osmosis concentrate was mainly composed by 3,4-DHPEA-EDA and verbascoside with very low amounts of, 3,4-DHPEA and *p*-HPEA, as a consequence of the freshness of OMWW processed. CPC was added to the olive pastes with concentrations of 10 and 5% v/w at the beginning of malaxation phase to examine the phenolic increase in the corresponding VOOs. The results showed an increase for all four Italian cultivars processed (Peranzana, Ogliarola, Coratina, and Moraiolo) in the phenolic content ranging between 27% and 44% higher than the controls (Fig. 11.3). This referred in particular to the main antioxidants 3,4-DHPEA-EDA and 3,4-DHPEA-EA that are strictly connected to oxidative stability, sensory properties (pungency and bitterness), and human health of a VOO.

This study is very interesting, as it was one of the first reports (to our knowledge) concerning the direct use of phenolic compounds, derived from the use of OMWW in a food process to improve the health properties of the product. This new line of research was followed by other authors involved in the olive sector. For instance, Lalas et al. (2011) dealt with the enrichment of table olives with polyphenols, accompanied by the possibility of increasing their nutritional value and the bitter taste without altering product's acceptability.

11.4.3 FORTIFICATION OF MILK BEVERAGES WITH POLYPHENOLS

Milk beverages were fortified with phenolic compounds extracted from olive vegetation water and fermented with lactic acid bacteria (*Lactobacillus plantarum* and *Lactobacillus paracasei*) to produce

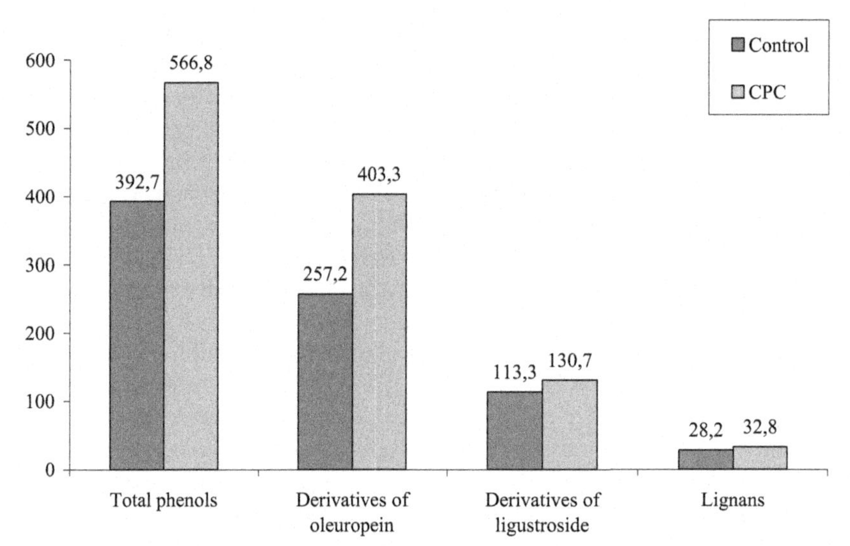

FIGURE 11.3 Phenolic Composition (mg/kg) of the EVOOs From Olive Pastes of cv. Moraiolo Malaxed With and Without the Crude Phenolic Concentrate (CPC) Addition

(Servili et al., 2011a)

a functional product. The phenolic fraction included in milk at different concentrations (100 and 200 mg/L) did not interfere with the fermentation process, and with the activities and survival of functional lactic acid bacteria (Servili et al., 2011b). The residual phenolic concentration after the production process was 77.2 and 172.5 mg/L, and 55.3 and 113.9 mg/L after 30 days of storage, respectively, for the two concentrations tested. All the phenolic compounds decreased, with the exception of hydroxytyrosol that increased, probably due to hydrolysis of 3,4-DHPEA-EDA. The presence of the phenolic fraction at the end of the storage period is another important opportunity to exploit OMWW and, most of all, the health benefits of these bioactive compounds that could be introduced into a human diet in the form of a functional milk beverage. In another scientific study, the impact of OMWW phenol compounds was evaluated on reactive carbonyl species and Maillard reaction end-products during UHT treatment of milk. The phenolic powder, added to milk at 0.1 and 0.05% w/v before UHT treatment, was able to control the Maillard reaction by trapping the reactive carbonyl species (hydroxycarbonyls and dicarbonyls), which led to Maillard-derived off-flavor formation. Phenolic compounds can be used as a functional ingredient to guarantee the improvement of nutritional and sensorial attributes of milk, without modifying its physical and sensory properties, even though the effects on the sensory acceptability of the product should be further investigated.

11.4.4 FAT REPLACEMENT WITH PECTIN DERIVED FROM OMWW IN MEATBALLS

Olive fruit contains an appreciable amount of fiber with promising functional properties like water holding and cation exchange capacity. The soluble polysaccharides present in OMWW are mainly composed of pectin material (Galanakis, 2011). Galanakis et al. (2010a) proposed a simplified method for the recovery of pectin from thermally concentrated OMWW. Extraction conducted using acid and solvent extraction, prior the isolation of the alcohol insoluble residue in a precipitation process. The water soluble fraction of the recovered material was able to form gels after a simple isolation and concentration procedure, despite the high ionic concentration, the low pectin content, and the high methylation degree (59%) of pectic polysaccharides. The method was further improved with regard to the role of endogenous pectin methyl esterase during thermal concentration step (Galanakis et al., 2010d) and utilized for the recovery of pectin containing material prior its application as fat replacement in meatballs (Galanakis et al., 2010c). According to the latter study, olive pectin was able to improve cooking properties of the product, by restricting the oil uptake during deep fat frying; thus giving rise to meatballs with sustained reduced fat content. Besides, this pectin containing extract was successfully clarified from the high ionic concentration using ultrafiltration, with the final purpose of optimizing the functional properties of pectin and improving the taste of the final product (Galanakis et al., 2010b).

11.5 ANTIMICROBIAL APPLICATIONS IN FOODS

11.5.1 ANTIMICROBIAL PROPERTIES OF OMWW-PHENOLS

Phenols of OMWW can be used as disinfectants in the food and chemical industries (Yangui et al., 2009) since they are known to exhibit antimicrobial action against *E. coli*, *P. aeruginosa*, *S. aureus*, and *B. subtilis* strains. This antimicrobial action is even greater than the respective activities induced by the individual phenolic compounds, indicating the synergistic action of the phenols contained in OMWW (Obied et al., 2007). For instance, singular phenolic compounds cannot inhibit the growth of the human pathogens *E. coli*, *K. pneumoniae*, *S.aureus*, and *S. pyogenes* when used at low concentrations, whereas OMWW is effective in the inhibition of Gram-positive and Gram-negative bacteria (Tafesh et al., 2011). In another study, Fasolato et al. (2015) tested a phenolic extract from OMWW using the minimum bactericidal concentration (MBC) on microtiter assay. The extract, containing 65% of phenols (mainly consisting of tyrosol, hydroxytyrosol, 3,4-DHPEA-EDA, and verbascoside) was recovered by applying three consecutive, membrane-filtration steps (with decreasing molecular cut-off values), and liquid–liquid extraction. The tested microorganisms were *Staphylococcus* (five strains), *Listeria* (four strains), *Escherichia* (two strains), *Salmonella* (one strain), *Pseudomonas* (three strains), *Lactobacillus* (two strains), and *Pediococcus* (one strain). *S. aureus* and *L. monocytogens* showed the lowest level of resistance to the phenolic extract (MBC = 1.5–3 mg/mL). In contrast, the Gram-negative strains (e.g., *S. typhimurium* and *Pseudomonas* spp.) were little or not affected by these levels of addition, with MBCs ranging between 6 and 12 mg/mL. Starter cultures were dramatically reduced in growth (e.g., *Staphylococcus xylosus*; 0.75 mg/mL MBC). As also shown in other studies (Tafesh et al., 2011), the maximal antimicrobial action of OMWW is performed by the natural mixture of phenols itself, instead of the single compounds. Taking into account the MBC results, the LAB, and *S. xylosus* were among the most sensitive bacteria.

In order to understand how the phenols act when come in contact with prokaryote cells, Carraro et al. (2014) investigated the primary, inhibitory effects of a purified extract of OMWW on the *E. coli* K-12 transcriptome. *E. coli* K-12 was chosen as a model organism, because it had been intensively investigated in relation to different stress responses (Jozefczuk et al., 2010). The concentration level used for the experiment (1 mg/mL) was much lower than the inhibition level shown in the in vitro test (6 mg/mL). According to the results, it showed a well-defined, inhibitory effect shortly after treatment, even though it did not affect *E. coli* growth on plates after 24 h. Genome-wide transcriptional analysis showed at least three effects on:

1. the involvement of genes linked to biofilm formation;
2. the repression of genes for flagellar synthesis; and
3. stress response.

Biofilm production is a stepwise process that begins when planktonic cells encounter a surface. Both cell surface adhesion and cell aggregation are essential to initiate bacterial biofilm formation (Zhang et al., 2007). The depressive effect of the purified extract from OMWW on the expression of several genes involved in the synthesis of fimbriae, curli, and exopolysaccharides suggests that phenols can modify the capacity of the bacteria to adhere on the surface. On the other hand, motility and chemotaxis are mechanisms used by bacteria in response to environmental stress. Exposure to the purified extract from OMWW caused a marked inhibition of the expression of the genes for flagellar synthesis and flagellar rotation. Since flagella are essential for both swimming and swarming, the results of the study of Carraro et al. (2014) showed that the concentration of 1 mg/mL was enough to significantly reduce swimming motility, but a higher concentration (at least 2 mg/mL) was required to significantly reduce swarming motility. The envelope is a known target for polyphenols. The purified extract from OMWW appears to be able to alter the envelope components by inducing an extracytoplasmic stress response. The upregulation of a gene that codes for (NADPH)-dependent reductase activity toward toxic aldehydes (Lee et al., 2010) means that purified extract OMWW causes oxidative stress, which promotes lipid peroxidation. The results shown by Carraro et al. (2014) agreed with the literature data reporting that, in addition to the protective effect against oxidative stress, polyphenols might act as prooxidants and induce DNA lesions. The manner of action on the cells (cytoprotection or cytotoxicity) has already been reported (Smirnova et al., 2009) and appears to depend on the cell type and the concentrations applied.

11.5.2 ANTIMICROBIAL PROPERTIES OF POLYPHENOLS IN FORTIFIED MILK BEVERAGES

In a functional milk beverage fortified with phenolic compounds, Servili et al. (2011b) did not find a significant reduction in LAB growth. Nevertheless, the level of inclusion was limited (100 and 200 mg/L). In other fermented products, the supplementation of higher levels of OMWW as a natural ingredient could reduce the performance or the quality of the end product. The MBC suggested that one strain of *Pediococcus pentosaceus* was able to grow at 12 mg/mL and, given that this microorganism has a widespread use in fermented food, the obtained data could be useful when scheduling the application of the purified extract from OMWW to food. These results highlighted the potential of the purified extract of OMWW to be used in food against some food-borne bacteria, especially against Gram-positive species. However, its potential antagonist effect on lactic acid bacteria that play a key role in the processing and preservation of fermented products must be taken into account. Therefore, phenols obtained from OMWW should be used after a preliminary check of the antibacterial spectrum (against pathogens, spoilers, and technological microorganisms). The aim would also be to define the optimal dose of use for the correct balance between the compatibility of their effects on the sensory properties of the foods (sometimes a slight modification of color and/or taste can be seen) and their preservative action.

11.5.3 ANTIMICROBIAL PROPERTIES OF POLYPHENOLS IN FORTIFIED BAKERY PRODUCTS

The efficacy of commercially available polyphenols recovered from OMWW (in a powder containing 10% polyphenols and 5% hydroxytyrosol) against other natural antioxidants (e.g., ascorbic acid, tocopherols mixture, and α-tocopherol) to inhibit microbial growth of bread and rusks during storage has been investigated (Galanakis, 2015b). Concentrations were selected in order to assure action against both oxidative deterioration and microbial spoilage. The produced bread and rusks were stored over a period of 20 days and 12 weeks, respectively, and assayed periodically to different microbiological assays (e.g., *Total coliforms*, *Yeasts-Moulds*, and *Bacillus spp.*). According to the results, antioxidants were able to induce antimicrobial properties in bakery products and subsequently prolong their shelf life. The optimal concentration of polyphenols was 200 mg/kg flour, as it extended the preservation of both bread and rusk samples. In addition, an emulsification of the powder enhanced the antimicrobial effect of polyphenols. Ascorbic acid and tocopherols mixture had no real effect to the overall bread preservation in both assayed concentrations (500 and 1000 mg/kg). Concerning rusk preservation, α-tocopherol, tocopherols mixture, and ascorbic acid were efficient at 600 mg/kg. The results of this study revealed the possibility of applying polyphenols from OMWW as an antmicrobial agent in bakery products.

11.5.4 ANTIMICROBIAL PROPERTIES OF POLYPHENOLS IN FORTIFIED MEAT PRODUCTS

L. monocytogenes is Gram-positive bacteria, able to cause food-borne disease especially in vulnerable people, such as the elderly, infants, pregnant women, and immunodeficient patients. Many kinds of food could be contaminated, due to the widespread diffusion of the bacteria, their ability to survive in the biofilm and grow at refrigeration temperatures. Ready-to-eat, with water activity ≥0.92 or pH ≥4.4, show a higher risk for *L. monocytogenes* than other foods. The antilisterial action of OMWW, previously verified in MBC trials, was tested in a challenge using salami as a food model. Three different batches were spiked with *L. monocytogenes* (at 100 CFU/g, as a mix of one commercial strain LMG 13303 NCTC 11994, together two wild type strains), one batch as control (ground meat, salt, and pepper), one as control plus nitrate (KNO_3 at 150 ppm) and another batch as control plus purified extract from OMWW (0.15%). In all the batches, a starter culture (the resistance of which to purified extract from OMWW had previously been verified) to corroborate the fermentation process was added. After an initial drying step, the salamis were ripened for up to 60 days. The water activity reached the critical value of 0.92 at 45 days of ripening, whereas the pH showed values between 5.2 (in the batches with phenols) and 5.7 (in the other batches) at 60 days of ripening. The addition of purified OMWW caused a substantial inhibition of *L. monocytogenes* growth, the values of which were always below 100 CFU/g (Fig. 11.4). The concentration of purified extract used in this preliminary trial did not determine any apparent sensorial modifications, regarding the flavor or the color of the salami (Fig. 11.5). This preliminary data gave useful information on the potential application of the purified extract from OMWW in the manufacturing of food to be commercialized as ready-to-eat, with the aim of improving food safety and public health.

In another study, the antimicrobial efficacy of a purified extract of OMWW for the purpose of improving the sensory characteristics and hygienic status of hamburgers made of white meat was investigated. Specifically, three doughs were prepared. One of which was the control, the second prepared

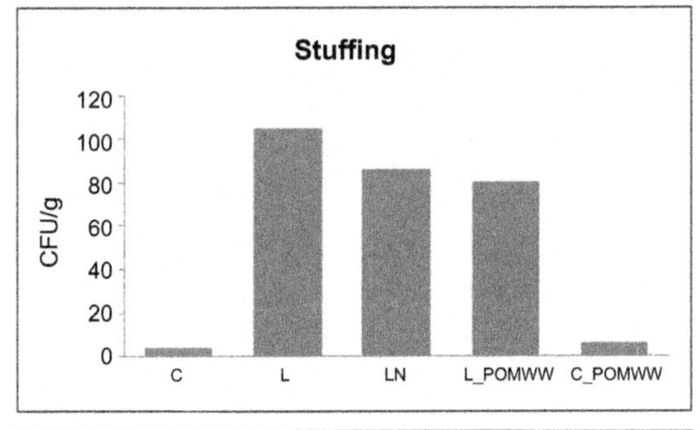

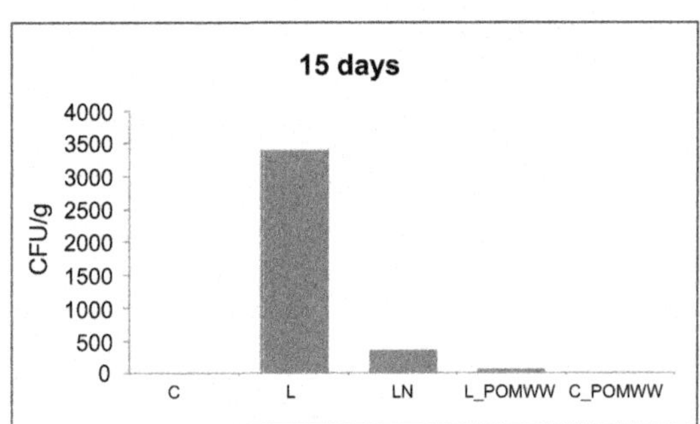

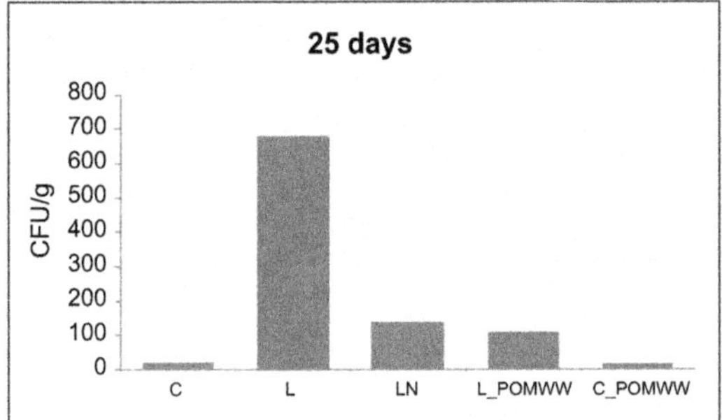

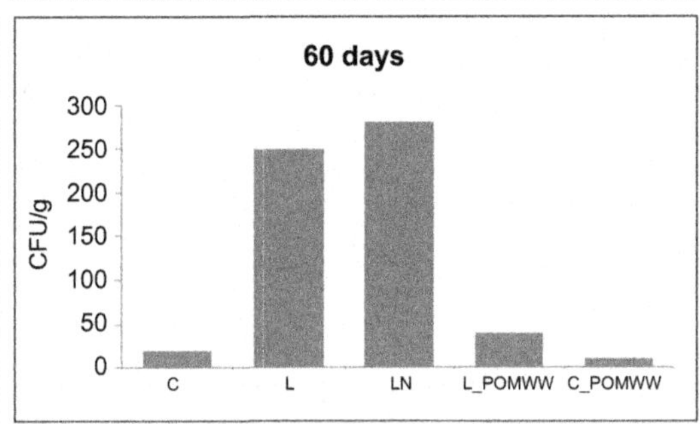

FIGURE 11.4 Challenge Test for *L. monocytogenes* in Salami

Control (C), Control + *L. monocytogenes* (L), Control + *L. monocytogenes* + nitrate (LN), Control + *L. monocytogenes* + purified OMWW (L_POMWW), Control + purified OMWW (C_POMWW).

with the addition of a purified extract of OMWW (0.75 g/kg of meat), and the third with the addition of a purified extract of OMWW at a higher concentration (1.50 g/kg of meat). No other ingredients or additives were added. Samples were packaged on a rigid tray, wrapped with PVC film (gas permeable) and stored at 4°C for up to 11 days. The dough containing added phenolic extract showed a certain delay in microbial growth when compared to the control sample. The effect was more evident in the dough at 1.5 g/kg compared to 0.75 g/kg (Fig. 11.6).

In practical terms, if the achievement of a bacterial load equal to 6 log can be considered the critical limit for the marketability of the product, it was noted that the dough at 1.5 g/kg reached 6 log a day later than the control one. Thus, it lengthens the self-life by 24 h, which is noteworthy from the commercial point of view. The sensory evaluation had almost always led to the discrimination of the hamburgers

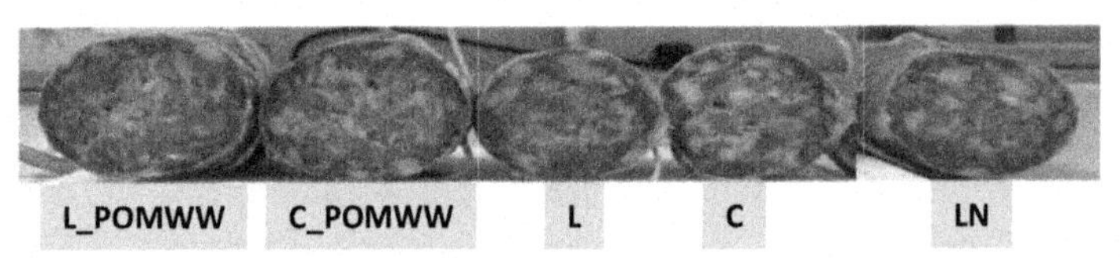

FIGURE 11.5 Challenge Test for *L. monocytogenes*, Salami at 60 days of Ripening

Control + *L. monocytogenes* + purified OMWW (L_POMWW), Control + purified OMWW (C_POMWW), Control + *L. monocytogenes* (L), Control (C), Control + *L. monocytogenes* + nitrate (LN).

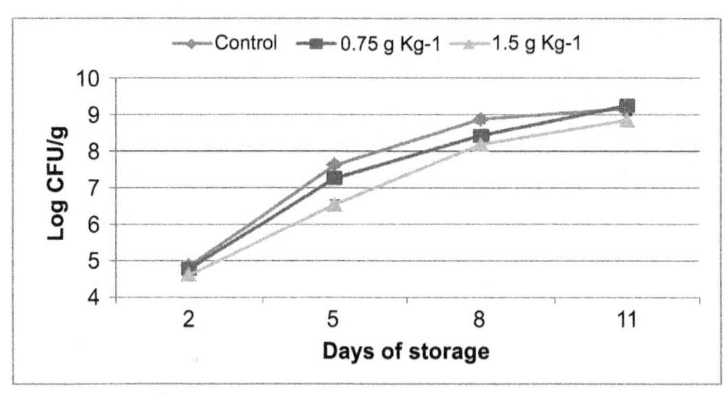

FIGURE 11.6 Hamburger of White Meat

Trial of addition of purified extract of OMWW at two final concentrations (0.75 g/kg and 1.5 g/kg). Total mesophilic count (log CFU/g, mean and standard deviation).

with added phenols compared to the control one, both in spite of color (darker in the case of the batch at 1.5 g/kg, Fig. 11.7) and sensory test, with particular reference to the flavor. The judges unanimously recognized a mild, but not unpleasant olive taste, differing a little from the control samples.

In order to test a natural alternative to counteract the growth of undesired moulds on the surface of fermented meat products, Chaves-López et al. (2015) dipped fresh sausages in a 2.5% aqueous solution of purified extract of OMWW for 1 min. During the following weeks of ripening, an effective reduction of fungal species, such as *Cladosporium cladosporioides*, *Penicillium aurantiogriseum*, *Penicillium commune*, and *Eurotium amstelodami* was shown in treated samples, when compared to the control batch, whereas the authors reported only a moderate antagonistic action over *Penicillium nalgiovense* and *Penicillium chrysogenum*. The same solution of phenols showed an effective inhibition of spore germination in all the

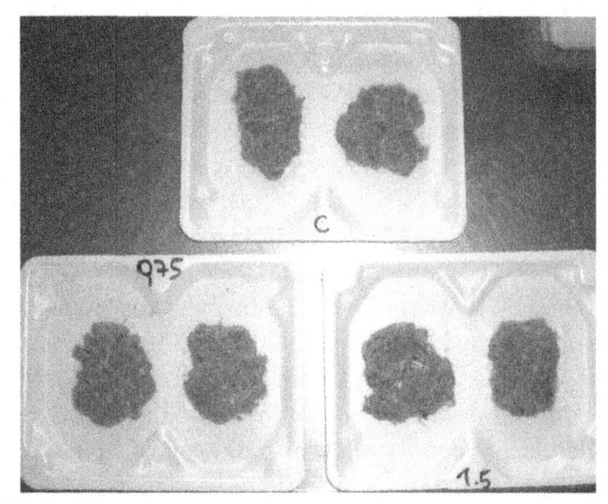

FIGURE 11.7 Hamburger of White Meat

Trial of addition of purified extract of OMWW at two final concentrations, 0.75 g/kg and 1.5 g/kg in comparison with a control batch (C).

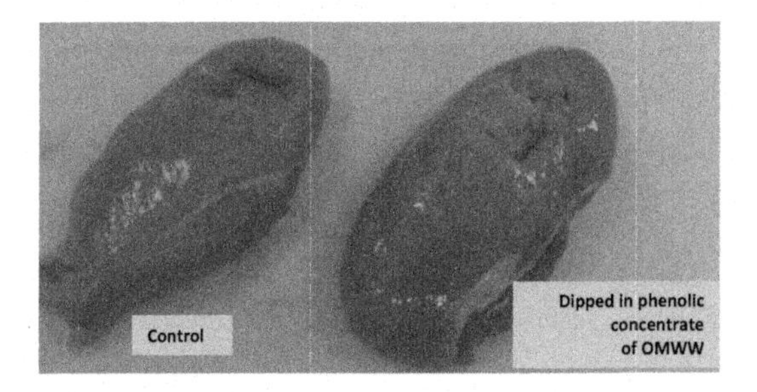

FIGURE 11.8

Chicken Breasts Submitted to Dipping into a Phenolic Concentrate of OMWW (right) and Control (left).

tested strains, with the exception of *A. parasiticus* and *A. flavus*. At the same time, a significant reduction in lipid oxidation and a moderate impact on sensory traits was shown without any undesired effects.

In another study, a phenolic concentrate from OMWW (Servili et al., 2011a) was used for the surface treatment of fresh breast of chicken in order to improve shelf-life at 4°C. The samples were dipped in the crude solution for 1 min., then wiped and packaged on a rigid tray and wrapped with a gas permeable pellicle (Fasolato et al., 2015). Either Enterobacteriaceae as *Pseudomonas* spp have shown a delay on growth in the dipped samples (OMWW) when compared to the control samples. The Enterobacteriaceae were nearly 4 log less at 7 days of storage in the treated samples, whereas *Pseudomonas* took 2 additional days to reach 7 log of CFU/g in the treated samples. The color of the surface of the dipped breasts tended toward yellow (Fig. 11.8) throughout the observational time and their odor was considered better in comparison to the control samples. The TBAR values were significantly lower in the samples subjected to dipping in the phenol solution.

11.6 ANTIMICROBIAL AND ANTIOXIDANT APPLICATIONS IN FEEDING

Gerasopoulos et al. (2015a) used a membrane retentate of OMWW as antioxidant ingredient for piglet feed (silage corn, formulated at a ratio corn: OMWW retentate 24:1 with a final OMWW retentate concentration in feed equal to 4%). The piglets fed for 30 days with OMWW retentate, showed a higher total antioxidant capacity, glutathione and catalase activity both in its blood and tissues, with lower oxidative and stress-induced damage to lipids and proteins, than the control group. With the same feed formula (silage corn: OMWW retentate or permeate 24:1), Gerasopoulos et al. (2015b) also tested the antioxidant action of the retentate and permeate of OMWW in broiler chickens, fed for 37 days. The groups of animals fed with the OMWW supplement showed lower values of protein and lipid oxidation and a higher total, antioxidant capacity in their plasma and tissues compared to the control group.

In another approach, a preliminary trial was conducted to test the antioxidant and antimicrobial effect of a crude phenolic concentrate from OMWW (obtained according to Servili et al., 2011a) as an ingredient in the formulation of feed for the finishing step of heavy pigs. The level of addition (~300 mg/kg feed, intended as total phenols and not as a solution as such, the concentration of which in the finished feed was 1.5%) was lower than concentration used by Gerasopoulos et al. (2015b) in the experiment with piglets. The crude phenolic concentrate had a phenol composition similar to that of the purified extract (tyrosol, hydroxytyrosol, 3,4-DHPEA-EDA, and verbascoside) with a concentration of approximately 2–3%. The loin muscle of the treated and control groups were subjected to a shelf-life trial after packaging in a protective atmosphere, using rigid trays and gas impermeable film (2 weeks of observational period at 4°C alternating exposure to darkness and light every 12 h). Surprisingly, the bacterial load with specific regard to *Pseudomonas* spp braked in the group of pigs that received the crude phenolic concentrate through the feedstuff, gaining at least 2 days in terms of shelf-life.

11.7 CONCLUSIONS

The scientific world and the whole olive oil sector are cooperating to valorize the OMWW with the common goal to reduce the by-product's environmental impact and to resolve the problem of its disposal cost for oil industry. The chemical composition of vegetation water, mainly for that concern the quantitative and qualitative phenolic concentration, was a good starting point toward an unavoidable and indispensable "green food processing." The well-known antioxidant properties of the main phenolic compounds of OMWW were utilized in different food applications with significant results: fortification of oils, preservation of fats, and production of functional beverages. On the other hand, the antimicrobial action of the phenols, which nevertheless exists and is widely referred in the literature through in vitro studies, it requires a greater enhancement in the knowledge also by tests that must be conducted directly on the food. The phenols obtained from the treatment and purification of the OMWW, can be directly added to the food products during their processing or that could be adsorbed on the packaging material (active packaging) can be a green and sustainable approach to ensure the food safety and the food quality. Up to now (June 2016), the olive polyphenols market is in its nascent stage, as the number of extraction plants is low and the demand for olive polyphenols has been limited by reduced market offer. However, this state of the art is about to change if more applications and sustainable solutions will be explored.

REFERENCES

Abenoza, M., Benito, M., Saldaña, G., Álvarez, I., Raso, J., Sánchez-Gimeno, A.C., 2013. Effects of pulsed electric field on yield extraction and quality of olive oil. Food Bioprocess Technol. 6, 1367–1373.

Alesci, A., Cicero, N., Salvo, A., Palombieri, D., Zaccone, D., Dugo, G., Bruno, M., Vadala, R., Lauriano, E.R., Pergolizzi, S., 2014. Extracts deriving from olive mill waste water and their effects on the liver of the goldfish Carassius auratus fed with hypercholesterolemic diet. Nat. Prod. Res. 28 (17), 1343–1349.

Angelino, D., Gennari, L., Blasa, M., Selvaggini, R., Urbani, S., Esposto, S., Servili, M., Ninfali, P., 2011. Chemical and cellular antioxidant activity of phytochemicals purified from olive mill waste waters. J. Agric. Food Chem. 59, 2011–2018.

Araújo, M., Pimentel, F.B., Alves, R.C., Oliveira, M.B.P.P., 2015. Phenolic compounds from olive mill wastes: health effects, analytical approach and application as food antioxidants. Trends Food Sci. Technol. 45 (2), 200–211.

Artajo, L.S., Romero, M.P., Morello, J.R., Motilva, M.J., 2006. Enrichment of refined olive oil with phenolic compounds: evaluation of their antioxidant activity and their effect on the bitter index. J. Agric. Food Chem. 54, 6079–6088.

Bahadoran, Z., Mirmiran, P., Azizi, F., 2013. Dietary polyphenols as potential nutraceuticals in management of diabetes: a review. J. Diabetes Metab. Disord. 12, 43–51.

Baldioli, M., Servili, M., Perretti, G., Montedoro, G.F., 1996. Antioxidant activity of tocopherols and phenolic compounds of virgin olive oil. J. Am. Oil Chem. Soc. 73, 1589–1593.

Barbera, A.C., Maucieri, C., Cavallaro, V., Ioppolo, A., Spagna, G., 2013. Effects of spreading olive mill wastewater on soil properties and crops, a review. Agric. Water Manag. 119, 43–53.

Ben Sassi, A., Boularbah, A., Jaouad, A., Walker, G., Boussaid, A., 2006. A comparison of olive oil mill wastewaters (OMW) from three different processes in Morocco. Process Biochem. 41 (1), 74–78.

Bianco, A., Buiarelli, F., Cartoni, G., Coccioli, F., Jasionowska, R., Margherita, P., 2003. Analysis by liquid chromatography-tandem mass spectrometry of biophenolic compounds in olives and vegetation waters, Part I. J. Sep. Sci. 26 (5), 409–416.

Bitler, C.M., Matt, K., Irving, M., Hook, G., Yusen, J., Eagar, F., Kirschner, K., Walker, B., Crea, R., 2007. Olive extract supplement decreases pain and improves daily activities in adults with osteoarthritis and decreases plasma homocysteine in those with rheumatoid arthritis. Nutr. Res. 27 (8), 470–477.

Brenes, M., Hidalgo, F.J., García, A., Rios, J.J., García, P., Zamora, R., Garrido, A., 2000. Pinoresinol and 1-acetoxypinoresinol, two new phenolic compounds identified in olive oil. J. Am. Oil Chem. Soc. 77, 715–720.

Cardinali, A., Linsalata, V., Lattanzio, V., Ferruzzi, M.G., 2011. Verbascosides from olive mill waste water: assessment of their bioaccessibility and intestinal uptake using an in vitro digestion/Caco-2 model system. J. Food Sci. 76 (2), H48–H54.

Cardoso, S.M., Falcão, S.I., Peres, A.M., Domingues, M.R.M., 2011. Oleuropein/ligstroside isomers and their derivatives in Portuguese olive mill wastewaters. Food Chem. 129 (2), 291–296.

Carofiglio, V.E., Romano, R., Servili, M., Goffredo, A., Alifano, P., Veneziani, G., Demitri, C., Centrone, D., Stufano, P., 2015. Complete valorization of olive mill wastewater through an integrated process for poly-3-hydroxybutyrate production. J. Life Sci. 9, 481–493.

Carraro, L., Fasolato, L., Montemurro, F., Martino, M.E., Balzan, S., Servili, M., Novelli, E., Cardazzo, B., 2014. Polyphenols from olive mill waste affect biofilm formation and motility in *Escherichia coli* K-12. Microb. Biotechnol. 7 (3), 265–275.

Carrasco-Pancorbo, A., Cerretani, L., Bendini, A., Segura-Carretero, A., del Carlo, M., Gallina-Toschi, T., Lercker, G., Compagnone, D., Fernández-Gutiérrez, A., 2005. Evaluation of the antioxidant capacity of individual phenolic compounds in virgin olive oil. J. Agric. Food Chem. 53, 8918–8925.

Casaburi, I., Puoci, F., Chimento, A., Sirianni, R., Ruggiero, C., Avena, P., Pezzi, V., 2013. Potential of olive oil phenols as chemopreventive and therapeutic agents against cancer: a review of in vitro studies. Mol. Nutr. Food Res. 57, 71–83.

Cases, J., Romain, C., Dallas, C., Gerbi, A., Cloarec, M., 2015. Regular consumption of Fiit-ns, a polyphenol extract from fruit and vegetables frequently consumed within the Mediterranean diet, improves metabolic ageing of obese volunteers: a randomized, double-blind, parallel trial. Int. J. Food Sci. Nutr. 66 (1), 120–125.

Ciriminna, R., Meneguzzo, F., Fidalgo, A., Ilharco, L.M., Pagliaro, M., 2016. Extraction, benefits and valorization of olive polyphenols. Eur. J. Lipid Sci. Technol. 118, 503–511.

Clodoveo, M.L., Durante, V., La Notte, D., 2013. Working towards the development of innovative ultrasound equipment for the extraction of virgin olive oil. Ultrason. Sonochem. 20, 1261–1270.

Chaves-López, C., Serio, A., Mazzarrino, G., Martuscelli, M., Scarpone, E., Paparella, A., 2015. Control of household mycoflora in fermented sausages using phenolic fractions from olive mill wastewaters. Int. J. Food Microbiol. 207, 49–56.

Chirumbolo, S., 2014. Dietary assumption of plant polyphenols and prevention of allergy. Curr. Pharm. Des. 20 (6), 811–839.

Comandini, P., Lerma-Garcia, M.J., Massanova, P., Simo-Alfonso, E.F., Toschi, T.G., 2015. Phenolic profiles of olive mill wastewaters treated by membrane filtration systems. J. Chem. Technol. Biotechnol. 90 (6), 1086–1093.

Dal Bosco, A., Mourvaki, E., Cardinali, R., Servili, M., Sebastiani, B., Ruggeri, S., Mattioli, S., Taticchi, A., Esposto, S., Castellini, C., 2012. Effect of dietary supplementation with olive pomaces on the performance and meat quality of growing rabbits. Meat Sci. 92, 783–788.

D'Antuono, I., Kontogianni, V.G., Kotsiou, K., Linsalata, V., Logrieco, A.F., Tasioula-Margari, M., Cardinali, A., 2014. Polyphenolic characterization of olive mill wastewaters, coming from Italian and Greek olive cultivars, after membrane technology. Food Res. Int. 65 (Part C), 301–310.

De Leonardis, A., Macciola, V., Lembo, G., Aretini, A., Ahindra, Nag., 2007. Studies on oxidative stabilisation of lard by natural antioxidants recovered from olive-oil mill wastewater. Food Chem. 100 (3), 998–1004.

De Marco, E., Savarese, M., Paduano, A., Sacchi, R., 2007. Characterization and fractionation of phenolic compounds extracted from olive oil mill wastewaters. Food Chem. 104 (2), 858–867.

EFSA, 2011. Scientific Opinion on the re-evaluation of butylated hydroxyanisole—BHA (E 320) as a food additive. EFSA J. 9 (10), 2392–2441.

EFSA, 2012. Scientific opinion on the re-evaluation of butylated hydroxytoluene BHT (E 321) as a food additive. EFSA J. 10 (3), 2588–2631.

EFSA, 2014. Scientific opinion on the re-evaluation of propyl gallate (E 310) as a food additive. EFSA J. 12 (4), 3642–3688.

EFSA, 2016. Statement on the refined exposure assessment of tertiary-butyl hydroquinone (E 319). EFSA J. 14 (1), 4363–4389.

Esposto, S., Taticchi, A., Di Maio, I., Urbani, S., Veneziani, G., Selvaggini, R., Sordini, B., Servili, M., 2015. Effect of an olive phenolic extract on the quality of vegetable oils during frying. Food Chem. 176, 184–192.

Estaun, J., Dosil, J., Al-Alami, A., Gimeno, A., De Vega, A., 2014. Effects of including olive cake in the diet on performance and rumen function of beef cattle. Anim. Prod. Sci. 54 (10), 1817–1821.

Fasolato, L., Cardazzo, B., Balzan, S., Carraro, L., Taticchi, A., Montemurro, F., Novelli, E., 2015. Minimum bactericidal concentration of phenols extracted from oil vegetation water on spoilers, starters and food-borne bacteria. Ital. J. Food Saf. 4, 75–77.

Fernández-Bolaños, J., Rodríguez, G., Rodríguez, R., Guillén, R., Jiménez, A., 2006. Extraction of interesting organic compounds from olive oil waste. Grasas Aceites 57, 95–106.

Food and Drug Administration (FDA), 2014. Available from: http://www.accessdata.fda.gov/scripts/fdcc/?set=GRASNotices&id=459

Fki, I., Allouche, N., Sayadi, S., 2005. The use of polyphenolic extract, purified hydroxytyrosol and 3,4-dihydroxyphenyl acetic acid from olive mill wastewater for the stabilization of refined oils: a potential alternative to synthetic antioxidants. Food Chem. 93 (2), 197–204.

Fiorentino, A., Gentili, A., Isidori, M., Monaco, P., Nardelli, A., Parrella, A., Temussi, F., 2003. Environmental effects caused by olive mill wastewaters: toxicity comparison of low-molecular-weight phenol components. J. Agric. Food Chem. 51 (4), 1005–1009.

Fuccelli, R., Sepporta, M.V., Rosignoli, P., Morozzi, G., Servili, M., Fabiani, R., 2014. Preventive activity of olive oil phenolic compounds on alkene epoxides induced oxidative DNA damage on human peripheral blood mononuclear cells. Nutr. Cancer 66 (8), 1322–1330.

Galanakis, C.M., Tornberg, E., Gekas, V., 2010a. A study of the recovery of the dietary fibres from olive mill wastewater and the gelling ability of the soluble fibre fraction. LWT Food Sci. Technol. 43, 1009–1017.

Galanakis, C.M., Tornberg, E., Gekas, V., 2010b. Clarification of high-added value products from olive mill wastewater. J. Food Eng. 99, 190–197.

Galanakis, C.M., Tornberg, E., Gekas, V., 2010c. Dietary fiber suspensions from olive mill wastewater as potential fat replacements in meatballs. LWT Food Sci. Technol. 43, 1018–1025.

Galanakis, C.M., Tornberg, E., Gekas, V., 2010d. The effect of heat processing on the functional properties of pectin contained in olive mill wastewater. LWT Food Sci. Technol. 43, 1001–1008.

Galanakis, C.M., 2011. Olive fruit and dietary fibers: components, recovery and applications. Trends Food Sci. Technol. 22, 175–184.

Galanakis, C.M., 2012. Recovery of high added-value components from food wastes: conventional, emerging technologies and commercialized applications. Trends Food Sci. Technol. 26, 68–87.

Galanakis, C.M., Schieber, A., 2014. Editorial. Special issue on recovery and utilization of valuable compounds from food processing by-products. Food Res. Int. 65, 299–300.

Galanakis, C.M., Martínez-Saez, N., del Castillo, M.D., Barba, F.J., Mitropoulou, V.S., 2015. Patented and commercialized applications. In: Galanakis, C.M. (Ed.), Food Waste Recovery: Processing Technologies and Industrial Techniques. Elsevier-Academic Press, USA, (Chapter 15).

Galanakis, C.M., 2015. Final report of "A study for the implementation of polyphenols from olive mill wastewater in foodstuff and cosmetics" (in Greek). Project "ΜΕΠΑΕ", General Secretariat for Research and Technology (GSRT) (Athens, Athens, Greece).

Garrido Fernández, A., Díez, M.J., Adams, M.R., 1997. In Table Olives: Production and Processing. Chapman & Hall, London.

Gerasopoulos, K., Stagos, D., Kokkas, S., Petrotos, K., Kantas, D., Goulas, P., Kouretas, D., 2015a. Feed supplemented with byproducts from olive oil mill wastewater processing increases antioxidant capacity in broiler chickens. Food Chem. Toxicol. 82, 42–49.

Gerasopoulos, K., Stagos, D., Petrotos, K., Kokkas, S., Kantas, D., Goulas, P., Kouretas, D., 2015b. Feed supplemented with polyphenolic byproduct from olive mill wastewater processing improves the redox status in blood and tissues of piglets. Food Chem. Toxicol. 86, 319–327.

Giacosa, A., Barale, R., Bavaresco, L., Gatenby, P., Gerbi, V., Janssens, J., Johnston, B., Kas, K., Vecchia, C., la Mainguet, P., Morazzoni, P., Negri, E., Pelucchi, C., Pezzotti, M., Rondanelli, M., 2013. Cancer prevention in Europe: the Mediterranean diet as a protective choice. Eur. J. Cancer Prev. 22 (1), 90–95.

Gravador, R.S., Serra, A., Luciano, G., Pennisi, P., Vasta, V., Mele, M., Pauselli, M., Priolo, A., 2015. Volatiles in raw and cooked meat from lambs fed olive cake and linseed. Animal 9 (4), 715–722.

Greco, Jr., G., Colarieti, M.L., Toscano, G., Iamarino, G., Rao, M.A., Gianfreda, L., 2006. Mitigation of olive mill wastewater toxicity. J. Agric. Food Chem. 54 (18), 6776–6782.

Hanifi, S., El-Hadrami, I., 2008. Phytotoxicity and fertilising potential of olive mill wastewaters for maize cultivation. Agron. Sustain. Dev. 28 (2), 313–319.

Hanifi, S., El Hadrami, I., 2009. Olive mill wastewaters: diversity of the fatal product in olive oil industry and its valorisation as agronomical amendment of poor soils: a review. J. Agron. 8 (1), 1–13.

Hassanzadeh, K., Keivan Akhtari, K., Hassanzadeh, H., Zarei, S.A., Fakhraei, N., Hassanzadeh, K., 2014. The role of structural CAH compared with phenolic OH sites on the antioxidant activity of oleuropein and its derivatives as a great non-flavonoid family of the olive components: a DFT study. Food Chem. 164, 251–258.

Karpouzas, D.G., Ntougias, S., Iskidou, E., Rousidou, C., Papadopoulou, K.K., Zervakis, G.I., Ehaliotis, C., 2010. Olive mill wastewater affects the structure of soil bacterial communities. Appl. Soil Ecol. 45 (2), 101–111.

Khan, H.Y., Hazeeb Zubair Faisal, M., Ullah, M.F., Mohd Farhan Sarkar, F.H., Ahmad, A., Hadi, S.M., 2014. Plant polyphenol induced cell death in human cancer cells involves mobilization of intracellular copper ions and reactive oxygen species generation: a mechanism for cancer chemopreventive action. Mol. Nutr. Food Res. 58 (3), 437–446.

Jiménez, A., Beltrán, G., Uceda, M., 2007. High-power ultrasound in olive paste pretreatment. Effect on process yield and virgin olive oil characteristics. Ultrason. Sonochem. 14, 725–731.

Jozefczuk, S., Klie, S., Catchpole, G., Szymanski, J., Cuadros-Inostroza, A., Steinhauser, D., Selbig, J., Willmitzer, L., 2010. Metabolomic and transcriptomic stress response of *Escherichia coli*. Mol. Syst. Biol. 6, 364.

Inglese, P., Famiani, F., Galvano, F., Servili, M., Esposto, S., Urbani, S., 2011. Factors affecting extra-virgin olive oil composition. Hortic. Rev. 38, 83–147.

Iriti, M., Varoni, E.M., 2014. Cardioprotective effects of moderate red wine consumption: polyphenols vs. ethanol. J. Appl. Biomed. 12 (4), 193–202.

Isemura, M., Timmermann, B.N., 2013. Special issue: plant polyphenols and health benefits. Curr. Pharm. Des. 19 (34), 6051–6225.

Lafka, T.I., Lazou, A.E., Sinanoglou, V.J., Lazos, E.S., 2011. Phenolic and antioxidant potential of olive oil mill wastes. Food Chem. 125 (1), 92–98.

Lalas, S., Athanasiadis, V., Gortzi, O., Bounitsi, M., Giovanoudis, I., Tsaknis, J., Bogiatzis, F., 2011. Enrichment of table olives with polyphenols extracted from olive leaves. Food Chem. 127 (4), 1521–1525.

Lampronti, I., Borgatti, M., Vertuani, S., Manfredini, S., Gambari, R., 2013. Modulation of the expression of the proinflammatory IL-8 gene in cystic fibrosis cells by extracts deriving from olive mill waste water. Evid. Based Complement. Alternat. Med. 2013, 11.

Laudadio, V., Ceci, E., Lastella, N.M.B., Tufarelli, V., 2015. Dietary high-polyphenols extra-virgin olive oil is effective in reducing cholesterol content in eggs. Lipids Health Dis. 14, 1–7.

Lee, C., Kim, I., Lee, J., Lee, K.L., Min, B., Park, C., 2010. Transcriptional activation of the aldehyde reductase YqhD by YqhC and its implication in glyoxal metabolism of *Escherichia coli* K-12. J. Bacteriol. 192, 4205–4214.

Leone, A., Tamborrino, A., Romaniello, R., Zagaria, R., Sabella, E., 2014. Specification and implementation of a continuous microwave assisted system for paste malaxation in an olive oil extraction plant. Biosyst. Eng. 125, 24–35.

Liu, J., Schalch, W., Wang-Schmidt, Y., Wertz, K., 2008. Novel use of hydroxytyrosol and olive extracts/concentrates containing it. WO Patent App. WO/2008/128552.

Lou-Bonafonte, J.M., Arnal, C., Navarro, M.A., Osada, J., 2012. Efficacy of bioactive compounds from extra virgin olive oil to modulate atherosclerosis development. Mol. Nutr. Food Res. 56 (7), 1043–1057.

Luciano, G., Pauselli, M., Servili, M., Mourvaki, E., Serra, A., Monahan, F.J., Lanza, M., Priolo, A., Zinnai, A., Mele, M., 2013. Dietary olive cake reduces the oxidation of lipids, including cholesterol, in lamb meat enriched in polyunsaturated fatty acids. Meat Sci. 93 (3), 703–714.

Manach, C., Scalbert, A., Morand, C., Rémésy, C., Jiménez, L., 2004. Polyphenols: food sources and bioavailability. Am. J. Clin. Nutr. 79, 727–747.

Marimoutou, M., Sage, F., le Smadja, J., d'Hellencourt, C.L., Gonthier, M.P., Robert-da Silva, C., 2015. Antioxidant polyphenol-rich extracts from the medicinal plants Antirhea borbonica, Doratoxylon apetalum and Gouania mauritiana protect 3T3-L1 preadipocytes against H2O2, TNF alpha and LPS inflammatory mediators by regulating the expression of superoxide dismutase and NF- kappa B genes. J. Inflam. 12 (10), 1–15.

Medina-Remon, A., Tresserra-Rimbau, A., Pons, A., Tur, J.A., Martorell, M., Ros, E., Buil-Cosiales, P., Sacanella, E., Covas, M.I., Corella, D., Salas-Salvado, J., Gomez-Gracia, E., Ruiz-Gutierrez, V., Ortega-Calvo, M., Garcia-Valdueza, M., Aros, F., Saez, G.T., Serra-Majem, L., Moreno Esteban, B., Lezcano Solis, D.A., 2015. Olive oil, a cornerstone of the Mediterranean diet. Olivae 121, 19–26.

Mele, M., Serra, A., Pauselli, M., Luciano, G., Lanza, M., Pennisi, P., Conte, G., Taticchi, A., Esposto, S., Morbidini, L., 2014. The use of stoned olive cake and rolled linseed in the diet of intensively reared lambs: effect on the intramuscular fatty-acid composition. Animal 8 (1), 152–162.

Mulinacci, N., Romani, A., Galardi, C., Pinelli, P., Giaccherini, C., Vincieri, F.F., 2001. Polyphenolic content in olive oil waste waters and related olive samples. J. Agric. Food Chem. 49 (8), 3510–3514.

Novelli, E., Fasolato, L., Cardazzo, B., Carraro, L., Taticchi, A., Balzan, S., 2014. Addition of phenols compounds to meat dough intended for salami manufacture and its antioxidant effect. Ital. J. Food Saf. 3, 154–156.

Niaounakis, M., Halvadakis, C.P., 2004. The Olive-Mill Waste Management: Literature Review and Patent Survey. Typothito-George Dardanos, Athens.

Obied, H.K., Allen, M.S., Bedgood, D.R., Prenzler, P.D., Robards, K., Stockmann, R., 2005. Bioactivity and analysis of biophenols recovered from olive mill waste. J. Agric. Food Chem. 53 (4), 823–837.

Obied, H.K., Karuso, P., Prenzler, P.D., Robards, K., 2007. Novel secoiridoids with antioxidant activity from Australian olive mill waste. J. Agric. Food Chem. 55 (8), 2848–2853.

Obied, H.K., Bedgood, D., Mailer, R., Prenzler, P.D., Robards, K., 2008a. Impact of cultivar, harvesting time, and seasonal variation on the content of biophenols in olive mill waste. J. Agric. Food Chem. 56 (19), 8851–8858.

Obied, H.K., Prenzler, P.D., Robards, K., 2008b. Potent antioxidant biophenols from olive mill waste. Food Chem. 111 (1), 171–178.

Pantano, D., Luccarini, I., Nardiello, P., Servili, M., Stefani, M., Casamenti, F., 2016. Oleuropein aglycone and polyphenols from olive mill waste water ameliorate cognitive deficits and neuropathology. Br. J. Clin. Pharmacol. doi: 10.1111/bcp.12993.

Pellegrini, N., Visioli, F., Buratti, S., Brighenti, F., 2001. Direct analysis of total antioxidant activity of olive oil and studies on the influence of heating. J. Agric. Food Chem. 49, 2532–2538.

Peixoto, F., Martins, F., Amaral, C., Gomes-Laranjo, J., Almeida, J., Palmeira, C.M., 2008. Evaluation of olive oil mill wastewater toxicity on themitochondrial bioenergetics after treatment with Candida oleophila. Ecotoxicol. Environ. Saf. 70 (2), 266–275.

Pandey, K.B., Rizvi, S.I., 2009. Plant polyphenols as dietary antioxidants in human health and disease. Oxid. Med. Cell. Longev. 2 (5), 270–278.

Pietta, P., Minoggio, M., Bramati, L., 2003. Plant polyphenols: structure, occurrence and bioactivity. Atta-ur-Rahman (Ed.), Studies in Natural Products Chemistry, vol. 28, Elsevier Science, Amsterdam, pp. 257–353.

Puértolas, E., Martínez de Marañón, I., 2015. Olive oil pilot-production assisted by pulsed electric field: impact on extraction yield, chemical parameters and sensory properties. Food Chem. 167, 497–502.

Ragni, M., Melodia, L., Bozzo, F., Colonna, M., A. Megna, V., Toteda, F., Vicenti, A., 2003. Use of a de-stoned olive pomace in feed for heavy lamb production. Ital. J. Anim. Sci. 2, 485–487.

Rahmanian, N., Jafari, S.M., Galanakis, C.M., 2014. Recovery and removal of phenolic compounds from olive mill wastewater. J. Am. Oil Chem. Soc. 91, 1–18.

Rovellini, P., Cortesi, N., Fedeli, E., 1997. Analysis of flavonoids from Olea europaea by HPLC-UV and HPLC-electrospray-MS. Riv. Ital. Sostanze Gr. 74, 273–279.

Saadi, I., Laor, Y., Raviv, M., Medina, S., 2007. Land spreading of olive mill wastewater: effects on soil microbial activity and potential phytotoxicity. Chemosphere 66 (1), 75–83.

Saez, L., Perez, J., Martinez, J., 1992. Low molecular weight phenolics attenuation during simulated treatment of wastewaters from olive oil mills in evaporation ponds. Water Res. 26 (9), 1261–1266.

Schaffer, S., Muller, W.E., Eckert, G.P., 2010. Cytoprotective effects of olive mill wastewater extract and its main constituent hydroxytyrosol in PC12 cells. Pharmacol. Res. 62 (4), 322–327.

Selvaggini, R., Esposto, S., Taticchi, A., Urbani, S., Veneziani, G., Di Maio, I., Sordini, B., Servili, M., 2014. Optimization of the temperature and oxygen concentration conditions in the malaxation during the oil mechanical extraction process of four Italian olive cultivars. J. Agric. Food Chem. 62, 3813–3822.

Servili, M., Baldioli, M., Selvaggini, R., Miniati, E., Macchioni, A., Montedoro, G.F., 1999. High-performance liquid chromatography evaluation of phenols in olive fruit, virgin olive oil, vegetation waters and pomace and 1D- and 2D-nuclear magnetic resonance characterization. J. Am. Oil Chem. Soc. 76, 873–882.

Servili, M., Selvaggini, R., Esposto, S., Taticchi, A., Montedoro, G., Morozzi, G., 2004. Health and sensory properties of virgin olive oil hydrophilic phenols: agronomic and technological aspects of production that affect their occurrence in the oil. J. Chromatogr. 1054, 113–127.

Servili, M., Esposto, S., Fabiani, R., Urbani, S., Taticchi, A., Mariucci, F., Selvaggini, R., Montedoro, G.F., 2009. Phenolic compounds in olive oil: antioxidant, health and organoleptic activities according to their chemical structure. Inflammopharmacology 17, 1–9.

Servili, M., Esposto, S., Veneziani, G., Urbani, S., Taticchi, A., Di Maio, I., Selvaggini, R., Sordini, B., Montedoro, G.F., 2011a. Improvement of bioactive phenol content in virgin olive oil with an olive-vegetation water concentrate produced by membrane treatment. Food Chem. 124 (4), 1308–1315.

Servili, M., Rizzello, C.G., Taticchi, A., Esposto, S., Urbani, S., Mazzacane, F., Di Maio, I., Selvaggini, R., Gobbetti, M., Di Cagno, R., 2011b. Functional milk beverage fortified with phenolic compounds extracted from olive vegetation water, and fermented with c-aminobutyric acid (GABA)- producing and potential probiotic lactic acid bacteria. Int. J. Food Microbiol. 147, 45–52.

Servili, M., Sordini, B., Esposto, S., Urbani, S., Veneziani, G., Di Maio, I., Selvaggini, R., Taticchi, A., 2014. Biological activities of phenolic compounds of extra virgin olive oil. Antioxidants 3, 1–23.

Shahidi, F., 1997. Natural antioxidants: chemistry, health effects and applications. AOCS press, Champaign.

Smirnova, G.V., Samoylova, Z.Y., Muzyka, N.G., Oktyabrsky, O.N., 2009. Influence of polyphenols on *Escherichia coli* resistance to oxidative stress. Free Rad. Biol. Med. 46, 759–768.

Tafesh, A., Najami, N., Jadoun, J., Halahlih, F., Riepl, H., Azaizeh, H., 2011. Synergistic antibacterial effects of polyphenolic compounds from olive mill wastewater. Evid. Based Complement. Alternat. Med. 2011, 9.

Tomas-Menor, L., Barrajon-Catalan, E., Segura-Carretero, A., Marti, N., Saura, D., Menendez, J.A., Joven, J., Micol, V., 2015. The promiscuous and synergic molecular interaction of polyphenols in bactericidal activity: an opportunity to improve the performance of antibiotics? Phytother. Res. 29 (3), 466–473.

Troise, A.D., Fiore, A., Colantuono, A., Kokkinidou, S., Peterson, D.G., Fogliano, V., 2014. Effect of olive mill wastewater phenol compounds on reactive carbonyl species and Maillard reaction end-products in ultrahigh-temperature-treated milk. J. Agric. Food Chem. 62 (41), 10092–10100.

Trujillo, M., Mateos, R., Collantesd, T., Espartero, J.L., Cert, R., Jover, M., Alcudia, F., Bautista, J., Cert, A., Parrado, J., 2006. Lipophilic hydroxytyrosyl esters. Antioxidant activity in lipid matrices and biological systems. J. Agric. Food Chem. 54, 3779–3785.

Tsimidou, M., Lytridou, M., Boskou, D., Pappa-Louisi, A., Kotsifaki, F., Petrakis, C., 1996. On the determination of minor phenolic acids of virgin olive oil by RP-HPLC. Grasas Aceites 47, 151–157.

Tufarelli, V., Introna, M., Cazzato, E., Mazzei, D., Laudadio, V., 2013. Suitability of partly destoned exhausted olive cake as by-product feed ingredient for lamb production. J. Anim. Sci. 91 (2), 872–877.

Uceda, M., Jiménez, A., Beltràn, G., 2006. Olive oil extraction and quality. Grasas Aceites 57 (1), 25–31.

Vasto, S., Buscemi, S., Barera, A., Carlo, M., di Accardi, G., Caruso, C., 2014. Mediterranean diet and healthy ageing: a Sicilian perspective. Gerontology 60 (6), 508–518.

Veneziani, G., Esposto, S., Taticchi, A., Selvaggini, R., Urbani, S., Di Maio, I., Sordini, B., Servili, M., 2015. Flash thermal conditioning of olive pastes during the oil mechanical extraction process: cultivar impact on the phenolic and volatile composition of virgin olive oil. J. Agric. Food Chem. 63, 6066–6074.

Visioli, F., Romani, A., Mulinacci, N., Zarini, S., Conte, D., Vincieri, F.F., Galli, C., 1999. Antioxidant and other biological activities of olive mill waste waters. J. Agric. Food Chem. 47, 3397–3401.

Vougogiannopoulou, K., Angelopoulou, M.T., Pratsinis, H., Grougnet, R., Halabalaki, M., Kletsas, D., Deguin, B., Skaltsounis, L.A., 2015. Chemical and biological investigation of olive mill waste water—OMWW secoiridoid lactones. Planta Med. 81 (12/13), 1205–1212.

Wang, G.-F., Shi, L.-P., Ren, Y.-D., Liu, Q.-F., Liu, H.-F., Zhang, R.-J., Li, Z., Zhu, F.-H., He, P.-L., Tang, W., Tao, P.-Z., Li, C., Zhao, W.-M., Zuo, J.-P., 2009. Anti-hepatitis B virus activity of chlorogenic acid, quinic acid and caffeic acid in vivo and in vitro. Antiviral Res. 83 (2), 186–190.

Yangui, T., Dhouib, A., Rhouma, A., Sayadi, S., 2009. Potential of hydroxytyrosol-rich composition from olive mill wastewater as a natural disinfectant and its effect on seeds vigour response. Food Chem. 117 (1), 1–8.

Yangui, T., Sayadi, S., Gargoubi, A., Dhouib, A., 2010. Fungicidal effect of hydroxytyrosol-rich preparations from olive mill wastewater against *Verticillium dahliae*. Crop Protect. 29 (10), 1208–1213.

Yangui, A., Abessi, M.H., Abderrabba, M., 2015. Antioxidant activity of olive mill wastewater extracts and its use as an effective antioxidant in olive oil; kinetic approach. J. Chem. Pharm. Res. 7 (3), 171–177.

Zahran, H.A., El-Kalyoubi, M.H., Khallaf, M.M., Abdel-Razek, A.G., 2015. Improving oils stability during deep-fat frying using natural antioxidants extracted from olive leaves using different methods. Middle East J. Appl. Sci. 5, 26–38.

Zhang, X.S., García-Contreras, R., Wood, T.K., 2007. YcfR (BhsA) influences *Escherichia coli* biofilm formation through stress response and surface hydrophobicity. J. Bacteriol. 189, 3051–3062.

CHAPTER

APPLICATIONS OF RECOVERED BIOACTIVE COMPOUNDS IN COSMETICS AND HEALTH CARE PRODUCTS

12

Francisca Rodrigues, Maria Antónia da Mota Nunes, Maria Beatriz Prior Pinto Oliveira
LAQV/Requimte, Faculty of Pharmacy, University of Porto, Rua Jorge Viterbo Ferreira, Porto, Portugal

12.1 INTRODUCTION

The global population has reached 7 billion and is projected to reach 9 billion by 2050 (Sahota, 2014). There is a general agreement that the planet resources cannot support with such arise in human population, especially at actual consumption rates (Sahota, 2014). Rising education levels and the World Wide Web are responsible for more refined consumers, leading them to questioning product origins, production methods, and ecological implications, as well as safety issues. A major impact on the cosmetics industry occurred due to this rise in ethical consumerism. A core component of industry responsibility is its commitment to sustainable development. Over the past decades, however, demand for natural resources has increased and that it is now widely considered a serious threat to the well-functioning of economies and societies due to the link to environmental problems, such as climate change, biodiversity loss, desertification, and ecosystem degradation (Behrens et al., 2007). Nowadays, the increasing ecoconsumerism focused end-user attention on all aspects of products from raw material origin, manufacturing, and leads to the disposal beauty products sector to be accused of numerous environmentally unfriendly practices (McPhee and Jain, 2015). According to Cosmetics Europe—The Personal Care Association, 450 million of Europeans daily use a variety of cosmetic products, such as soap, shampoo, hair conditioner, toothpaste, deodorant, shaving cream, skin care, perfume, and make-up (COLIPA, 2015). According to several studies, at consumer level the highest environmental impact of cosmetic products ranged from unsustainable sourcing of raw materials, pollution both in the manufacturing phase and disposal of packaging and products, to animal testing (McPhee and Jain, 2015). In response, companies have attempted to address these issues holistically by focusing on sustainable raw material sourcing and greener formulations, as well as through packaging reductions (McPhee and Jain, 2015). A widely accepted definition of the Brundtland Commission of the United Nations refers to sustainability in the concept of sustainable development: "meeting the needs of the present without compromising the ability of future generations to meet their own needs" (Drexhage and Murphy, 2010).

However, these laudable practices face both consumer skepticism that often views these activities as "green washing" and the reality of the market, where in a global economy, there is little appetite for premium pricing to cover the increased costs of these products compared to traditional one (McPhee and Jain, 2015).

Olive Mill Waste. http://dx.doi.org/10.1016/B978-0-12-805314-0.00012-1

Cosmetic application could be a solution to reuse by-products discarded by several agro industries. As it is well known, human skin is a complex organ that regulates body heat and water loss, while preventing the entry of toxic substances and microorganisms (Rodrigues et al., 2015c). Natural ingredients, phytonutrients, microbial metabolites, dairy derived actives, minerals, and animal protein components are believed to benefit healthy skin ageing (Prakash and Majeed, 2009). Olive oil has been used on the skin for thousands of years but most of the mechanisms underlying these beneficial effects remain unclear (Badiu et al., 2010). According to Badiu et al. (2010) olive oil has antioxidant properties and contains several essential fatty acids required for the production of phospholipids, being the case of alpha-linolenic acid and gamma-linolenic acid. However, few studies reported the possible uses of agroindustrial wastes on cosmetic industry (Martinez-Saez et al., 2014; Mussatto et al., 2012). Indeed, from a sustainable point of view, this new application could provide, in the near future, a way of recover added value products for olive companies, developing cost-effective processing methods, decreasing the negative impacts of wastes on the environment, and providing other economical advantages for companies. But, how could these new ingredients act on skin? Which compounds of OMW could be reused? Are there any recent studies on this field? Is it possible to develop new cosmetic ingredients based on agroindustrial by-products? With growing scarcity of resources and rising ethical consumerism, how can the food industry, associated with the cosmetic industry, become more sustainable? What are the best ways to improve the use of this food by-product? Are they safe? The aim of this chapter is to address such questions and evaluate OMW possible application as active ingredients in the cosmetic field. In this review, the challenging applications of OMW as active ingredients for skin care products were discussed, analyzing the main compounds responsible for their activity, and highlighting their effects on skin.

12.2 SKIN STRUCTURE

The human skin is the largest organ of human body, forming the outermost biological barrier between the human body and the external environment. Among the multiple, complex functions of mammalian skin, one of its major roles is to prevent invasion of the organism, by acting as a defensive barrier to threats from the external environment (Prow et al., 2011). In the 1960s the tremendous influence of the thickness of the *stratum corneum* (SC) became obvious (Schäfer-Korting et al., 2007). It consists of different layers exhibiting individual composition and physiology (Franzen and Windbergs, 2015). The barrier function of the skin is realized by a cornified layer of protein-rich dead cells (corneocytes) embedded in a lipid matrix, namely SC (Mathes et al., 2014). The corneocytes are held together by corneodesmosomes, which help to form a tough outer layer by maintaining cellular shape and regular packing (Prow et al., 2011). The lipid-enriched component consists predominantly of ceramides (about 50% by weight), as well as cholesterol and free fatty acids (Schurer and Elias, 1991). This layer is formed at the end of a balanced differentiation process, beginning at the basal layer of the epidermis, where progenitor cells divide and newly emerged keratinocytes are pushed toward the apical side of the epidermis (Mathes et al., 2014). Keratinocytes undergo a process of keratinization, in which the cell differentiates and moves upward from the basal layer (*stratum basale*), through the *stratum spinosum* and *stratum granulosum*, to the outermost layer, the SC, where they anucleate and flattened and are eventually sloughed off (Prow et al., 2011). Interspersed among the keratinocytes in the viable epidermis are cells with roles, such as melanin production (melanocytes), sensory perception (Merkel cells), and immunological function (Langerhans and other cells) (Prow et al., 2011). The water content at normal skin surface is low (10–15%) (Schäfer-Korting et al., 2007). Deeper skin layers are constituted by the viable epidermis (50–100 μm) with the basal membrane (consisting of at least one member of the protein family of laminin, type IV collagen and nidogen, and the proteoglycan perlecan) as well as by the dermis (1–2 mm), where appendices, such as sweat glands and hair follicle are located (Franzen and Windbergs, 2015; Schäfer-Korting et al., 2007). The dermis is divided into an upper papillary layer containing loosely arranged collagen fibers and a reticular layer with dense collagen fibers arranged in parallel to the surface of the skin. As well as collagen, the dermal matrix comprises a high amount of elastin, to provide the elastic properties of the skin (Mathes et al., 2014). This matrix is produced by fibroblasts, which are the main cell type of the dermis. Dermis is pervaded by blood and lymph vessels. Nerves sweat and sebaceous glands as well as hair follicles and shafts are embedded in the dermis. Beneath the dermis lies the subcutis, also known as the hypodermis. The subcutis functions as both an insulator, conserving the body's heat, and as a shock-absorber. Next to fibroblasts, adipocytes are the most prominent cell type in this compartment (Mathes et al., 2014). In addition to the structured cellular components of skin, there are appendages including the pilosebaceous units (hair follicles and associated sebaceous glands), apocrine and eccrine sweat glands (Prow et al., 2011). Fig. 12.1 summarizes the skin structure.

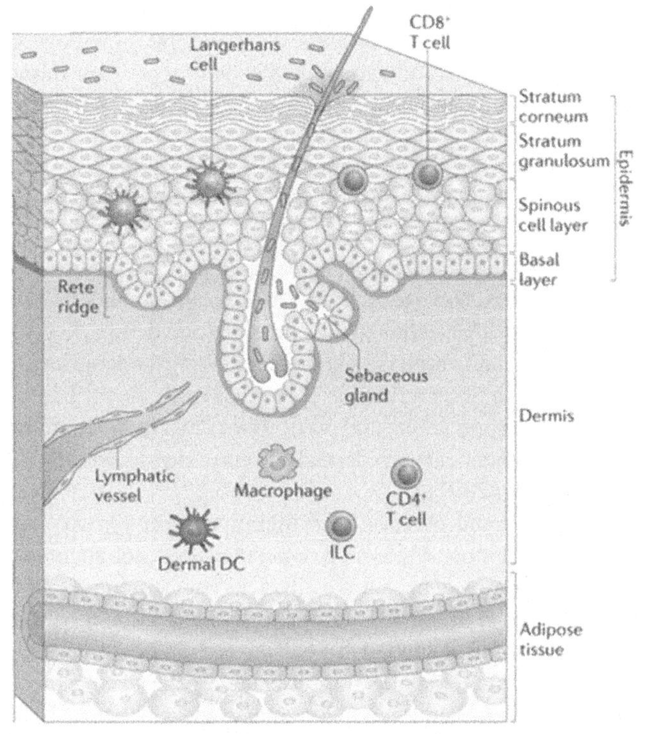

FIGURE 12.1 Schematic Picture of the Native Skin That is Subclassified Into Three Main Compartments: Epidermis, Dermis, and Subcutis (Hypodermis)

Adapted by Pasparakis, M., Haase, I., Nestle, F.O., 2014. Mechanisms regulating skin immunity and inflammation. Nat. Rev. Immunol. 14, 289–301, 3792130776332 (2016), with permission from Macmillan Publishers Ltd.

The surface of the skin has long been recognized to be acidic, with a pH of 4.2–5.6, being described as the acid mantle (Schmid-Wendtner and Korting, 2006). It has a number of diverse functions, such as antimicrobial defense, maintenance of the permeability barrier by effects on extracellular lipid organization and processing, preservation of optimal corneocyte integrity and cohesion, regulated by pH-sensitive proteolytic enzymes, and restriction of inflammation by inhibiting the release of proinflammatory cytokines (Honari and Maibach, 2014).

The transport of substances across the SC mainly occurs by passive diffusion and based on the dual-compartment bricks and mortar structure of the SC, interrupted by appendages. Three possible routes can be considered: the transcellular, the intercellular and the appendageal (Prow et al., 2011). Polar solutes normally take the transcellular route and more lipophilic ones go via the intercellular lipids (Prow et al., 2011). The delivery of drugs or particles via the appendageal route is regarded as a realistic alternative to delivery across the SC, covering only 0.1% of the human skin surface area.

Skin ageing is thought to be a complex biological process traditionally classified as intrinsic and extrinsic ageing, affected by a variety of internal, such as, genetic predisposition, hormonal disorders, vitamin deficiencies and external factors, such as, ultraviolet (UV) radiation, environmental pollution, improper care (Dzwigałowska et al., 2013; Farage et al., 2007, 2008; Longo et al., 2013). This complex and multifactorial process, whose baseline rate is genetically determined, may be accelerated by environmental, mechanical, or socioeconomic factors (Farage et al., 2007). Traditionally, the clinical signs of skin ageing are thinning skin with exaggerated expression lines, wrinkles, age spots and actinic keratoses (Longo et al., 2013). The changes undergone by skin as it ages occur throughout the epidermis, dermis, and subcutaneous tissue and can manifest in discrete and broad alterations in skin topography (Baumann, 2007).

12.2.1 EPIDERMIS

Epidermis manifests some important changes related to ageing. Some studies related that aged skin is characterized by a thinner epidermis, however is common the concept that SC does not change with age (Baumann, 2007). In aged skin, the intersection of the epidermis and dermis, known as dermal-epidermal junction, is altered, being flatter and diminishing the connecting area surface (Baumann, 2007). A decrease in the cell turnover is also observed in the epidermal layer. According to Kligman (1979) the SC transit time is of 20 days in young adults and 30 or more days in older adults. This slow turnover leads to a protracted SC replacement rate, epidermal atrophy, slower wound healing, and often less effective desquamation (Baumann, 2007).

On a cellular level, apoptosis and decreased proliferative capacity, two key features of ageing, affect epidermal structure and function, while on a molecular level, telomere shortening and reactive oxygen species (ROS), respectively, account for the characteristic structural and functional changes of chronologically aged skin (Fisher et al., 2002). Telomeres, which shorten with chronological ageing, are repetitive DNA sequences, in all mammals TTAGGG and its complement, that cap the ends of chromosomes (Thiele et al., 2006). Telomeres do not encode genes, but rather protect the proximal genes and regulatory sequences in several ways (Krutmann and Gilchrest, 2006). Destabilization of the loop structure by progressive depletion with chronological ageing, exposes the telomeric repeat sequence. Then, an as-yet-unidentified sensing mechanism interacts with the overhang to initiate a signal cascade that can lead to cell cycle arrest, premature senescence or apoptosis (Thiele et al., 2006). Fig. 12.2 summarizes the epidermal organization.

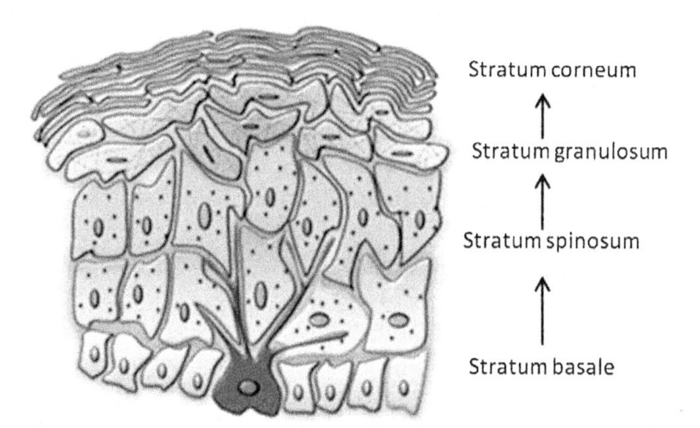

FIGURE 12.2 Epidermal Organization

Adapted by Natarajan, V.T., Ganju, P., Ramkumar, A., Grover, R., Gokhale, R.S., 2014. Multifaceted pathways protect human skin from UV radiation. Nat. Chem. Biol. 10, 542–551, with permission from Macmillan Publishers Ltd.

12.2.2 DERMIS

About 20% of dermal thickness disappears as people become elderly (Baumann, 2007). Even more striking changes occur in the dermis: massive elastosis (deposition of abnormal elastic fibers), collagen degeneration, and twisted, dilated microvasculature (Gilchrest, 1996). As the three primary structural components of the dermis, collagen, elastin, and glycosaminoglycans (GAGs) have been the subjects of the majority of antiageing research pertaining to the skin (Baumann, 2007).

Collagen is the primary structural component of the dermis and the most abundant protein found in humans, comprising about 25–35% of total body protein content, being responsible for strength and support of human skin (Thompson and Maibach, 2010). This protein comprises approximately 70–80% of the dry weight of the dermis (Waller and Maibach, 2009). Alterations in collagen play an integral role in the ageing process. This, in turn, partly explains the popularity of collagen-containing products intended for "antiageing" purposes (Baumann, 2007). The ratio of collagen types found in human skin also changes with age (Baumann, 2007). During ageing process, a marked loss of fibrillin-positive structures as well as a reduced content of collagen VII is observed, weakening the bond between dermis and epidermis (Contet et al., 1999; Watson et al., 2001). In older skin, collagen looks irregular and disorganized (Ganceviciene et al., 2012) and the overall collagen content per unit area of the skin surface is known to decline approximately 1% per year (Shuster et al., 1975). Significantly lower levels of collagen IV have been identified at the base of wrinkles.

Dermal fibroblasts make precursor molecules called procollagen, which is converted into collagen. Fibroblast collapse, due to the accumulation of degraded collagen fibers that prohibit construction of a healthy collagen matrix, causes the ratio of collagen synthesis to collagen degradation to become deranged in a self-perpetuating cycle (Fisher et al., 2008). There are two important regulators of collagen production: (1) transforming growth factor (TGF)-β, a cytokine that promotes collagen production, and (2) activator protein (AP)-1, a transcription factor that inhibits collagen production and upregulates collagen breakdown by upregulating enzymes called matrix metalloproteinases (MMPs) (Kang et al., 1997). In aged skin an increase of AP-1 occurred comparatively to young skin (Chung et al., 2000). MMP activity is increased in aged human skin, being associated with dramatic higher levels of degraded collagen (Fisher et al., 2002). This combination results in an overall decrease in collagen levels in the dermis.

MMP are a family of ubiquitous endopeptidases playing a role in many different physiological and pathological processes in the skin, including cutaneous ageing (Nelson et al., 2000; Sárdy, 2009). This zinc-containing proteinases family is responsible for degradation of extracellular matrix (ECM) proteins, which form skin dermal connective tissue. Together with collagen, elastin and elastic fibers, MMP are essential for strengthening of muscles, tendons, and joints (Pimple and Badole, 2014). Degradation of these vital tissues and/or oxidative damage of DNA lead to loosening of the skin and eventual formation of wrinkles (Pimple and Badole, 2014).

Functional elastin also declines in the dermis with age, as elastin becomes calcified in aged skin and elastin fibers degrade (Boss and Seegmiller, 1981). The cell turnover also declines. GAGs, the primary dermal skin matrix constituents assisting in binding water and an important contributor to the skin structure, declines with age (Ganceviciene et al., 2012). The same occurs with hyaluronic acid produced by fibroblasts and the interfibrillary ground substance, also a component of a healthy dermal matrix (Sudel et al., 2005).

12.2.3 HYPODERMIS

The overall volume of subcutaneous fat typically diminishes with age, as well as fat distribution. The physiological significance may be to increase thermoregulatory function by further insulating internal organs (Farage et al., 2010).

The efficacy of active ingredients is related to their diffusion through the barrier and their concentration in the formulation (Rawlings and Matts, 2005). There are numerous synthetic skincare formulations containing active ingredients, such as alpha-hydroxyl acids, hyaluronic acid, cohesive polydensified matrix, etc., that have adverse reactions; for example, allergic contact dermatitis, irritant contact dermatitis, phototoxic, and photoallergic reactions (Pimple and Badole, 2014). However, plant-derived polyphenolic substances, such as alloin, catechin, epicatechin, curcumin, myricetin, quercetin, etc., are beneficial as antiageing ingredients (Pimple and Badole, 2014). Many herbal polyphenolic substances have found to be effective in reducing the rate and intensity of wrinkle formation. But, how this ingredients work? How can they reach the final function desired in ageing?

12.3 APPLICATION OF OMW AS ACTIVE INGREDIENT FOR COSMETIC PROPOSES

Olive oil is one of the most well-known natural ingredients of the world. Global production of olive oil has doubled in the last 20 years with a positive increasing socioeconomic impact in Europe, especially in Mediterranean countries (Meksi et al., 2012). At the same time, as the production of olive oil is increasing the olive by-products discharge is also escalating (Rahmanian et al., 2014). Olive pomace and OMW, both by-products of olive oil processing, have high phytotoxicity related to environmental problems when discarded without any treatment (Rusan et al., 2015; Roselló-Soto et al., 2015). OMW is a dark-colored liquid and represents the liquid fraction of residues in three phase's extraction systems and traditional mills, being constituted by the water used in the different stages of olive oil extraction and the vegetable water of the fruit (Roig et al., 2006).

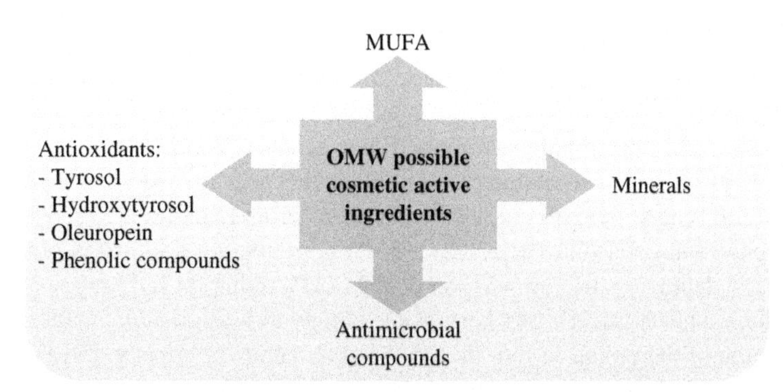

FIGURE 12.3 Major Interesting Compounds for Cosmetic Industry Detected in OMW

Consumers are demanding natural ingredients and rapidly a wide range of products that have emerged in the market last year (Skaltsounis et al., 2015). These features anticipate new opportunities to recover functional ingredients of OMW for further applications, namely for cosmetics. A multiplicity of organic compounds can be found in OMW, such as sugars, phenolic compounds, polyalcohols, lipids, and pectins suspended in a relatively stable emulsion (Asfi et al., 2012). Fig. 12.3 summarizes different compounds detected in OMW that present interesting characteristics for cosmetic purposes.

Moreover, they have also been recognized as beneficial to human health and well-being, due to their antioxidant properties (Roselló-Soto et al., 2015).

12.3.1 BIOACTIVE COMPONENTS OF OMW

OMW presents a huge potential source of bioactive compounds for many biotechnological applications. Phenolic compounds, such as hydroxytyrosol, tyrosol, oleuropein, and phenolic acids which have major antioxidant properties, are targeted for cosmetics applications (Takac and Karakaya, 2009; Yener, 2015). Furthermore, other valuable compounds with potential interest for cosmetic industry are present in OMW, such as fatty acids [monounsaturated fatty acids (MUFA) specifically] and minerals, as Fig. 12.2 resumes (Meksi et al., 2012). The question is: How they act on skin?

12.3.1.1 Antioxidants

As previously explained, both chronologic and environment independent processes accelerate skin ageing. Physiological processes, such as ageing have been associated with an imbalance between the action of ROS and antioxidants (Bulotta et al., 2014). The cumulative effect of intrinsic age (biological ageing) and photo ageing contributes to a progressive loss of structural integrity and physiological function of the skin. When in excess, ROS have deleterious effects as lipid oxidation, protein oxidation, DNA strand break, and modulation of gene expression (Esteve et al., 2015). The antioxidant activity is an excellent example of a functional benefit that plant extracts can deliver. Cosmetic treatments of the latest generation, developed against wrinkles, rely on antioxidant properties of some ingredients, especially those derived from plants (Rona et al., 2004). Antioxidants can mitigate the effects of the skin ageing process by limiting the biochemical consequences of oxidation (Rodrigues et al., 2015c). Plants are known to contain a variety of natural antioxidants that protect and preserve themselves (Angerhofer et al., 2009). The exposure to UV radiation is the initial step to several skin harmful effects as wrinkling, dryness, pigment abnormalities, and in last instance skin cancer (Nichols and Katiyar, 2010). Notwithstanding, a topical application of antioxidant formulas also seems to be an interesting strategy to protect skin against oxidative stress related with several extrinsic agents (Leonardi et al., 2013). Skin exposure to UVA radiation (320–400 nm) can damage DNA molecules and other lipid components. Also UVB radiation (280–320 nm) cause oxidative stress resulting in promutagenic DNA lesions which induce mutations in the epidermal cells (Guo et al., 2010). Photoprotection comprises human body mechanisms to minimize the damages suffered when exposed to UV radiation (D'Andria et al., 2013). Plants secondary metabolites act, such as UV blockers. Some compounds as phenolic acids and flavonoids can prevent the penetration of the radiation into the skin (Saewan and Jimtaisong, 2015).

Once OMW is regarded as a potent source of natural antioxidants, several methodologies have been applied to its recovery. Recently, Goula and Lazarides (2015) developed an integrated process that can turn OMW and pomegranate wastes into ingredients for further applications (Goula and Lazarides, 2015). These researchers applied technological processes as fermentation, spray drying, and encapsulation to OMW in order to produce olive paste spread or olive powder and encapsulated phenols. Therefore, OMW powder obtained will be able to be used in a wide variety of food and pharmaceutical applications and in cosmetic products, such as creams, balms, shampoos, or hair conditioners (Goula and Lazarides, 2015).

Due to the sensitivity of natural compounds to the presence of factors like light or oxygen, encapsulation is a promising technology that can be used to improve bioactive compounds stability. Several encapsulation techniques have been studied (Soto et al., 2015). However, spray-drying is one of the most used (Munin and Edwards-Lévy, 2011). In order to obtain active ingredients, it should be considered some particle properties such as particle size and density, kinetics, release and degradation mechanisms. However, although the particles obtained have generally a mean size range of 10-100 μm, they could have a large size distribution related to the variety of droplet sizes in the spray (Fang and Bhandari, 2010; Munin and Edwards-Lévy, 2011). In cosmetic formulations it must be assured that the particles obtained are suitable to skin permeation through the skin barrier, in particular *stratum corneum* (Rawlings and Matts, 2005). Therefore, it is advisable to exploit encapsulation techniques specific for an application field (in this case, cosmetics), according to what is required: improve compounds stability, control their release, avoid incompatibility of substances, reduce odor of compounds, and assure compounds active function (Soto et al., 2015).

In these conditions, encapsulation techniques can be viable in cosmetics. The successful development of products for topical bioactive compounds delivery is based on understanding skin permeation to design an effective formulation. The development of standard analytical methods to evaluate the accordance of OMW-based products with its content and skin delivery kinetics is necessary and an innovative field. For instance, for the determination of hydroxytyrosol and tyrosol (two major phenolic compounds) in different types of olive extract raw materials and cosmetic cream samples, an analytical method has been developed. This method besides having environment friendly features, it allows control the whole process from olive extracts until the final product. In last instance, it aims to assure the quality of the finished cosmetic products (Miralles et al., 2015).

Hydroxytyrosol, 2-hydroxytyrosol, tyrosol, oleanolic acid, flavonoids, anthocyanins, and tannins are natural antioxidants found in OMW with potential added value. Hydroxytyrosol and tyrosol are the major representative hydrophilic phenols identified, respectively ≈18 and ≈4 g/L (Bazoti et al., 2006).

Moreover, hydroxytyrosol, the main product of oleuropein degradation, is identified as the olive phenolic compound with the strongest in vitro antioxidant properties (Romero et al., 2004). It has been demonstrated that hydroxytyrosol can have a chemo-protective effect against UVB-induced DNA impairment in a human skin keratinocyte cell line, using the comet assay. Hydroxytyrosol (100 μM) reduced intracellular ROS formation and attenuated the expression of p53 and NF-κB in a concentration dependent manner, proposing a positive effect as topical use (Guo et al., 2010). Additionally, hydroxytyrosol have a good solubility in oil and aqueous media which predicts its successful application in cosmetic products (Bouzid et al., 2005).

Caffeic and ferulic acids are two hydroxycinnamic acids present in olives and therefore in OMW. Recently, in albino Swiss mice, caffeic acid administration (15 mg per kg of body weight) before UVB daily exposition decreased lipid peroxidation, inflammatory markers expression, and enhanced antioxidant status in mice skin (Balupillai et al., 2015).

Oleuropein belongs to the secoiridoids group and is an ester of 2-(3, 4-dihydroxyphenyl)-ethanol (hydroxytyrosol). The oleuropein content of olive fruit varies according to the stage of maturation (higher in early stages and declining along maturation). Along with the decline of oleuropein content other oleosides as ligustroside also decline, while other phenolic compounds, such as verbascoside increase (Omar, 2010). Oleuropein have a potent scavenging effect demonstrated against oxidative substances. It effectively inhibits copper sulfate-induced low-density lipoproteins (LDL) oxidation (Visioli et al., 2006); scavenge nitric oxide and increase the inducible nitric oxide synthase (de la Puerta et al., 2001); scavenge hypochlorous acid (produced by neutrophil myeloperoxidase) (Visioli et al., 2002); and scavenge oxygen radicals and inhibits lipid peroxidation (Umeno et al., 2015). In order to study topical applications of oleuropein formulas, Perugini et al. (2008) evaluated the effects of an emulsion and emulgel (100 μL) with oleuropein on UVB-induced erythema in healthy female volunteers with 20–30 years old ($n = 10$). Results showed that oleuropein formulations mitigated erythema, transepidermal water loss, and blood flow (Perugini et al., 2008).

Squalene, a polyunsaturated triterpene wide spread in nature, is present in the non saponifiable fraction of olive oil. The squalene content in olive oil is approximately 1% (Ghanbari et al., 2012). In humans, it is secreted by the sebaceous glands, being quantified in high amounts in sebum (≈12%) (Viola and Viola, 2009). In cosmetic formulations, it is already used as moisturizer or emollient agent. Due to its features it can be a high valued asset in creams, skincare or massage oils and hair serums attributting soft properties to these products (Rodrigues et al., 2015d). Nevertheless, the UV radiation effect on sebum and in its components, as squalene, has been mainly focused on its antioxidant properties (Viola and Viola, 2009). Squalene inhibits the lipid peroxidation induced by UV radiation acting as scavenger of singlet oxygen (Ghanbari et al., 2012). The oxidized squalene derivatives (result of UV radiation) at low level, can act as endogenous signaling molecules, which have an important function in the initiation of the protective and immunological response against radiation.

Therefore, sebaceous lipids, upon oxidation by UV radiation, can have a positive role in the skin defense through contributing to the regulation of specific mediators. If this immunomodulatory system is disrupted by, for example, prolonged or intense sun exposure, tissue damage can occur. Also, induced inflammatory responses in normal sun exposed skin needs further research (Oyewole and Birch-Machin, 2015). Olive oil squalene is a well-known and studied compound (Ghanbari et al., 2012). However its occurrence in OMW and stability needs particular attention. In future, OMW squalene content and its relevant importance in the development of OMW-based new cosmetic products is advisable.

Antioxidants are frequently used in sunscreen formulations to complement UV filter photoprotection since they are able to prevent the damage induced by the free radicals generated by solar irradiation previously mentioned (Chaudhuri, 2005). Sunscreen compounds (e.g., benzophenon derivatives, parabens, etc.) include typically single or multiple aromatic structures, conjugated or not with carbon-carbon double bonds or carbonyl moieties, being able to attenuate the transmission of energetic solar photons (Giokas et al., 2007). Indeed, they absorb light in both UVA and UVB wavelength regions and are widely used as active substances in sunscreen products. Nowadays, as UV filters of cosmetics and particularly sunscreens are becoming widespread available, questions have been raised concerning their long-term usage and the resulted skin damage in the presence of UV radiation (Gasparro et al., 1998; Nohynek and Schaefer, 2001). For instance, synthetic UV filters may penetrate the skin resulting in systemic exposure to potentially harmful xenobiotic and allergic chemicals as well as estrogenic effects (Boehm et al., 1995; Nohynek and Schaefer, 2001). Moreover, in vitro studies have shown that some sunscreen ingredients cause DNA damage (inflected by free radicals) and subsequently may be carcinogenic (McHugh and Knowland, 1997). In this way, antioxidants are becoming a challenge in sunscreens, being typically used to complement UV filter protection. Natural antioxidants found in fruits, vegetables, and relevant bioresources include compounds, such as phenols, ascorbate, and tocopherols (Galanakis et al., 2010). Possessing similar structures, natural phenols can also act with the same mechanism as synthetic sunscreen agents. Subsequently, they have been proposed for their utilization as active compounds in cosmetics (Fang and Bhandari, 2010) and are frequently used in sunscreen formulations to complement UV filter (Damiani et al., 2006). For instance, quercetin (a typical flavonol) has been reported to enhance photo-stability of two common UVA and UVB filters: methoxydibenzoylmethane and octyl methoxycinnamate, respectively (Scalia and Mezzena, 2010). Hydroxytyrosol has also shown a significant protective ability against UVB-induced DNA damage (Guo et al., 2010) and against UVA-induced protein damage (D'Angelo et al., 2005). Besides, natural phenolic extracts could eventually replace other sunscreen formulations that are designated to protect topical cultivations, such as pomegranate fruits (Weerakkody et al., 2010). Indeed, at least four companies around the world produce phenol rich products from OMW with the final purpose of merchandising them as natural preservatives or bioactive additives in foodstuff and cosmetics (Galanakis and Schieber, 2014). For instance, Afonso et al. (2014) referred that antioxidants like vitamin C (ascorbic acid), vitamin E (α-tocopherol), and coenzyme Q10 (ubiquinone) decreased UV-related skin damage by increasing the photostability and in vitro SPF of avobenzone. Natural phenols have also been proposed for their valorization as active agents in sunscreen formulations (Fang and Bhandari, 2010) since they have similar structures with chemical UV filters and thus could act with the same mechanism. A typical example is quercetin, which has been referred to enhance the photostability of methoxy dibenzoylmethane (UVA) and octyl methoxy cinnamate (UVB) filters, respectively (Scalia and Mezzena, 2010). Likewise, hydroxytyrosol has been referred to have a protective role against UVA-induced protein (D'Angelo et al., 2005) and UVB-induced DNA (Guo et al., 2010) damages. Galanakis et al. (2010) demonstrated that phenols recovered from OMW are more active UV filters than ascorbic acid and α-tocopherol in both UVB and UVA regions, probably due to their higher antioxidant capacity. The fact that olive phenols absorb in both UV regions of interest allowed their application as UV booster in particular cases, that is, to enhance the absorption of titanium dioxide solutions in both UVB and UVA region, the absorption of bisoctrizole and diethylamino hydroxybenzoyl hexyl benzoate solutions in the UVB region and the absorption of Octocrylene, octyl-dimethyl-PABA, and octyl methoxy cinnamate in the UVA region.

The efficacy of bioactive compounds in cosmetics is related to their diffusion through the skin. Small soluble molecules with lipophilic and hydrophilic properties have greater ability to cross the stratum corneum than polymers or highly lipophilic substances (Rawlings and Matts, 2005). However, if they are very soluble in water, there is always the problem of being removed from the skin during seawater immersion. This is a common problem with the applications of phenols. Following these considerations, researchers are investigating the implementation of phenols as UV booster in cosmetics using different concentrations. In addition, the encapsulation of olive phenols prior their emulsification in cosmetics was investigated with the final purpose of increasing their water resistance.

Natural or seminatural cosmetics can be used to answer to consumers' current preferences and demands (natural ingredients) and also be a way of preventing skin damage. However, due to the natural instability of antioxidants, these can be easily oxidized before reaching its site of action. Considering the development of new cosmetic formulas, it is important to ensure the stability of bioactive compounds present in the final product. According to some authors and as previously discussed, the encapsulation of polyphenols could ensure some stability (Fang and Bhandari, 2010). However, it should be highlight that a desirable value of particle size for skin permeation is established as inferior to 500 nm, in order to penetrate the skin epithelium, which is not as easy as expected (Kohli and Alpar, 2004).

Besides antioxidant capacity, OMW also presents interesting pharmacological properties, such as antimicrobial activity, due to phenolic content (Leouifoudi et al., 2015; Tafesh et al., 2011). Most studies on antimicrobial activity of OMW have been performed under an ecological/agronomical perspective focusing, for example, the microbial diversity in soil. Therefore, antimicrobial activity of OMW phenols related has been scarcely studied (Leouifoudi et al., 2015).

Studies on bioactive compounds derived from OMW showed that single phenolic compounds or their combination was effective in growth inhibition of different microorganisms (Tafesh et al., 2011; Vagelas et al., 2009). OMW compounds that presented antibacterial activity are hydroxytyrosol, oleuropein, 4-hydroxybenzoic acid, vanillic acid, and *p*-coumaric acid (Capasso et al., 1995; Soler-Rivas et al., 2000; Sousa et al., 2006). The in vitro growth inhibition of some bacteria, such as *Escherichia coli*, *Klebsiella peneumoniae,* and *Bacillus cereus,* by oleuropein, vanillic, and *p*-coumaric acids has been reported (Aziz et al., 1997; Ghanbari et al., 2012). Although natural antimicrobials appear as promising ingredients in topical formulations against pathogenic microorganisms, some caution is necessary in extrapolation of results (Medina-Martínez et al., 2015). Nevertheless, the potential contribute as natural preservative, despite synthetic ones, in cosmetic products, ought to be explored.

12.3.1.2 Fatty acids

In skin, fatty acids are integrated into complex lipids or are present in a free form (Rodrigues et al., 2015c). They can be synthesized by skin cells or taken up of exogenous sources in a protein mediated process (Lin and Khnykin, 2014). Their presence is crucial in epidermis physiology with particular impact as constituents of the lipid film previously described that protects skin surface. During lifetime, the skin fatty acids content decrease resulting in a less moisture and emollient skin. Topical applications of cosmetics rich in fatty acids can reach epidermal and dermal layers of tissue promoting beneficial effects (Viola and Viola, 2009). This effect is also probably due to the fatty acids enhancer potential. In fact, fatty acids have been studied as potent skin penetration enhancers for the development of successful topical and transdermal delivery systems, particularly unsaturated fatty acids

(Aungst, 1989; Babu et al., 2015). Structurally, fatty acids consist of an aliphatic hydrocarbon chain and a terminal carboxyl group. According to Aungst (1989), this mechanism involves the disruption of the densely packed lipids that fill the extracellular spaces of SC (Aungst, 1989).

Olive oil is well-known as a key ingredient in several cosmetics (shampoos, soaps, body lotions, etc.) due to its content in phenolic compounds and fatty acids, particularly monounsaturated (MUFA) (Asfi et al., 2012; Barbulova et al., 2015). The total lipid content can vary between 2% and 14% according to different authors (Gonçalves et al., 2012; Zbakh and El Abbassi, 2012). Among fatty acids, MUFA content ranged from 71% to 78% (Asfi et al., 2012; Gonçalves et al., 2012).

12.3.1.3 Minerals

The mineral fraction represents 0.5–2% of OMW total organic and inorganic load (Asses et al., 2009; Pérez et al., 1998). The major minerals present are sodium (Na), magnesium (Mg), iron (Fe), copper (Cu), manganese (Mn), potassium (K), and zinc (Zn) (Amaral et al., 2008; Roig et al., 2006). According to Arienzo and Capasso (2000), K is the predominant mineral in OMW (≈17 g/L), followed in decreasing order by Mg (≈3 g/L), Ca (≈2 g/L) Na (≈0.4 g/L), Fe (≈0.129 g/L), Zn (≈0.063 g/L), Mn (≈0.015 g/L), and Cu (≈0.009 g/L) (Arienzo and Capasso, 2000).

The presence of water is crucial to the maintenance of skin elasticity and integrity. The loss of moisture, which leads to dry skin, is closely linked to age-related skin condition (Thompson and Maibach, 2010). For example, the topical and systemic estrogen therapies probably preserve skin moisture based in the increase of acid mucopolysaccharides and hyaluronic acid in the dermis (high water holding capacity), higher sebum levels and water-holding capacity of the SC, and changes in the corneocytes surface area (Thompson and Maibach, 2010). The natural moisturizing factor (NMF) is responsible for maintaining proper moisture levels in the SC and for appropriate barrier homeostasis, desquamation, and plasticity (Robinson et al., 2010). NMF represents 10% of the corneocyte mass and generates an osmotic force that pulls water in the corneocyte, acting as a humectant and a plasticizer in the SC (Roberts et al., 2007). NMF consists in a mixture of amino acids and derivatives, and specific salts. Particularly, K is an important mineral component of the NMF correlated with the state of hydration, stiffness, and pH in the SC (Rawlings and Matts, 2005). As OMW are rich in minerals, namely K, the recovery and purification of these compounds can be strategic to add-value to cosmetic products with hydration finality.

Table 12.1 summarizes the skin effects of the different compounds previously described.

12.4 OTHER APPLICATIONS AND FUTURE PERSPECTIVES

Considering the OMW richness in minerals, fatty acids and phenols, another interesting application could be spa treatments, with some similarity to wine therapy spas, as the moisturizer and antiageing properties of bioactive compounds are beneficial to skin health.

Dyes industry is another field for application of OMW. Synthetic dyes are used in several industries, including cosmetic industry. OMW, a colored olive oil processing by-product, can be a valuable source of natural dyes to replace the relatively toxic synthetic ones. Some researchers have been done to evaluate the optimum conditions to extract dye from olive pomace with promissory results. Additionally, other studies were performed in order to valorize OMW as possible dyebath for textile sector (Elksibi et al., 2014; Meksi et al., 2012).

Table 12.1 Skin Effect of the Different Components of OMW

	Compound	Characteristics	Skin Effect
Antioxidants	Hydroxytyrosol	Main product of oleuropein degradation	Chemo-protective effect Antiageing
	Oleuropein	Oleuropein content vary according to the stage maturation (higher in early stages and declining along maturation)	↓ Tumors growth in exposed skin Skin protecting Antiageing Antiinflammatory Antimicrobial ↓ Intracellular levels of ROS ↓ Erythema ↓ TEWL
	Fatty acids	Total lipid 2–14% MUFA content is 71–78%	↓ Permeability barrier Acidification of SC ↑ Hydratation ↑ Softness ↑ Elasticity ↑ Protective barrier
	Minerals	0.5–2% of OMW K the predominant metal K> Mg> Ca> Na> Fe> Zn> Mn> Cu	↑ Hydratation ↑ Stiffness pH control

Due to its versatility, OMW could be considered as a big challenge to become an active ingredient, whether used whole or in extract form, in pharmaceutical, cosmetics, or even food supplements industries. However, as other agroindustry by-products that have been explored and recently introduced in the market as cosmetic ingredients, the development of this active ingredient from OMW depends on the industries interest (Rodrigues et al., 2015a,b,d). Indeed, it is imperative to stimulate research from different areas to enhance the content of compounds of interest, but also to explain their potentialities to olive farmers (Rodrigues et al., 2015c). Without any doubt, the positive points of OMW as cosmetic ingredient is the increasing interest of consumers for natural products as well as the multiplicity of active ingredients that can be found and isolated from OMW. Nevertheless, the safety of this olive by-product and its well documented composition and biological activities should not be forgotten. The involvement of olive farmers and suppliers have also to guarantee the regular supply of interested industries, which should have in mind the equipments and treatments involved for its use. Moreover, specific extraction methods based on sustainable concepts should be developed considering solvents toxicity.

12.5 CONCLUSIONS

In this chapter, it was well documented that OMW had a large amount of interesting compounds with promising application in cosmetic field. Most of these are still undervalued, even though their chemical composition reveals high potential as a source of natural compounds, including antioxidants, fatty acids, and mineral that can be used as functional and technical ingredients.

OMW is a dynamic and complex matrix with enzymes, organic acids, metals, and substrates (proteins, polysaccharides, and phenolic compounds) in a reaction medium (water) (Tafesh et al., 2011). Most studies are focused in individual bioactive compounds. Additionally, inside OMW matrix chemical reactions can occur allowing a synergistic effect.

The valorization of OMW for cosmetic industry could be promoted based on some important points: the recovery of bioactive compounds allows adding value to this by-product; the environmental burden can be diminished; benefic bioactive compounds could be incorporated in skin care products. In the development of innovative cosmetics, using agroindustrial by-products, it is important to guarantee low-level of metals that can be transferred during the handling and processing of olive oil products and by-products. Sustainable and environment friendly extraction techniques have been successfully used to recover such compounds, which can be converted into high value added natural ingredients for cosmetic formulations.

The importance of antiageing cosmetics is actually well established. Many molecules are promising "active" ingredients to slowdown the different symptoms. Presently, the best therapy against ageing is to limit the DNA damage with antioxidants, vitamins, and minerals. As well documented throughout this chapter, OMW is extremely rich in bioactive compounds, such as antioxidants, fatty acids, and minerals. Indeed, they can become a source of possible active ingredients for cosmetic products with different claims, such as antiageing or hydration.

ACKNOWLEDGMENTS

Francisca Rodrigues is thankful to Foundation for Science and Technology (Portugal) for the PhD grant SFRH/BDE/51385/2011 financed by POPH-QREN and subsidized by European Science Foundation. M. Antónia Nunes is thankful for the research grant from project UID/QUI/500006/2013. This work received financial support from the European Union (FEDER funds through COMPETE) and National Funds (FCT, Foundation for Science and Technology) through project LAQV/UID/QUI/50006/2013 The work also received financial support from the European Union (FEDER funds) under the framework of QREN through Project NORTE-07-0124-FEDER-000069. To all financing sources the authors are greatly indebted.

REFERENCES

Afonso, S., Horita, K., Sousa e Silva, J.P., Almeida, I.F., Amaral, M.H., Lobao, P.A., Costa, P.C., Miranda, M.S., Esteves da Silva, J.C.G., Sousa Lobo, J.M., 2014. Photodegradation of avobenzone: stabilization effect of antioxidants. J. Photochem. Photobiol. B 140, 36–40.

Amaral, C., Lucas, M.S., Coutinho, J., Crespí, A.L., do Rosário Anjos, M., Pais, C., 2008. Microbiological and physicochemical characterization of olive mill wastewaters from a continuous olive mill in Northeastern Portugal. Biores. Technol. 99, 7215–7223.

Angerhofer, C.K., Maes, D., Giacomoni, P.U., 2009. The use of natural compounds and botanicals in the development of anti-aging skin care products. In: Nava, D. (Ed.), Skin Aging Handbook. William Andrew Publishing, Norwich, pp. 205–263.

Arienzo, M., Capasso, R., 2000. Analysis of metal cations and inorganic anions in olive oil mill waste waters by atomic absorption spectroscopy and ion chromatography. Detection of metals bound mainly to the organic polymeric fraction. J. Agric. Food Chem. 48, 1405–1410.

Asfi, M., Ouzounidou, G., Panajiotidis, S., Therios, I., Moustakas, M., 2012. Toxicity effects of olive-mill wastewater on growth, photosynthesis and pollen morphology of spinach plants. Ecotoxicol. Environ. Saf. 80, 69–75.

Asses, N., Ayed, L., Bouallagui, H., Sayadi, S., Hamdi, M., 2009. Biodegradation of different molecular-mass polyphenols derived from olive mill wastewaters by *Geotrichum candidum*. Int. Biodeterior. Biodegradation 63, 407–413.

Aungst, B., 1989. Structure/effect studies of fatty acid isomers as skin penetration enhancers and skin irritants. Pharm. Res. 6, 244–247.

Aziz, N., Farag, S., Mousa, L., Abo-Zaid, M., 1997. Comparative antibacterial and antifungal effects of some phenolic compounds. Microbios 93, 43–54.

Babu, R.J., Chen, L., Kanikkannan, N., 2015. Fatty alcohols, fatty acids, and fatty acid esters as penetration enhancers. In: Dragicevic, N., Maibach, H.I. (Eds.), Percutaneous Penetration Enhancers Chemical Methods in Penetration Enhancement—Modification of the Stratum Corneum. Springer, Berlin, pp. 133–150.

Badiu, D., Luque, R., Rajendram, R., 2010. Effect of olive oil on the skin. In: Preedy, V.R., Watson, R.R. (Eds.), Olives and Olive Oil in Health and Disease Prevention. Academic Press, San Diego, pp. 1125–1132.

Balupillai, A., Prasad, R.N., Ramasamy, K., Muthusamy, G., Shanmugham, M., Govindasamy, K., Gunaseelan, S., 2015. Caffeic acid inhibits UVB-induced inflammation and photocarcinogenesis through activation of peroxisome proliferator-activated receptor-γ in mouse skin. Photochem. Photobiol. 91, 1458–1468.

Barbulova, A., Colucci, G., Apone, F., 2015. New trends in cosmetics: by-products of plant origin and their potential use as cosmetic active ingredients. Cosmetics 2, 82–92.

Baumann, L., 2007. Skin ageing and its treatment. The Journal of Pathology 211, 241–251.

Bazoti, F.N., Gikas, E., Skaltsounis, A.L., Tsarbopoulos, A., 2006. Development of a liquid chromatography-electrospray ionization tandem mass spectrometry (LC-ESI MS/MS) method for the quantification of bioactive substances present in olive oil mill wastewaters. Analytica chimica acta 573–574, 258–266.

Behrens, A., Giljum, S., Kovanda, J., Niza, S., 2007. The material basis of the global economy: worldwide patterns of natural resource extraction and their implications for sustainable resource use policies. Ecol. Econ. 64, 444–453.

Boehm, F., Meffert, H., Tribius, S., 1995. Hazards of sunscreens—cytotoxic properties of micronized titandioxide. In: Holick, M.F., Jung, E.G. (Eds.), Biological Effects of Light. Walter de Gruyter, Berlin, pp. 446–451.

Boss, G.R., Seegmiller, J.E., 1981. Age-related physiological changes and their clinical significance. Western J. Med. 135, 434–440.

Bouzid, O., Navarro, D., Roche, M., Asther, M., Haon, M., Delattre, M., Lorquin, J., Labat, M., Asther, M., Lesage-Meessen, L., 2005. Fungal enzymes as a powerful tool to release simple phenolic compounds from olive oil by-product. Process Biochem. 40, 1855–1862.

Bulotta, S., Celano, M., Lepore, S.M., Montalcini, T., Pujia, A., Russo, D., 2014. Beneficial effects of the olive oil phenolic components oleuropein and hydroxytyrosol: focus on protection against cardiovascular and metabolic diseases. J. Transl. Med. 12, 219.

Capasso, R., Evidente, A., Schivo, L., Orru, G., Marcialis, M.A., Cristinzio, G., 1995. Antibacterial polyphenols from olive oil mill waste waters. J. Appl. Bacteriol. 79, 393–398.

Chaudhuri, R.K., 2005. Role of antioxidants in sun care products. In: Shaath, N. (Ed.), Sunscreens. Taylor Francis Group, Boca Raton, FL, pp. 603–618.

Chung, J.H., Kang, S., Varani, J., Lin, J., Fisher, G.J., Voorhees, J.J., 2000. Decreased extracellular-signal-regulated kinase and increased stress-activated MAP kinase activities in aged human skin in vivo. J. Invest. Dermatol. 115, 177–182.

COLIPA, 2015. Cosmetics Europe—The Personal Care Association.

Contet, A., Jeanmaire, Pauly, 1999. A histological study of human wrinkle structures: comparison between sun-exposed areas of the face, with or without wrinkles, and sun-protected areas. Br. J. Dermatol. 140, 1038–1047.

D'Andria, R., Di Salle, A., Petillo, O., Sorrentino, G., Peluso, G., 2013. Nutraceutical, cosmetic, health products derived from olive. In: Arcas, N., Arroyo López, F.N., Caballero, J., D'Andria, R., Fernández, M., Fernandez, Escobar, R., Garrido, A., López-Miranda, J., Msallem, M., Parras, M., Rallo, L., Zanoli, R., (Eds.), Present and future of the Mediterranean olive sector, Zaragoza, Spain. pp. 153–161.

D'Angelo, S., Ingrosso, D., Migliardi, V., Sorrentino, A., Donnarumma, G., Baroni, A., Masella, L., Tufano, M.A., Zappia, M., Galletti, P., 2005. Hydroxytyrosol, a natural antioxidant from olive oil, prevents protein damage induced by long-wave ultraviolet radiation in melanoma cells. Free Rad. Biol. Med. 38, 908–919.

Damiani, E., Rosati, L., Castagna, R., Carloni, P., Greci, L., 2006. Changes in ultraviolet absorbance and hence in protective efficacy against lipid peroxidation of organic sunscreens after UVA irradiation. J. Photochem. Photobiol. 82, 204–213.

de la Puerta, R., Domínguez, M.E.M., Ruíz-Gutíerrez, V., Flavill, J.A., Hoult, J.R.S., 2001. Effects of virgin olive oil phenolics on scavenging of reactive nitrogen species and upon nitrergic neurotransmission. Life Sci. 69, 1213–1222.

Drexhage, J., Murphy, D., 2010. Sustainable Development: From Brundtland to Rio 2012. United Nations Headquarters, New York, p. 26.

Dzwigałowska, A., Sołyga-Żurek, A., Dębowska, R.M., Eris, I., 2013. Preliminary study in the evaluation of anti-aging cosmetic treatment using two complementary methods for assessing skin surface. Skin Res. Technol. 19, 155–161.

Elksibi, I., Haddar, W., Ben Ticha, M., Gharbi, R., Mhenni, M.F., 2014. Development and optimisation of a non conventional extraction process of natural dye from olive solid waste using response surface methodology (RSM). Food Chem. 161, 345–352.

Esteve, C., Marina, M.L., García, M.C., 2015. Novel strategy for the revalorization of olive (*Olea europaea*) residues based on the extraction of bioactive peptides. Food Chem. 167, 272–280.

Fang, Z., Bhandari, B., 2010. Encapsulation of polyphenols—a review. Trends Food Sci. Technol. 21 (10), 510–523.

Farage, M., Miller, K., Maibach, H., 2010. Degenerative changes in aging skin. In: Farage, M., Miller, K., Maibach, H. (Eds.), Textbook of Aging Skin. Springer, Berlin, pp. 25–35.

Farage, M.A., Miller, K.W., Elsner, P., Maibach, H.I., 2007. Structural characteristics of the aging skin: a review. Cutan. Ocul. Toxicol. 26, 343–357.

Farage, M.A., Miller, K.W., Elsner, P., Maibach, H.I., 2008. Intrinsic and extrinsic factors in skin ageing: a review. Int. J. Cosmet. Sci. 30, 87–95.

Fisher, G.J., Kang, S., Varani, J., et al., 2002. Mechanisms of photoaging and chronological skin aging. Arch. Dermatol. 138, 1462–1470.

Fisher, G.J., Varani, J., Voorhees, J.J., 2008. Looking older: fibroblast collapse and therapeutic implications. Arch. Dermatol. 144, 666–672.

Franzen, L., Windbergs, M., 2015. Applications of Raman spectroscopy in skin research—from skin physiology and diagnosis up to risk assessment and dermal drug delivery. Adv. Drug Deliv. Rev. 89, 91–104.

Galanakis, C.M., Tornberg, E., Gekas, V., 2010. Recovery and preservation of phenols from olive waste in ethanolic extracts. J. Chem. Technol. Biotechnol. 85, 1148–1155.

Galanakis, C.M., Schieber, A., 2014. Editorial. Special issue on recovery and utilization of valuable compounds from food processing by-products. Food Res. Int. 65, 299–300.

Ganceviciene, R., Liakou, A.I., Theodoridis, A., Makrantonaki, E., Zouboulis, C.C., 2012. Skin anti-aging strategies. Dermatoendocrinol. 4, 308–319.

Gasparro, F.P., Mitchnick, M., Nash, J.F., 1998. A review of sunscreen safety and efficacy. Photochem. Photobiol. 68, 243–256.

Ghanbari, R., Anwar, F., Alkharfy, K.M., Gilani, A.-H., Saari, N., 2012. Valuable nutrients and functional bioactives in different parts of olive (*Olea europaea* L.)—a review. Int. J. Mol. Sci. 13, 3291–3340.

Gilchrest, B.A., 1996. A review of skin ageing and its medical therapy. Br. J. Dermatol. 135, 867–875.

Giokas, D.L., Salvador, A., Chisvest, A., 2007. UV filters: from sunscreens to human body and the environment. Trac-Trends Anal. Chem. 26, 360–374.

Gonçalves, M.R., Costa, J.C., Marques, I.P., Alves, M.M., 2012. Strategies for lipids and phenolics degradation in the anaerobic treatment of olive mill wastewater. Water Res. 46, 1684–1692.

Goula, A.M., Lazarides, H.N., 2015. Integrated processes can turn industrial food waste into valuable food by-products and/or ingredients: the cases of olive mill and pomegranate wastes. J. Food Eng. 167 (Part A), 45–50.

Guo, W., An, Y., Jiang, L., Geng, C., Zhong, L., 2010. The protective effects of hydroxytyrosol against UVB-induced DNA damage in HaCaT cells. Phytother. Res. 24, 352–359.

Honari, G., Maibach, H., 2014. Skin structure and function. In: Honari, H.M. (Ed.), Applied Dermatotoxicology. Academic Press, Boston, pp. 1–10.

Kang, S., Fisher, G.J., Voorhees, J.J., 1997. Photoaging and topical tretinoin: therapy, pathogenesis, and prevention. Arch. Dermatol. 133, 1280–1284.

Kligman, A.M., 1979. Perspectives and problems in cutaneous gerontology. J. Invest. Dermatol. 73, 39–46.

Kohli, A., Alpar, H.O., 2004. Potential use of nanoparticles for transcutaneous vaccine delivery: effect of particle size and charge. Int. J. Pharm. 275, 13–17.

Krutmann, J., Gilchrest, B.A., 2006. Photoaging of skin. In: Barbara, A., Gilchrest, J.K. (Eds.), Skin Aging. Springer, Berlin, pp. 33–44.

Leonardi, G.R., Moreno, I., Melo, P.S., Alencar, S.M., Monteio e Silva, S.A., 2013. Application of phenolic acids in gels: new antioxidants perspective for antiaging effect. J. Am. Acad. Dermatol. 68, 21.

Leouifoudi, I., Harnafi, H., Zyad, A., 2015. Olive mill waste extracts: polyphenols content, antioxidant, and antimicrobial activities. Adv. Pharmacol. Sci. 2015, 11.

Lin, M.-H., Khnykin, D., 2014. Fatty acid transporters in skin development, function and disease. Biochimica et Biophysica Acta 1841, 362–368.

Longo, C., Casari, A., Beretti, F., Cesinaro, A.M., Pellacani, G., 2013. Skin aging: in vivo microscopic assessment of epidermal and dermal changes by means of confocal microscopy. J. Am. Acad. Dermatol. 68, 73–82.

Martinez-Saez, N., Ullate, M., Martín-Cabrejas, M.A., Martorell, P., Genovés, S., Ramón, D., Castillo, M.D., 2014. A novel antioxidant beverage for body weight control based on coffee silverskin. Food Chem. 150, 227–234.

Mathes, S.H., Ruffner, H., Graf-Hausner, U., 2014. The use of skin models in drug development. Adv. Drug Deliv. Rev. 69–70, 81–102.

McHugh, P.J., Knowland, J., 1997. Characterization of DNA damage inflicted by free radicals from a mutagenic sunscreen ingredient and its location using an in vitro genetic reversion assay. Photochem. Photobiol. 66, 276–281.

McPhee, D., Jain, R., 2015. Price, performance, supply and sustainability all-in-one—The Amyris case. Household Personal Care Today 10, 71–72.

Medina-Martínez, M.S., Truchado, P., Castro-Ibáñez, I., Allende, A., 2015. Antimicrobial activity of hydroxytyrosol: a current controversy. Biosci. Biotechnol. Biochem., 1–10.

Meksi, N., Haddar, W., Hammami, S., Mhenni, M.F., 2012. Olive mill wastewater: a potential source of natural dyes for textile dyeing. Ind. Crops Prod. 40, 103–109.

Miralles, P., Chisvert, A., Salvador, A., 2015. Determination of hydroxytyrosol and tyrosol by liquid chromatography for the quality control of cosmetic products based on olive extracts. J. Pharm. Biomed. Anal. 102, 157–161.

Munin, A., Edwards-Lévy, F., 2011. Encapsulation of natural polyphenolic compounds: a review. Pharmaceutics 3, 793–829.

Mussatto, S.I., Machado, E.M.S., Carneiro, L.M., Teixeira, J.A., 2012. Sugars metabolism and ethanol production by different yeast strains from coffee industry wastes hydrolysates. Appl. Energy 92, 763–768.

Nelson, A.R., Fingleton, B., Rothenberg, M.L., Matrisian, L.M., 2000. Matrix metalloproteinases: biologic activity and clinical implications. J. Clin. Oncol. 18, 1135.

Nichols, J., Katiyar, S., 2010. Skin photoprotection by natural polyphenols: anti-inflammatory, antioxidant and DNA repair mechanisms. Arch. Dermatol. Res. 302, 71–83.

Nohynek, G.J., Schaefer, H., 2001. Benefit and risk of organic ultraviolet filters. Regul. Toxicol. Pharmacol. 33, 285–299.

Omar, S.H., 2010. Oleuropein in olive and its pharmacological effects. Scientia pharmaceutica 78, 133.

Oyewole, A.O., Birch-Machin, M.A., 2015. Sebum, inflammasomes and the skin: current concepts and future perspective. Exp. Dermatol. 24, 651–654.

Pérez, J., de la Rubia, T., Hamman, O.B., Martínez, J., 1998. Phanerochaete flavido-alba laccase induction and modification of manganese peroxidase isoenzyme pattern in decolorized olive oil mill wastewaters. Appl. Environ. Microbiol. 64, 2726–2729.

Perugini, P., Vettor, M., Rona, C., Troisi, L., Villanova, L., Genta, I., Conti, B., Pavanetto, F., 2008. Efficacy of oleuropein against UVB irradiation: preliminary evaluation. Int. J. Cosmet. Sci. 30, 113–120.

Pimple, B.P., Badole, S.L., 2014. Polyphenols: a remedy for skin wrinkles. In: Watson, R.R., Preedy, V.R., Zibadi, S. (Eds.), Polyphenols in Human Health and Disease. Academic Press, San Diego, pp. 861–869.

Prakash, L., Majeed, M., 2009. Natural ingredients for anti-ageing skin care. Household Personal Care Today 2, 44–46.

Prow, T.W., Grice, J.E., Lin, L.L., Faye, R., Butler, M., Becker, W., Wurm, E.M.T., Yoong, C., Robertson, T.A., Soyer, H.P., Roberts, M.S., 2011. Nanoparticles and microparticles for skin drug delivery. Adv. Drug Deliv. Rev. 63, 470–491.

Rahmanian, N., Jafari, S., Galanakis, C., 2014. Recovery and removal of phenolic compounds from olive mill wastewater. J. Am. Oil Chem. Soc. 91, 1–18.

Rawlings, A.V., Matts, P.J., 2005. Stratum corneum moisturization at the molecular level: an update in relation to the dry skin cycle. J. Invest. Dermatol. 124, 1099–1110.

Roberts, M.S., Bouwstra, J., Pirot, F., Falson, F., 2007. Skin hydration—a key determinant in topical absorption. In: Walters, K.A., Roberts, M.S. (Eds.), Dermatologic, Cosmeceutic, and Cosmetic Development: Therapeutic and Novel Approaches. CRS Press, London, pp. 115–128.

Robinson, M., Visscher, M., LaRuffa, A., Wickett, R., 2010. Natural moisturizing factors (NMF) in the stratum corneum (SC). Effects of lipid extraction and soaking. J. Cosmet. Sci. 61, 13.

Rodrigues, F., Gaspar, C., Palmeira-de-Oliveira, A., Sarmento, B., Amaral, M.H., Oliveira, M.B., 2015a. Application of coffee silverskin in cosmetic formulations: physical/antioxidant stability studies and cytotoxicity effects. Drug Dev. Ind. Pharm. 22, 1–8.

Rodrigues, F., Pereira, C., Pimentel, F.B., Alves, R.C., Ferreira, M., Sarmento, B., Amaral, M.H., Oliveira, M.B., 2015b. Are coffee silverskin extracts safe for topical use? An in vitro and in vivo approach. Ind. Crops Prod. 63, 167–174.

Rodrigues, F., Pimentel, F.B., Oliveira, M.B., 2015c. Olive by-products: challenge application in cosmetic industry. Ind. Crops Prod. 70, 116–124.

Rodrigues, F., Sarmento, B., Amaral, M.H., Oliveira, M.B., 2015d. Exploring the antioxidant potentiality of two food by-products into a topical cream: stability, in vitro and in vivo evaluation. Drug Dev. Ind. Pharm. 42 (6), 880–889.

Roig, A., Cayuela, M.L., Sánchez-Monedero, M.A., 2006. An overview on olive mill wastes and their valorisation methods. Waste Manag. 26, 960–969.

Romero, C., Brenes, M., Yousfi, K., García, P., García, A., Garrido, A., 2004. Effect of cultivar and processing method on the contents of polyphenols in table olives. J. Agric. Food Chem. 52, 479–484.

Rona, C., Vailati, F., Berardesca, E., 2004. The cosmetic treatment of wrinkles. J. Cosmet. Dermatol. 3, 26–34.

Roselló-Soto, E., Koubaa, M., Moubarik, A., Lopes, R.P., Saraiva, J.A., Boussetta, N., Grimi, N., Barba, F.J., 2015. Emerging opportunities for the effective valorization of wastes and by-products generated during olive oil production process: non-conventional methods for the recovery of high-added value compounds. Trends Food Sci. Technol. 45, 296–310.

Rusan, M.M., Albalasmeh, A., Zuraiqi, S., Bashabsheh, M., 2015. Evaluation of phytotoxicity effect of olive mill wastewater treated by different technologies on seed germination of barley (*Hordeum vulgare* L.). Environ. Sci. Pollut. Res. 22, 9127–9135.

Saewan, N., Jimtaisong, A., 2015. Natural products as photoprotection. J. Cosmet. Dermatol. 14, 47–63.

Sahota, A., 2014. Introduction to sustainability. In: Sahota, A. (Ed.), Sustainability: How the Cosmetics Industry is Greening Up. Wiley, London, pp. 1–16.

Sárdy, M., 2009. Role of matrix metalloproteinases in skin ageing. Connect. Tissue Res. 50, 132–138.

Scalia, S., Mezzena, M., 2010. Photostabilization effect of quercetin on the UV filter combination, butyl methoxydibenzoymethane-octyl methoxycinnamate. Photochem. Photobiol. 86, 273–278.

Schäfer-Korting, M., Mehnert, W., Korting, H.-C., 2007. Lipid nanoparticles for improved topical application of drugs for skin diseases. Adv. Drug Deliv. Rev. 59, 427–443.

Schmid-Wendtner, M.H., Korting, H.C., 2006. The pH of the skin surface and its impact on the barrier function. Skin Pharmacol. Physiol. 19, 296–302.

Schurer, N.Y., Elias, P.M., 1991. The biochemistry and function of stratum corneum lipids. Adv. Lipid Res. 24, 27–56.

Shuster, S.A.M., Black, M.M., McVitie, E.V.A., 1975. The influence of age and sex on skin thickness, skin collagen and density. Br. J. Dermatol. 93, 639–643.

Skaltsounis, A.-L., Argyropoulou, A., Aligiannis, N., Xynos, N., 2015. Recovery of high added value compounds from olive tree products and olive processing byproducts. In: Bouskou, D. (Ed.), Olive and Olive Oil Bioactive Constituents. AOCS Press, Illinois, pp. 333–356.

Soler-Rivas, C., Espín, J.C., Wichers, H.J., 2000. Oleuropein and related compounds. J. Sci. Food Agric. 80, 1013–1023.

Sousa, A., Ferreira, I.C.F.R., Calhelha, R., Andrade, P.B., Valentão, P., Seabra, R., Estevinho, L., Bento, A., Pereira, J.A., 2006. Phenolics and antimicrobial activity of traditional stoned table olives 'alcaparra'. Bioorg. Med. Chem. 14, 8533–8538.

Soto, M.L., Falqué, E., Domínguez, H., 2015. Relevance of natural phenolics from grape and derivative products in the formulation of cosmetics. Cosmetics 2, 259–276.

Sudel, K.M., Venzke, K., Mielke, H., Breitenbach, U., Mundt, C., Jaspers, S., Koop, U., Sauermann, K., Knussman-Hartig, E., Moll, I., Gercken, G., Young, A.R., Stab, F., Wenck, H., Gallinat, S., 2005. Novel aspects of intrinsic and extrinsic aging of human skin: beneficial effects of soy extract. Photochem. Photobiol. 81, 581–587.

Tafesh, A., Najami, N., Jadoun, J., Halahlih, F., Riepl, H., Azaizeh, H., 2011. Synergistic antibacterial effects of polyphenolic compounds from olive mill wastewater. Evid. Based Complement. Alternat. Med. 2011, 431021.

Takac, S., Karakaya, A., 2009. Recovery of phenolic antioxidants from olive mill wastewater. Recent Patents Chem. Eng. 2, 230–237.

Thiele, J., Barland, C.O., Ghadially, R., Elias, P.M., 2006. Permeability and antioxidant barriers in aged epidermis. In: Barbara, A., Gilchrest, J.K. (Eds.), Skin Aging. Springer, New York, pp. 65–79.

Thompson, Z., Maibach, H.I., 2010. Biological effects of estrogen on skin. In: Miranda, A., Farage, K.W.M., Howard, I., Maibach (Eds.), Textbook of Aging Skin. Springer, Berlin, pp. 361–367.

Umeno, A., Takashima, M., Murotomi, K., Nakajima, Y., Koike, T., Matsuo, T., Yoshida, Y., 2015. Radical-scavenging activity and antioxidative effects of olive leaf components oleuropein and hydroxytyrosol in comparison with homovanillic alcohol. J. Oleo Sci. 64, 793–800.

Vagelas, I., Kalorizou, H., Papachatzis, A., Botu, M., 2009. Bioactivity of olive oil mill wastewater against plant pathogens and post-harvest diseases. Biotechnol. Biotechnol. Equip. 23, 1217–1219.

Viola, P., Viola, M., 2009. Virgin olive oil as a fundamental nutritional component and skin protector. Clin. Dermatol. 27, 159–165.

Visioli, F., Bogani, P., Galli, C., 2006. Healthful properties of olive oil minor components. In: Boskou, D. (Ed.), Olive Oil, Chemistry and Technology. AOCS Press, Illinois, pp. 173–190.

Visioli, F., Galli, C., Galli, G., Caruso, D., 2002. Biological activities and metabolic fate of olive oil phenols. Eur. J. Lipid Sci. Technol. 104, 677–684.

Waller, J.M., Maibach, H.I., 2009. A quantitative approach to age and skin structure and function: protein, glycosaminoglycan, water, and lipid content and structure. In: André, O., Barel, M.P., Howard, I., Maibach (Eds.), Handbook of Cosmetic Science and Technology. Informa Healthcare USA, Inc, New York, pp. 243–260.

Watson, R.E.B., Craven, N.M., Kang, S., Jones, C.J.P., Kielty, C.M., Griffiths, C.E.M., 2001. A short-term screening protocol, using fibrillin-1 as a reporter molecule, for photoaging repair agents. J. Invest. Dermatol. 116, 672–678.

Weerakkody, P., Jobling, J., Infante, M.M.V., Rogers, G., 2010. The effect of maturity, sunburn and the application of sunscreen on the internal and external qualities of pomegranate fruit grown in Australia. Scientia Horticulturae 124, 57–61.

Yener, M.E., 2015. Supercritical fluid processing for the recovery of bioactive compounds from food industry by-products. In: Fornari, T., Stateva, R.P. (Eds.), High Pressure Fluid Technology for Green Food Processing. Springer, Berlin, pp. 305–355.

Zbakh, H., El Abbassi, A., 2012. Potential use of olive mill wastewater in the preparation of functional beverages: a review. J. Funct. Foods 4, 53–65.

Index

D

E

F

G

H

P

T

V

Printed in the United States
By Bookmasters